AF614724

Methods in Molecular Biology™

For further volumes:
http://www.springer.com/series/7651

Multiple Sequence Alignment Methods

Edited by

David J. Russell

Department of Electrical Engineering, University of Nebraska–Lincoln, Lincoln, NE, USA

Editor
David J. Russell
Department of Electrical Engineering
University of Nebraska–Lincoln
Lincoln, NE, USA

ISSN 1064-3745 ISSN 1940-6029 (electronic)
ISBN 978-1-62703-645-0 ISBN 978-1-62703-646-7 (eBook)
DOI 10.1007/978-1-62703-646-7
Springer New York Heidelberg Dordrecht London

Library of Congress Control Number: 2013947475

Printed on acid-free paper

Humana Press is a brand of Springer
Springer is part of Springer Science+Business Media (www.springer.com)

Preface

Introduction

Multiple sequence alignment has become one of the indispensable tools of bioinformatics, in fact of biology, as scientists try to make sense of the rapidly increasing flood of sequence information. Multiple sequence alignments are fundamental to tasks such as homology searches, genomic annotation, protein structure prediction, and the areas of computational evolutionary biology, gene regulation networks, and functional genomics. Over the last 25 years, and increasingly over the last 10 years, there has been the development of a number of different multiple sequence alignment algorithms and implementations. As many of these implementations are well on their way to becoming standard laboratory tools, there was a need for a single source that would provide an in-depth introduction and analysis of the various algorithms being used. And who best to describe these algorithms and their nuances than the people who developed these algorithms—hence this handbook.

Who Might Find This Handbook of Use

This volume is intended as both a multiple sequence alignment textbook and as a reference book; it begins at a level suitable for those with no previous exposure to the problem of performing sequence alignment and carries the reader through to a reasonable degree of proficiency at understanding how most industry-standard alignment algorithms achieve their results. The people who might find this handbook of use are computational biologists in general, people involved in tasks and areas that use multiple sequence alignments in particular, and students embarking on a study of computational biology.

For novices to the field, the chapters presented in the first part of the handbook introduce the fundamental concepts necessary for understanding how and why sequence alignment algorithms function the way they do. Because the results of multiple sequence alignments have such a direct impact on our understanding and interpretation of the information contained in biological sequences, it is important to understand the workings and limitations of different algorithms. This is especially true for practitioners in the field. Treating multiple sequence alignment as a black box which delivers results to be used unquestioningly could be a recipe for disaster. The chapters in the second part of the handbook describe detailed practical procedures for the most commonly used multiple sequence alignment algorithms available today. Additionally, extensive practical detail for each algorithm's implementation is provided such that a competent scientist who is unfamiliar with the method can carry out the technique successfully by simply following the detailed, practical procedures presented.

Organization

The first set of five chapters deals with issues common to all multiple sequence alignment algorithms. At the heart of many multiple sequence alignment schemes is the idea of dynamic programming. This is a solution strategy in which the problem to be solved is broken up into overlapping subproblems which are solved and then combined to provide a solution to the overall problem. Chapter 1 provides a thorough description of the dynamic programming approach and its application to the pairwise sequence alignment problem. Because the generation of multiple sequence alignments is an NP-complete problem, there is a need for heuristic strategies. Chapter 2 details the various heuristic approaches currently used to generate multiple alignments. When using a heuristic approach, it is necessary to find objective scoring techniques to select between possible multiple alignments. Chapter 3 provides a survey of different scoring schemes that can be used during multiple sequence alignment. Given the number of different multiple alignment algorithms available, an important issue is performance comparison. This is usually done using benchmarks. Different benchmarking strategies are studied in Chapter 4 considering desirable properties of benchmarks. Multiple sequence alignments assume that the sequences being aligned are homologous. The process of selecting homologous sequences using BLAST and FASTA packages is detailed in Chapter 5.

Each of the 13 chapters in the second set deal with a particular multiple sequence alignment algorithm or package. The most widely used algorithm for multiple sequence alignments has been the *Clustal* algorithms. The latest version of Clustal, *Clustal Omega*, is described in detail in Chapter 6. Almost as well known as the *Clustal* algorithms are the *T-Coffee* (Tree-Based Consistency Objective Function for Alignment Evaluation) algorithms. As the authors point out, the *T-Coffee* algorithms incorporate structural, evolutionary, and experimental evidence to reach a more meaningful and accurate multiple sequence alignment. Chapter 7 provides a practical overview of various *T-Coffee* implementations. Both the *Clustal* and *T-Coffee* algorithms are progressive algorithms. The *MAFFT* algorithm, which has been gaining in popularity in recent years, uses an iterative refinement approach to provide a fast alignment algorithm. The *MAFFT* algorithm is described in Chapter 8 along with the *MUSCLE* algorithm. The chapter contains detailed instructions in the use of several different popular options in the *MAFFT* package. *Probcons* is a well-known example of an algorithm that uses Hidden Markov Models (HMMs) to provide sequence alignment. *Probcons* and *Probalign*, which uses a partition function approach, are described in Chapter 9. One of the primary applications of multiple sequence alignment is in phylogenetic analysis. *PRANK* is a phylogeny-aware alignment algorithm which uses phylogenetic information to distinguish between alignment gaps caused by insertions and deletions. This determination can be used to provide the inferred ancestral sequences and mark the alignment gaps differently depending on their origin in insertion or deletion events. Chapter 10 provides a detailed description of *PRANK* and provides practical advice for using *PRANK* for evolutionary analysis. Chapter 11 describes *GramAlign*, an alignment algorithm that uses a grammar-based relative complexity distance metric to determine the alignment order, the benefit being a computationally efficient and scalable program useful for managing the increasing amount and size of biological data made available due to the continuing advancements in sequencing technology. Detection of local homologies is another major application of multiple sequence alignment. The *DIALIGN* algorithms construct multiple alignments from local pairwise sequence similarities thus making them particularly useful for discovering conserved functional regions in sequences that share only local homologies but are otherwise unrelated. The different *DIALIGN* algorithms are described in Chapter 12. Another algorithm that focuses on local similarities is *PicXAA*, a nonprogressive, greedy algorithm that uses regions of high local similarity to drive the initial alignment which can then be iteratively refined. The *PicXAA* algorithm, as well as its implementation and usage, is described in Chapter 13. The computational cost of multiple sequence alignments can be defrayed in part by the intelligent use of the multiple cores with which most current computers are equipped. *MSAprobs*, described in Chapter 14, is a progressive alignment method which, along with various other improvements, has been parallelized using multithreading for use on multicore CPUs. Phylogeny inference often includes a paradox in which the accuracy of an inferred phylogeny depends on the accuracy of a multiple sequence alignment which depends on the accuracy of the inter-sequence distance metric. Many alignment techniques use a phylogeny to specify these distances, and so each inference relies on the accuracy of the other. Chapter 15 presents *SATé*, an iterative method for simultaneously estimating accurate multiple sequence alignments and phylogenetic trees. The *PRALINE* toolkit is described in Chapter 16. The algorithms in *PRALINE* use progressive alignment; however, instead of using a pre-determined guide tree, they continuously reevaluate at each stage which alignment will be optimal, thus generating an adaptive guide tree on the fly. As reflected in its name (Profile ALIgNmEnt), *PRALINE* uses various global, local, and homology-extended profile preprocessing protocols to address the problems caused by the greediness of a progressive alignment method. The algorithms described in the last two chapters, *PROMALS3D* and *MSACompro*, both focus on protein sequences. The *PROMALS3D* algorithm, described in Chapter 17, uses a multipronged approach including fast sequence alignment and the utilization of side information. Fast sequence alignment methods align similar sequences while additional information such as structure-based constraints from alignments of three-dimensional structures, for relatively dissimilar sequences, is used to construct multiple sequence alignments. The *MSACompro* algorithm described in Chapter 18 makes use of predicted secondary structure, relative solvent accessibility, and residue–residue contact information to improve the accuracy of multiple sequence alignments, deriving the structural information from the sequence itself, rather than from an external database.

The various multiple sequence alignment algorithms presented in this handbook give a flavor of the broad range of choices available for multiple sequence alignment generation. Their diversity is a reflection of the complexity of the multiple sequence alignment problem and the amount of information that can be obtained from multiple sequence alignments. Each of these chapters not only describes the algorithm it covers but also presents instructions and tips on using their implementations. This handbook will hopefully provide a readily available resource which will allow practitioners to experiment with different algorithms and find the particular algorithm that is of most use in their application.

Lincoln, NE, USA ***David J. Russell***

Contents

Contributors

PUNTO BAWONO • *Centre for Integrative Bioinformatics (IBIVU), VU University Amsterdam, Amsterdam, The Netherlands; Netherlands Bioinformatics Centre (NBIC), Nijmegen, The Netherlands*

GIOVANNI BUSSOTTI • *European Bioinformatics Institute (EBI), Wellcome Trust Genome Campus, Cambridge, UK*

JIA-MING CHANG • *Bioinformatics and Genomics Programme, Centre for Genomic Regulation (CRG), Universitat Pompeu Fabra, Barcelona, Spain*

JIANLIN CHENG • *Computer Science Department, Life Science Center, Informatics Institute, University of Missouri, Columbia, MO, USA*

XIN DENG • *Computer Science Department, University of Missouri, Columbia, MO, USA*

CHRISTOPHE DESSIMOZ • *EMBL-European Bioinformatics Institute, Cambridge, UK*

PAOLO DI TOMMASO • *Bioinformatics and Genomics Programme, Centre for Genomic Regulation (CRG), Universitat Pompeu Fabra, Barcelona, Spain*

HALUK DOĞAN • *Department of Genetics and Bioengineering, Istanbul Bilgi University, Istanbul, Turkey*

IONAS ERB • *Bioinformatics and Genomics Programme, Centre for Genomic Regulation (CRG), Universitat Pompeu Fabra, Barcelona, Spain*

JOSÉ ESPINOSA-CARRASCO • *Bioinformatics and Genomics Programme, Centre for Genomic Regulation (CRG), Universitat Pompeu Fabra, Barcelona, Spain*

MANUEL GIL • *Max F. Perutz Laboratories, Center for Integrative Bioinformatics Vienna, Medical University Vienna, University of Vienna, Vienna, Austria*

NICK GOLDMAN • *EMBL-European Bioinformatics Institute, Cambridge, UK*

KEVIN GORI • *EMBL-European Bioinformatics Institute, Cambridge, UK*

OSAMU GOTOH • *Computational Biology Research Center (CBRC), Tokyo, Japan; National Institute of Advanced Industrial Science and Technology (AIST), Tokyo, Japan*

NICK V. GRISHIN • *Department of Biophysics, Howard Hughes Medical Institute, Dallas, TX, USA; Department of Biochemistry, University of Texas Southwestern Medical Center, Dallas, TX, USA*

JAAP HERINGA • *Centre for Integrative Bioinformatics (IBIVU), Amsterdam Institute for Molecules, Medicines and Systems (AIMMS), VU University Amsterdam, Amsterdam, The Netherlands; Netherlands Bioinformatics Centre (NBIC), Nijmegen, The Netherlands*

DESMOND G. HIGGINS • *Conway Institute, University College Dublin, Dublin, Ireland*

STEFANO IANTORNO • *Wellcome Trust Sanger Institute, Cambridge, UK; National Institute of Allergy and Infectious Diseases, National Institutes of Health, Bethesda, MD, USA*

KAZUTAKA KATOH • *Immunology Frontier Research Center, Osaka University, Suita, Japan; Computational Biology Research Center, The National Institute of Advanced Industrial Science and Technology (AIST), Tokyo, Japan*

KEVIN LIU • *Department of Computer Science, Rice University, Houston, TX, USA*

YONGCHAO LIU • *Institut für Informatik, Johannes Gutenberg Universitat Mainz, Mainz, Germany*
ARI LÖYTYNOJA • *Institute of Biotechnology, University of Helsinki, Helsinki, Finland*
CEDRIK MAGIS • *Bioinformatics and Genomics Programme, Centre for Genomic Regulation (CRG), Universitat Pompeu Fabra, Barcelona, Spain*
BURKHARD MORGENSTERN • *Abteilung für Bioinformatik (IMG), Universitat Göttingen, Göttingen, Germany*
Ö. UFUK NALBANTOĞLU • *Department of Electrical Engineering, University of Nebraska-Lincoln, Lincoln, NE, USA*
CEDRIC NOTREDAME • *Bioinformatics and Genomics Programme, Centre for Genomic Regulation (CRG), Universitat Pompeu Fabra, Barcelona, Spain*
HASAN H. OTU • *Department of Genetics and Bioengineering, Istanbul Bilgi University, Istanbul, Turkey*
WILLIAM R. PEARSON • *Department of Biochemistry and Molecular Genetics, University of Virginia School of Medicine, Charlottesville, VA, USA*
JIMIN PEI • *Howard Hughes Medical Institute, University of Texas Southwestern Medical Center, Dallas, TX, USA*
USMAN ROSHAN • *Department of Computer Science, New Jersey Institute of Technology, Newark, NJ, USA*
DAVID J. RUSSELL • *Department of Electrical Engineering, University of Nebraska-Lincoln, Lincoln, NE, USA*
SAYED MOHAMMAD EBRAHIM SAHRAEIAN • *Department of Plant and Microbial Biology, University of California, Berkeley, CA, USA*
BERTIL SCHMIDT • *Institut für Informatik, Johannes Gutenberg Universitat Mainz, Mainz, Germany*
FABIAN SIEVERS • *Conway Institute, University College Dublin, Dublin, Ireland*
DARON M. STANDLEY • *Immunology Frontier Research Center, Osaka University, Suita, Japan*
JEAN-FRANÇOIS TALY • *Bioinformatics Core Facility, Centre for Genomic Regulation (CRG), Barcelona, Spain*
TANDY WARNOW • *Department of Computer Science, The University of Texas at Austin, Austin, TX, USA*
BYUNG-JUN YOON • *Department of Electrical and Computer Engineering, Texas A&M University, College Station, TX, USA*

Part I

Theory

Chapter 1

Dynamic Programming

Ö. Ufuk Nalbantoğlu

Abstract

Independent scoring of the aligned sections to determine the quality of biological sequence alignments enables recursive definitions of the overall alignment score. This property is not only biologically meaningful but it also provides the opportunity to find the optimal alignments using dynamic programming-based algorithms. Dynamic programming is an efficient problem solving technique for a class of problems that can be solved by dividing into overlapping subproblems. Pairwise sequence alignment techniques such as Needleman–Wunsch and Smith–Waterman algorithms are applications of dynamic programming on pairwise sequence alignment problems. These algorithms offer polynomial time and space solutions. In this chapter, we introduce the basic dynamic programming solutions for global, semi-global, and local alignment problems. Algorithmic improvements offering quadratic-time and linear-space programs and approximate solutions with space-reduction and seeding heuristics are discussed. We finally introduce the application of these techniques on multiple sequence alignment briefly.

Key words Dynamic programming, Needleman–Wunsch algorithm, Smith–Waterman algorithm, Affine gap penalties, Hirschberg's algorithm, Banded dynamic programming, Bounded dynamic programming, Seeding

1 Introduction

Biological sequence alignment is undoubtedly one of the major techniques used in several areas of computational biology. Several tasks such as inferring phylogenetic relationships, homology search of functional elements, classification of proteins, designing detection markers require an extensive amount of sequence alignments. This volume can change between the multiple alignment of a few gene-size sequences to an extensive molecular database search of large queries. Considering the fast rate of increase in the database volumes and the number of sequences to be aligned, we can have an idea about the fact that the sequence alignment programs employed are quite successful in handling the problem. Here we introduce the basic methodology behind the success of the alignment programs, namely dynamic programming.

David J. Russell (ed.), *Multiple Sequence Alignment Methods*, Methods in Molecular Biology, vol. 1079,
DOI 10.1007/978-1-62703-646-7_1, © Springer Science+Business Media, LLC 2014

Efficient algorithms employed in biological sequence alignment have been around since the early 1970s. Over the decades, many improvements and modifications have been achieved. However, the main methodology enabling sequence alignment has stayed as the kernel of the programs. Dynamic programming is an efficient computation scheme applied on a class of problems that can be solved recursively. Defining the alignment problem in such manner enables the computation of optimal alignments in polynomial time and memory for a pair of sequences.

Dynamic programming offers a feasible optimal solution for pairwise alignments. However, an exact application becomes impractical when multiple sequences are considered. A similar practical issue is observed when queries are needed to be aligned against large databases. Several heuristics are proposed for approximate solutions of the corresponding problems. Yet, the idea behind most of the popular alignment heuristics is to approximate the solution by dividing the problem into subpairwise alignment problems. In that sense, it is possible to say the idea of aligning two sequences using dynamic programming is a main building block of the multiple sequence alignment and database search algorithms.

In this chapter, we first review the pairwise sequence alignment problem and the polynomial time solution offered by dynamic programming. The minor variations on the algorithms to find global, semi-global, local, and near-optimal alignments are introduced. We discuss the fundamental algorithmic complexity reductions and heuristics to provide fast approximate solutions. The final issue we consider is the relation of multiple sequence alignment and dynamic programming.

2 The Pairwise Alignment Problem

To pose the problem and the solution program, perhaps the introduction of pairwise sequence alignment problem is a preferable starting point because of a couple of reasons. Firstly, the multiple sequence alignment is generally defined as a direct generalization of pairwise sequence alignment. Therefore, the pairwise solution offered can also be generalized to the multiple sequence case. Secondly, the practical applications of multiple sequence alignment approximate the solution considering the pairwise alignment of sequences. A multiple alignment is generally met by extending the pairwise alignments to a consensus of multiple alignment.

An alignment of two biological sequences is simply an ordered mapping. Let's say X and Y are two sequences and each element of a sequence is called a residue. Assume the ith residue of X, X_i, is paired with Y_j. Then, a residue coming after X_i cannot match up with a residue of Y coming before Y_j. Moreover, a residue can be matched with at most one residue and the matchings are not bijective. If a residue is not matched with another residue, it is referred as to be aligned with a gap. A popular visual representation of sequence alignments is to print the sequence couples as two lines, in which the matching residues are printed in the same column. The gaps, corresponding to the unmatched residues, are usually represented with dashes "–." Since the matchings are ordered, the primary structure of the sequences, which is the permutation of the letters, is preserved. In other words, it is possible to edit sequence X by substituting, inserting, and deleting residues and converting it to Υ. An alignment offers a map indicating which residues are conserved, substituted, deleted, and inserted. As an example, assume two DNA fragments $X = GACAT$ and $\Upsilon = GAGACAT$. An alignment A of X and Υ is

GAGACAT
GACA– –T.

Here, according to the alignment the first two, the fourth, and the last residues of Υ are conserved, the third one is substituted to a C, and fifth and the sixth ones are deleted.

This definition of sequence alignment does not impose any restriction on the matching positions. The two extreme cases are the alignment that all the residues of X are aligned to gaps placed before the first residue of Υ, and the one that all X residues are matched with gaps after the last residue of Υ. The intermediate alignments, ranging between these extremes are all valid alignments. To represent the alignments graphically, a rectangular grid shown in Fig. 1 has been used traditionally as it provides a simple understanding of the alignment space, as well as the dynamic programming solutions proposed. The rows of the grid correspond to the residues of a sequence, where the columns represent the residues of the second sequence to be aligned. In this structure, a cell or vertex $c(i, j)$, $(0 \leq i \leq |X|, 0 \leq j \leq |\Upsilon|)$ is connected to its surrounding neighbors, $(i-1, j), (i-1, j-1), (i, j-1)$, with incoming directed edges, and to $(i+1, j), (i+1, j+1)$, $(i, j+1)$, with outgoing directed edges. We can show that each alignment of X and Υ corresponds to a path from the upper-left corner to the lower-right corner of this graph by the following assignments. A diagonal edge from the vertex $c(i-1, j-1)$ to $c(i, j)$ corresponds to the aligned pair X_i, Υ_j, a horizontal edge from $c(i, j-1)$ to $c(i, j)$ corresponds to the pairing of Υ_j with a gap, and a vertical edge from $c(i-1, j)$ to $c(i, j)$ corresponds to the pairing of X_i with a gap. We can represent any alignment A by reading the pairings of an alignment from left to right in order, and converting it to an edge sequence E. The adjacent pairs of A must involve at least one incrementation of i and j since gap-to-gap pairings are not valid. Therefore, the adjacent edges in E are also adjacent in the graph, i.e., two adjacent edges in E are incoming and outgoing edges of the same vertex in the grid. This consecutive placement implies that E is a path from the upper-left to the

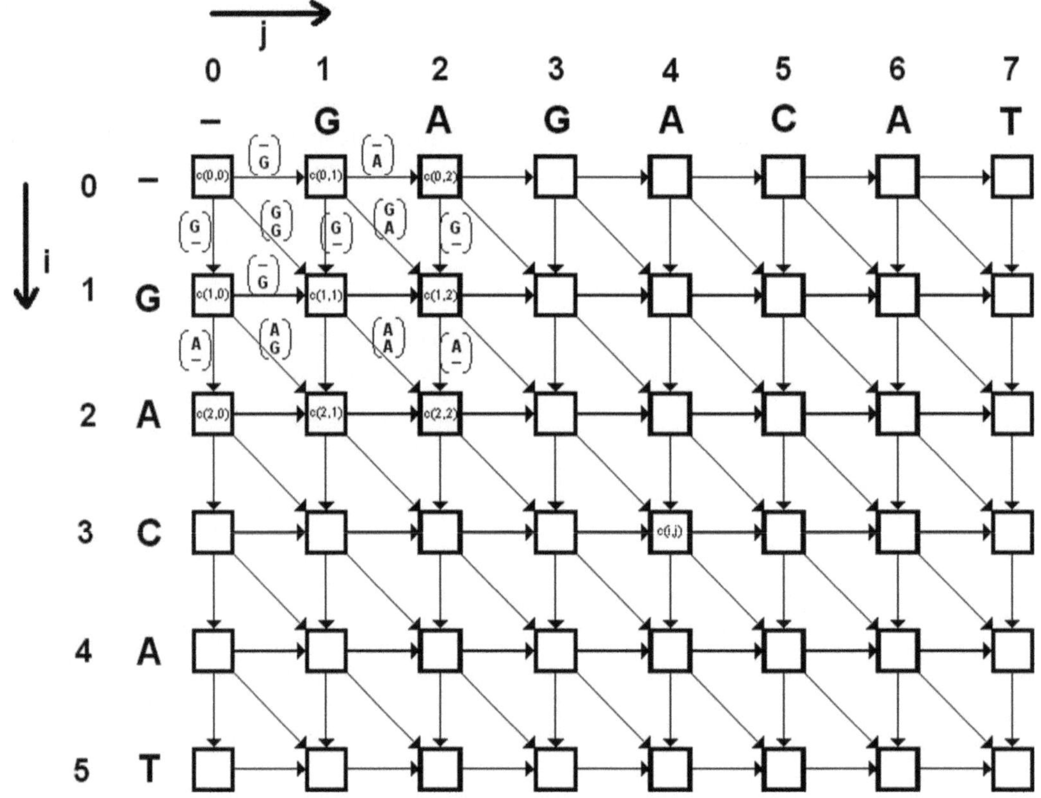

Fig. 1 Pairwise sequence alignment as a directed graph

lower-right of the grid. Each alignment can be represented by a unique path, conversely, each path corresponds to a unique alignment. Thus, the set of the alignments and the set of the described paths in the graph have one-to-one mapping and can be used interchangeably for practical purposes.

A sequence alignment program should search for the alignment which makes the most biological sense. Practically, this is achieved by assigning alignment scores to each of the alignments by a defined computation procedure, and searching for the best scores among the valid alignments. Every alignment has to be considered as a candidate and evaluated in a search procedure. According to this, all alignment paths in the alignment grid have to be considered in the search space. However, the cardinality of this search set grows rapidly with the length of the sequences. The number of possible alignments can be computed using the function f of Waterman [1]. This function considers the cases given that k residues of X and Y match and the rest are aligned with gaps. In this case, $(|X| - k)$ residues of X and $(|Y| - k)$ residues of Y are aligned with a gap. The total length of the alignment is then $(|X| - k) + (|Y| - k) + k = |X| + |Y| - k$. Out of $(|X| + |Y| - k)!$ permutations, picking the valid ones with the correct ordering of X and Y residues,

$$\#\ alignments\ with\ k\ residue\ matches = \frac{(|X| + |\Upsilon| - k)!}{k!(|X| - k)!(|\Upsilon| - k)!}. \quad (1)$$

Summing up for the all possibilities of k matches gives the total number of alignments

$$\#\ alignments = \sum_{k=0}^{\min(|X|,|\Upsilon|)} \frac{(|X| + |\Upsilon| - k)!}{k!(|X| - k)!(|\Upsilon| - k)!}. \quad (2)$$

Waterman function attains large numbers very rapidly. For the example above, with $|X| = 7$ and $|\Upsilon| = 5$, there are 7,183 possible alignments. This exceeds a million when we have sequences of ten residues. As a real-life example, we can assume aligning the β_2 microglobulin proteins of human and mouse (both with 119 residues). If we had a computer capable of performing one zetta-operations (10^{21} operations per second), it would take around 2×10^{61} years to evaluate and find the best alignments.

2.1 The Optimum Path Problem and Dynamic Programming

Fortunately, best scoring alignments can be found without computing the score of each alignment individually. Currently accepted alignment scoring schemes can be formulated recursively. Dynamic programming exploits this formation and attempts to solve the problem by dividing the problem into subproblems [2]. The computational simplification comes from the fact that a subproblem is needed to be solved several times to compute the total problem. The program basically "remembers" the solutions used several times and calls it from the memory instead of evaluating them over and over again.

Before going into the details of evaluating sequence alignments using dynamic programming, we can briefly see how the methodology handles recursive computations efficiently. In computer science terms, a recursive function is a finite step procedure that contains itself in its definition and different argument instances are drawn from an argument sequence for each time the function is called. A minimal set of function values has to be known by the computer to evaluate other instances. According to the top-down computation procedure, the function is called for a given argument. Another instance of the same function is called in the function due to the definition. This recursive callings terminate when a known instance of the function is met, and each function visited previously is computed traversing back to the desired instance of the function, where the recursion started.

Computation of Fibonacci numbers is a simple yet instructive example to observe the top-down evaluation of recursive functions. The nth Fibonacci number is defined as

$$F(n) = F(n-1) + F(n-2), \quad F(0) = 0, \quad F(1) = 1. \quad (3)$$

Figure 2a represents the top-down computation of $F(5)$. By the number definition $F(5) = F(4) + F(3)$, so the functions $F(4)$

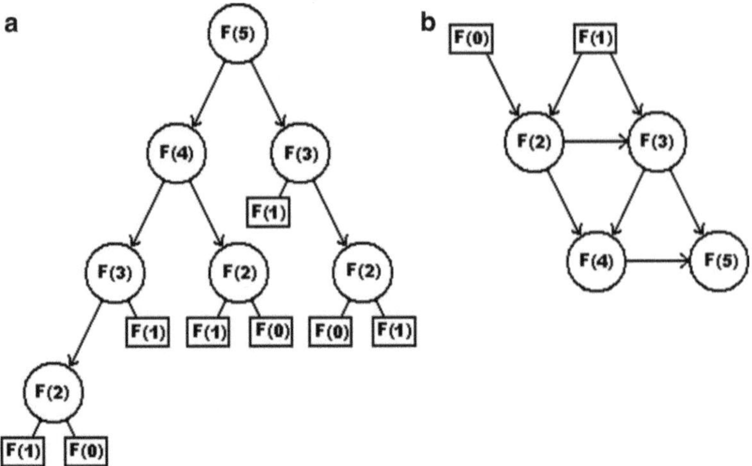

Fig. 2 (**a**) Top-down and (**b**) bottom-up recursive computations of Fibonacci numbers

and $F(3)$ are called recursively. Each function is called recursively until they meet $F(0)$ and $F(1)$ as shown in the graphical representation. Adding up these functions and traversing back to the top node $F(5)$ will return the fifth Fibonacci number. From this tree structured graph, it can be seen that $F(2)$ and $F(3)$ are computed three times and two times, respectively, by different function callings. A total of seven summation operations are performed. However, if we operated considering a space-time trade-off by saving $F(2)$ and $F(3)$ as they were evaluated, less number of evaluations would be required. This can be achieved by following a bottom-up recursive approach. According to the bottom-up scheme, the computation starts from the known function values and $F(2) = F(1) + F(0)$ is evaluated and saved. Calling previously memorized function values as they are required in the recursive function, the next Fibonacci number is evaluated iteratively until the desired number is met. We can see the bottom-up recursion in Fig. 2b. Here, the graph structure forms a trellis instead of a tree structure as in the top-down approach. $F(2)$ is evaluated once and called from the memory for the computation of $F(3)$ and $F(4)$. Same is true for $F(3)$ which is used in the computation of $F(4)$ and $F(5)$. A total of four summation are performed. When this scheme is generalized, the top-down approach will require $F(n+1)-1$ summations for the computation of $F(n)$, where the bottom-up approach requires $n-1$ summations for the same computation. The introduction of memory and single computation of duplicated function instances by bottom-up recursion is the fundamental methodology of dynamic programming. Similar to the Fibonacci number computation, dynamic programming can provide significant reduction in computational complexity for similar recursive problems where overlapping subproblems are observed.

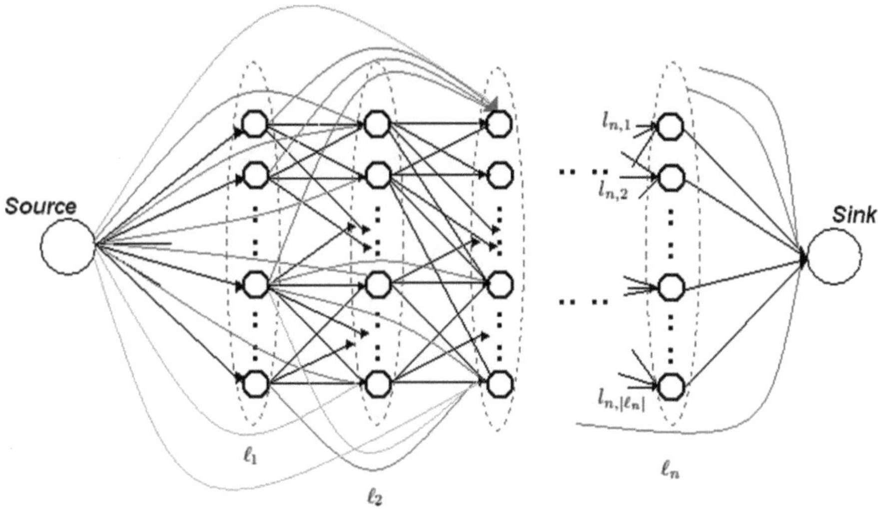

Fig. 3 Feedforward multilayer network with directed edges

The single-source optimal path problem [3] is a well-known computer science problem that is addressed by dynamic programming. Finding the optimal path in a directed multilayered network has a direct correspondence with the sequence alignment problem as we will see. Let's assume a multilayered network with a source node, a sink node and n middle layers (Fig. 3). Each middle layer ℓ_k has a finite positive number of nodes $\{l_{k,1}, l_{k,2}, \ldots, l_{k,|\ell_k|}\}$. Directed edges exist from the previous layers ℓ_m, $m < k$ to ℓ_k, and the vertices in a layer are not connected. Each edge has a corresponding score, and the optimal path problem is to find the highest scoring path from the source to the sink where the score of a path is the sum of the edges in that path. This problem can be divided into subproblems and solved recursively as follows.

Assume that the optimal path visits node $l_{k,\,i}$. The set of the paths from the source to $l_{k,\,i}$ consists of the edges between the layers ℓ_m, $m \leq k$. On the other hand, the paths from $l_{k,\,i}$ to the sink consists of the edges between the layers ℓ_l, $l \geq k$. As a result, the paths from the source to $l_{k,\,i}$ and from $l_{k,\,i}$ to the sink are non-overlapping sets, and these two problems can be solved separately. Thus

$$L(source, sink) = L(source, l_{k,i}) + L(l_{k,i}, sink), \qquad (4)$$

where $L(x, y)$ denotes the optimal path from node x to node y. Based on this divisibility to subproblems property, the optimal path to $l_{k,\,i}$ can be defined recursively by dividing the optimal paths from the source to the parent nodes of $l_{k,\,i}$ ($\widehat{l_{k,j}}$) and from $\widehat{l_{k,j}}$ to $l_{k,\,i}$:

$$L(source, l_{k,i}) = \max_j \left(L(source, \widehat{l_{k,j}}) + L(\widehat{l_{k,j}}, l_{k,i})\right), \quad 1 \leq j \leq |\widehat{\ell_k}|. \qquad (5)$$

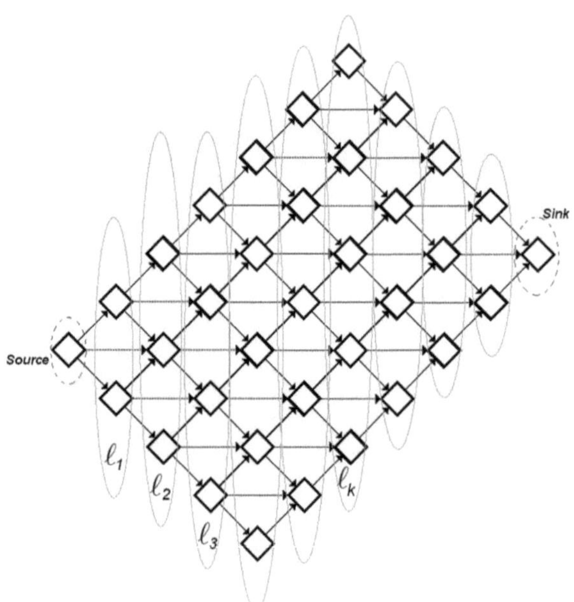

Fig. 4 Topological equivalence of multilayer networks and alignment graphs

Using the bottom-up recursion methodology of dynamic programming, we can start with calculating the score from the source to its children (with zero initials), and save these paths and their scores to be used in the next iterations. The computation terminates when the sink layer is reached and the optimal path is found. This dynamic programming algorithm lets the computer to deal with only the path combinations of adjacent nodes in each iteration instead of computing the scores of every possible path. Clearly we can see that addition of layers increases the number of computations nearly in linear fashion given a constant average connectivity, whereas the search space increases in exponential fashion. This is the implication of the reduction in computational complexity.

The graphical representation of pairwise sequence alignment has the same multilayered network structure defined above (Fig. 4). We can see the topological equivalence by assigning the top-left cell as the source, bottom-right cell as the sink, and the cells $\{c(i,j) : i + j = k\}$ as the middle layer ℓ_k.

A property has to hold in order to satisfy the equality of optimal path finding and sequence alignment problems. This is the additive property of the path scores. In terms of sequence alignment, each pairing in the alignment (the scores of individual edges) has to be independent of each other, and the total alignment score is obtained by summing the up scores of each pair. Fortunately, such a scoring scheme is biologically plausible and the currently preferred alignment score calculation method. An alignment is accepted to be corresponding to a parsimonious molecular evolution scenario when the conserved sites are aligned together. According to this, conserved residue pairings are rewarded with positive scores and less likely substitutions, insertions and deletions are usually punished with negative scores. In this naive form of scoring alignments, the model of scoring is applied independently to each pairing and summed up together. Therefore the problem reduces to a maximum path finding problem on the alignment graph and it can be solved using dynamic programming.

3 Pairwise Sequence Alignment Algorithms

The application of the optimal path finding algorithm to the sequence alignment problem provides polynomial time solution for finding the optimal alignments with the defined scoring schemes. Various versions of these algorithms have been proposed to date, with different objectives of alignment preferences and performance specifications. Yet, the main idea behind is preserved and it is based on the algorithm proposed by Needleman and Wunsch [4].

Although in its original article it is not referenced that 1970 Needleman–Wunsch algorithm is based on dynamic programming, it is an exact application of the optimum path finding solution we have mentioned. The algorithm states the methodology of finding the similarities between two protein sequences. With appropriate scoring modifications, this can be generalized to DNA/RNA sequences.

Needleman–Wunsch algorithm assigns positive scores to the pairing residues of the same kind and zero to the residues of different kinds. The penalties introduced by gaps depend on the length of a gap. This is because a novel insertion or detection is a significant event and consecutive insertions and deletions are more likely to happen once the initial event occurs. The penalty for a consecutive gap is usually defined as a concave function of the length of the gap [5]. Concave scoring reflects the relatively decreasing importance of a new adjacent gap introduced. As a result, variable length of gap introductions has to be considered individually in the computation of the alignment scores. This corresponds to a modification in the graphical structure in Fig. 1. In the graph, only the transitions of single gaps have directed edges, and they are represented by the horizontal and the vertical edges. The consecutive gaps of length k can be represented as jumping horizontal (vertical) edges between the cells $c(i, j)$ and $c(i + k, j)$ ($c(i, j)$ and $c(i, j + k)$). This modification enables the representation of any alignment as a path and its corresponding score as the additive score of this path, so the equivalence of the optimal path and the sequence alignment problems is preserved. Note that this network structure is in the class of the multilayer networks we have covered, which satisfies the recursion relation in Eq. 5. In this form, the dynamic programming is formulated as follows. The optimal path to the cell $c(i, j)$ is the best path among the combinations of the optimal path from the upper-left cell to the parents of $c(i, j)$ and the edge between the parent and $c(i, j)$. From the alignment graph, we can see that the parents of the cell $c(i, j)$ are the cell $c(i-1, j-1)$ and the cells $c(p, j)$, $p < i$, and $c(i, q)$, $q < j$. The recursive function turns out to be

$$S(i,j) = \max \begin{cases} S(i-1,j-1) + e_{i,j}(i-1,j-1) \\ \max\limits_{p} S(i-p,j) + e_{i,j}(i-p,j) & 0 < p \leq i \\ \max\limits_{q} S(i,j-q) + e_{i,j}(i,j-q) & 0 < q \leq j \end{cases} \quad (6)$$

$S(i, j)$ denotes the maximal path score from the upper-left corner to the cell $c(i, j)$ and $e_{i,j}(x, y)$ is the score of the edge from the parent cell $c(x, y)$ to $c(i, j)$.

A standard approach to calculate the best alignment score is to start the recursion from the upper left corner and calculate $S(i, j)$ by scanning the matrix line by line, i.e., incrementing i and j in a double loop. At the end of $|X| \times |\Upsilon|$ (the length of the sequences) iterations, $S(|X|, |\Upsilon|)$ is calculated as the optimal alignment score. This matrix scan constitutes the first phase of the algorithm. In a second reverse scan, starting from the cell $(|X|, |\Upsilon|)$, the edges leading to the cell $c(i, j)$ are traced back by finding the parent cell using the score change. The reverse-scan procedure records the sequence of the edges in the optimal path and the alignment is recovered from this sequence.

As an example, we can apply Needleman–Wunsch algorithm to the DNA sequences $X = GACAT$ and $\Upsilon = GAGACAT$ with the scoring scheme in which the same residues are awarded with +2 score, different residues are punished by $-$ 0.5 score and the gaps have the penalty function $\{-1, -1.1, -1.11, -1.111, \ldots\}$. Figure 5 shows the resulting cell scores calculated by the procedure, and the corresponding path recovered by tracing back from the lower-right cell.

It is notable that starting the bottom-up recursive algorithm from the top-left corner, or from the bottom-right corner using the dual graph finds the same alignment as the directionality will not alter the problem. In fact in the original Needleman–Wunsch algorithm the recursion was applied in the reverse direction.

More than one optimal alignment might exist as multiple paths on the grid can have the same score. In multiple optimal alignment case, the reverse-scan phase might select one of the paths randomly as it detects multiple parents leading to the same score, or an extra backtracing thread can be generated at each such instance, resulting in exploring all optimal alignments.

Needleman–Wunsch algorithm requires $|X| \times |\Upsilon|$ ($\mathcal{O}(n^2)$) space since it records the optimal alignment score for every cell.

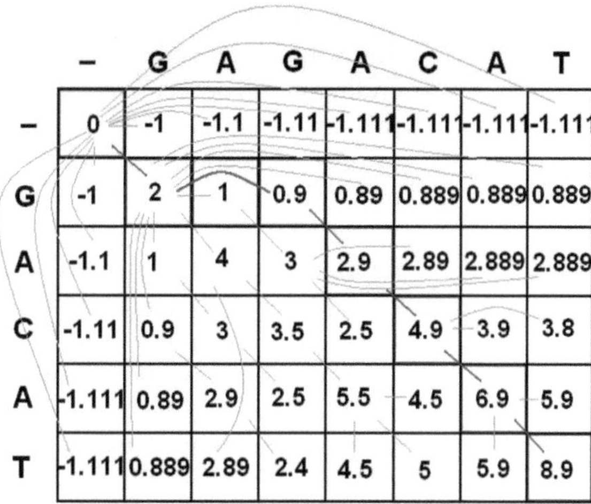

	–	G	A	G	A	C	A	T
–	0	-1	-1.1	-1.11	-1.111	-1.111	-1.111	-1.111
G	-1	2	1	0.9	0.89	0.889	0.889	0.889
A	-1.1	1	4	3	2.9	2.89	2.889	2.889
C	-1.11	0.9	3	3.5	2.5	4.9	3.9	3.8
A	-1.111	0.89	2.9	2.5	5.5	4.5	6.9	5.9
T	-1.111	0.889	2.89	2.4	4.5	5	5.9	8.9

Fig. 5 Needleman–Wunsch algorithm example

The space required for the reverse-scan phase is the same as the length of the alignment that can be neglected. For the computation of $S(i, j)$, the operation number equals the number of the parents of the cell $c(i, j)$, $i + j + 1$. Then the total number of computations is

$$\sum_{i}^{|X|} \sum_{j}^{|\Upsilon|} i + j = 0.5(|X|^2|\Upsilon| + |X||\Upsilon|^2 + 4|X||\Upsilon|) \qquad (7)$$

so that the complexity is $\mathcal{O}(n^3)$. Similar to the space requirement, the number of operations performed at the reverse-scan phase equals the alignment size and it can be included in the cubic complexity. Therefore, the original Needleman–Wunsch algorithm runs in cubic time with quadratic space requirements.

3.1 Extension to Different Alignment Problems

The alignment algorithm discussed addresses the global alignment in which a pair of sequences needs to be aligned entirely from the first residues to the last ones. However in certain cases, the similarity searched between two sequences might not be global. For example, if we are searching for the location of a gene in a genome, the gene sequence has to be aligned globally only in a local region of a genome sequence. Overlapping fragment regions are also the main interest in DNA fragment assembly problems. In both problems, the flanking regions outside the locally aligned parts are not of interest and they should be filled with gaps in the alignment. This problem is referred as the semi-global alignment problem. A variation of the Needleman–Wunsch algorithm can address this problem by behaving liberal on the gaps that are at the beginning or at the end of an alignment. The simple modification is not punishing these gaps by assigning zero scores to the edges at the boundaries of the alignment grid (i.e., $e_{0,p}(0, q) =$

$e_{p,0}(q,0) = e_{|X|,p}(|X|,q) = e_{p,|Y|}(q,|Y|) = 0)$. Because the paths following these paths are not punished, the alignments staring or ending with consecutive gaps are more likely to be found, which is the desired solution in semi-global alignment problems.

An application of the pairwise alignment algorithm is proposed by Waterman and Smith [6] and it is known as the Smith–Waterman algorithm. The main objective is finding the similar regions of two sequences instead of aligning them globally. In this sense, Smith–Waterman algorithm searches for local alignments. Local alignments are crucial in finding the homologous regions between proteins and in classifying newly discovered biological elements. Clearly what we look for in the alignment graph is a high-scoring path section among all available paths. An algorithm searching for the highest scoring local alignment should search for the maximal path that the score difference of the beginning and the end of the segment is the greatest among all segments (i.e., highest scoring subpath). Another requirement for such maximal paths is that a cell visited by the path cannot have a score smaller than the score of the starting cell. This situation would be contradict with the definition of a high-scoring alignment, because the section of the path between the initial cell and the lower scoring cell contributes a negative score. This motivation alters the scoring scheme to a direction that negatively contributing trends are avoided and they should not be included in an optimal path.

To satisfy this property, the optimal alignment scores are computed using the same recursive procedure of the Needleman--Wunsch algorithm is applied with a modification. During the recursive computation, when the score of a path becomes negative, the cell associated is set to zero score, to exclude it from optimal path candidates. In this case, starting from zero-scored cells, an optimal alignment will attain a high score and the path section between the maximum score and the zero-scored cell will be the best-conserved subsequences of the pair of input sequences. The scoring function is simply modified to

$$S(i,j) = \max \begin{cases} 0 \\ S(i-1,j-1) + e_{i,j}(i-1,j-1) \\ \max_p S(i-p,j) + e_{i,j}(i-p,j) & 0 < p \leq i \\ \max_q S(i,j-q) + e_{i,j}(i,j-q) & 0 < q \leq j \end{cases} \tag{8}$$

and the forward-scan phase is the same as the Needleman–Wunsch algorithm. In the reverse-scan phase we look for the best scoring path section. This is found by locating the highest scoring cell and tracing back to a zero-scoring cell. In Fig. 6, we can see the resulting local alignment when the modified algorithm is run on the same example used in the Needleman–Wunsch algorithm.

	-	G	A	G	A	C	A	T
-	0	0	0	0	0	0	0	0
G	0	2	1	2	1	0.9	0.89	0.889
A	0	1	4	3	4	3	2.9	2.89
C	0	0.9	3	3.5	3	6	5	4.9
A	0	0.89	2.9	2.5	5.5	5	8	7
T	0	0.889	2.89	2.4	4.5	5	7	10

Fig. 6 Smith–Waterman algorithm example

Another modification on the dynamic programming algorithm addresses the problem of finding all near-optimal alignments. The optimal global alignment provided by the Needleman–Wunsch algorithm is sensitive on the scoring function defined, and it might not reflect a biologically meaningful alignment perfectly. Another alignment with a slightly lower score might be more interesting; however, it is not found by the original algorithm. Waterman [7] suggested an algorithm to find not just the optimal alignment but all of the near-optimal alignments performing at the same space and time complexity with the Needleman–Wunsch algorithm.

The near-optimal alignment finding algorithm is based on one observation on the problem separability of alignments as shown in Eq. 4. The same optimal path and score is found whether the recursion is performed from top-left to bottom-right or vice versa (i.e., directionality does not change the solution). At the first phase, two score matrices are generated: one computing the recursion in forward direction (i.e., top-left to bottom-right) and one in the reverse direction. For the forward computation the $S_F(i, j)$ has the optimal path score from the top-left cell to the cell $c(i, j)$, and $S_R(i, j)$ has the optimal path score from the bottom-right cell to the cell $c(i, j)$. Assume a path attains score s. For any edge connecting cell $c(i, j)$ and its parent (p, q) in this path, $S_F(p, q) + e_{i,j}(p, q) + S_R(i, j) = s$, according to Eq. 4. Using this observation, given an alignment score s, starting from the bottom-right (or top-left) corner, a path with the score s can be traced back. The proposed algorithm performs a reverse-scan phase, by adding the detected edges iteratively to a list. When multiple edges are found in an iteration, multiple paths are traced separately to find every satisfying alignments. The original algorithm sets s to a range of values close to the optimal score, in order to find all near-optimal alignments.

3.2 Algorithmic Improvements on the Dynamic Programming Methods

The dynamic programming scheme and its modifications we have introduced to solve global, semi-global, local and near-optimal sequence alignments require quadratic space and cubic time as a function of sequence length. It has been shown that with an appropriate change in the alignment scoring method the complexity can be reduced to quadratic time. Additional to this improvement, the problem can be solved exactly in linear space. Such a space-time reduction is a very significant gain in practical terms. Considering the length of the biological sequences, the improvement means about 100–1,000 folds speed ups, and the same rate of less memory usage for pairwise alignments. The gain is more dramatic for multiple sequence alignment programs that could range between tens of thousands to million-fold. Many modern sequence alignment programs use the modified version of the algorithm based on the same improvement principles.

Here, we will briefly see two fundamental performance improvement modifications, one reducing the time complexity from cubic to quadratic and the other providing linear-space solutions. Gotoh showed that changing the gap penalty rules to affine gap penalties, the score computation can be performed in $\mathcal{O}(n^2)$ [8]. Later, by incorporating Hirschberg's method [9] to the sequence alignment algorithm, it was shown that saving the full matrix of optimal path scores is not necessary in order to perform the same computation.

3.2.1 Quadratic-Time Alignment by Affine Gap Penalties

The concave gap penalty functions used in the original Needleman–Wunsch algorithm assign an individual score to each consecutive gap of different length. Therefore a cell $c(i, j)$ has $i + j$ parents and the score of each one has to be evaluated and considered in computing $S(i, j)$ (e.g., Eq. 6). This concave function can be defined in the form of a piecewise linear function $g(k) = a + b(k-1)$, $0 < b < a$, where a is the penalty for an introduction of new gap and b is the penalty for each residue extension of a gap. In this form of constant score increase for consecutive gaps, the total gap penalty can be computed recursively instead of performing a gap score computation for each parent. Consider the second and the third computation lines in the recursive function Eq. 6. The gap penalty scores $e_{i,j}(i-p, j) = e_{i,j}(i, j-p) = g(p) = a + b(p-1)$ by definition. The scores of the vertical and the horizontal edges are then

$$V(i,j) = \max_p S(i-p,j) + g(p), \quad 0 < p \leq i \tag{9}$$

$$H(i,j) = \max_q S(i,j-q) + g(q), \quad 0 < q \leq j. \tag{10}$$

The vertical score can be expressed in recursive form as

$$\begin{aligned} V(i,j) &= \max_{p} S(i-p,j) + g(p), \quad 0 < p \le i \\ &= \max\{S(i-1,j) + a, \max_{p} S(i-p,j) + g(p)\}, \quad 1 < p \le i \\ &= \max\{S(i-1,j) + a, \max_{p'} S(i-1-p',j) + g(p'+1)\}, \\ &\quad 0 < p' \le i-1,\ p' = p-1 \\ &= \max\{S(i-1,j) + a, \max_{p'} S(i-1-p',j) + g(p') + b\}, \\ &\quad 0 < p' \le i-1,\ p' = p-1 \\ &= \max\{S(i-1,j) + a, V(i-1,j) + b\}, \quad 0 < p' \le i-1. \end{aligned} \tag{11}$$

The same recursion applies for $H(i, j)$. The new recursion relation turns out to be

$$\begin{aligned} S(i,j) &= \max\{S(i-1,j-1) + e_{i,j}(i-1,j-1), V(i,j), H(i,j)\} \\ V(i,j) &= \max\{S(i-1,j) + a, V(i-1,j) + b\} \\ H(i,j) &= \max\{S(i,j-1) + a, H(i,j-1) + b\}. \end{aligned} \tag{12}$$

The modified recursive function implies that if we save the three scoring matrices for S, V, and H scores, only five score evaluations have to be performed for each cell resulting in around $5\ |X|\ |\Upsilon|$ calculations, which has the time complexity of $\mathcal{O}(n^2)$.

3.2.2 Sequence Alignment in Linear-Space

As we can see from the recursive function equation 12 to be used in dynamic programming, only the values in cells $c(i-1, j-1)$, $c(i-1, j)$, and $c(i, j-1)$ are called from the memory for calculating the optimal path score of the cell $c(i, j)$. In this case, recording all $\mathcal{O}(n^2)$ scores is not necessary for computing the optimal alignment score. For example, when the scores are computed by scanning each line starting from the top-left, the active memory can only hold the remaining required values of the previous line (i.e., scores of cells $c(i-1, p)$, $j-1 \le p \le |\Upsilon|$), and the values computed in the current line (i.e., scores of cells (i, q), $0 \le q \le j-1$). In this fashion, usage of memory equal to the length of one of the sequences is sufficient to compute $S(i, j)$ in each iteration. However, the scores computed for the optimal subpaths are lost and the trace back phase cannot be conducted to recover an optimal path. Hirschberg's algorithm offers a recursive procedure to recover the optimal path for finding the longest common subsequence of two strings. This problem is solved by the same dynamic programming algorithm we cover, so Hirschberg's algorithm also applies to sequence alignment [10].

Hirschberg's algorithm searches for an intermediate edge in an optimal path. Once the edge is found, the optimal path problem is

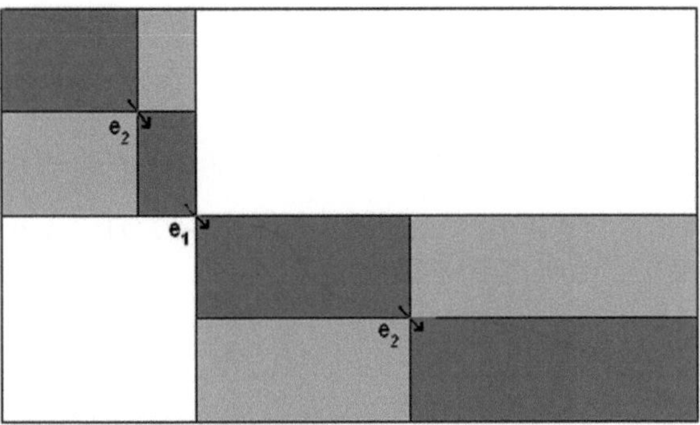

Fig. 7 Subproblem division in Hirschberg's algorithm

divided into subproblems as the search for the optimal path section before this edge and the search for the section after it. Recursive repetition of the same procedure for the intermediate edges in the subsections is performed until the all intermediate edges are found. A typical application is as follows. Assume the procedure attempting to find the edge contained in the optimal path, where the path is crossing the median row of the alignment grid. To search for this edge, similar forward and backward score computations are performed as in finding the near-optimal paths described previously. In this case the $S_F(i, j)$, $0 \leq i < \lfloor |X| /2 \rfloor$, $0 \leq j \leq |\Upsilon|$, and $S_R(i, j)$, $\lfloor |X|/2 \rfloor \leq i \leq |X|$, $0 \leq j \leq |\Upsilon|$ are computed. The maximal path score is found by looking for the highest score $S_F(\lfloor |X|/2 \rfloor - 1, q) + e_{\lfloor |X|/2 \rfloor, j}(\lfloor |X|/2 \rfloor, q) + S_R(\lfloor |X|/2 \rfloor, j)$ and the corresponding edge belongs to the optimal path. The same procedure is employed for the median of the two subproblems (i.e., to find the optimal path edges crossing rows $\lfloor |X|/4 \rfloor$, and $\lfloor 3\ |X|/4 \rfloor$). All non-horizontal paths are discovered following this binary fashion, and finally the optimal path is completed by remaining horizontal edges. At kth recursion, 2^k subproblems are defined. From Fig. 7, it can be seen that each subproblem has half number of rows (and columns) of the previous one (i.e., the upper-left submatrix and the lower-right submatrix have half of the cells contained in the current matrix). Due to quadratic complexity, a subproblem requires $(2^k)^2$ fold less computation than the original problem. Therefore, the number of total computations required in the recursive procedure is at most

$$\begin{aligned} \mathbb{S}_{\text{Hirschberg}} &= \mathbb{S} + 2\frac{\mathbb{S}}{4} + \cdots + 2^k \frac{\mathbb{S}}{4^k} + \cdots \quad 0 \leq k \\ &= 2\mathbb{S}, \end{aligned} \tag{13}$$

where $\mathbb{S}$ is the amount computation required for the original problem. While the computation is multiplied by a constant factor, and the algorithm still has quadratic complexity, only the scores of the median lines are required to find the median edges. As we know that the scores can be computed in sequence-length-size memory, Hirschberg's algorithm can be run in linear space.

At the implementation level, different versions of the linear-space algorithm are available [11, 12]. As a variation, the median edge can be found by only using a forward scan of score computation [13]. In this version, the forward score computation continues when the median line is reached. For each cell in the $\lfloor |X|/2 \rfloor + 1$th line, $S_F(\lfloor |X|/2 \rfloor + 1, j)$ are computed and the median edges resulting in each one of these scores are recorded. Clearly an additional sequence-length memory space is needed for this ancestral records. At iteration k, for each cell in the $(\lfloor |X|/2 \rfloor + k)$th line, the cell visited by the optimal path in the previous line is determined by the score computation. Therefore, the median edge for the current cell can be updated at each iteration. At the final iteration, the median edge assigned to the lower-right cell is determined to be the median edge belonging to the solution path.

The linear-space algorithm using median edge candidate records provides a space-time trade-off flexibility. The algorithm can be generalized to m splits of subproblems instead of binary splits at each recursion. Clearly, keeping the records of the edges crossing the $\lfloor |X|/m \rfloor$th line, we can detect the edge belonging to the optimal path at that portion. In a forward score computation pass, we can keep records of such edges for every $\lfloor |X|/m \rfloor$ lines $(2 \leq m \leq |X|)$. This would require m sequence size memory spaces. At each iteration m mid-edges are discovered, and m subproblems are defined. For iteration k, m^k subproblems are defined and each problem requires $(m^k)^2$ fold less computation than the original problem. In this case, the amount of computation is $\mathbb{S}_{\text{Hirschberg}} = \sum_{k=0}^{\infty} \mathbb{S}/3^k = \mathbb{S} + \mathbb{S}/(k-1)$. This space-time trade-off might be important in practical applications.

As we have seen the $\mathcal{O}(n^2)$ time and $\mathcal{O}(n)$ space algorithms are available by affine gap penalties and Hirschberg's algorithm for global alignments, the semi-global alignment extension is straightforward. Initializing the first row and columns of S, V, and H matrices and defining $g(p) = 0$ for the last rows and columns is sufficient. On the other hand, additional requirements are needed to perform local alignments in linear time [14, 15]. For calculating the scores, the same recursive function equation 12 with the Smith–Waterman modification is used. Here the scoring modification is setting the negative scores to zero. The linear-space scoring memory is the same as the global alignment case, and the main difference is a root cell is recorded for computed cell $c(i, j)$. Recall that the Smith–Waterman algorithm terminates a local alignment when a zero-scored ending cell is met during the backward scan phase. The root of $c(i, j)$ is simply the zero-scored starting point of the optimal path passing through $c(i, j)$. The root is empty when $S(i, j) = 0$ and it is initialized when a cell attains a positive score with an edge coming from a zero-scored parent. This root is updated at each iteration with the similar edge-updating procedure in the global alignment algorithm. The current highest scoring cell and its root is saved during the process. Once the forward score computation is finished, the highest scoring cell and its root is already known. As the second phase, calling the global alignment procedure with Hirschberg's algorithm assigning the root as the top-left corner and the highest scoring cell as the bottom-right corner is sufficient to find the local alignment. This linear-space local alignment algorithm runs in $\mathbb{S} + 2\ell \leq 3\mathbb{S}$ time where ℓ is the time required for aligning the local segment by the quadratic space Needleman–Wunsch algorithm. More sophisticated strategies based on Hirschberg's divide-and-conquer principle have also been applied to the problem of finding near-optimal alignments [16].

3.3 Heuristical Improvements on Sequence Alignment

The quadratic-time solution offered by dynamic programming is the optimal solution to the sequence alignment problem when an exact solution is required. As a result, the entire set of alignment candidates are searched in a smart way. Several tasks of computational biology require alignment of high number of sequences or of relatively long sequences as a natural result of the length of the molecular sequences found in life. The required computational burden is high when exact dynamic programming solutions are considered. For that reason, researches have proposed fast approximate solutions with lower costs. The most popular approach to an approximate solution is defining heuristics to constrain the search space and employing alignment algorithms in a reduced search space. One might think that skipping potential alignment candidates would bring a great statistical risk of finding solutions far away from a set of desired solutions. While this is true, in a sense that the optimal solution cannot be guaranteed with approximating heuristics, in practice such approaches are quite successful. Usually biological sequences are not random with a great statistical complexity, and they are expected to share significant homologies with other sequences. Therefore, some regions of the dynamic programming matrices are more likely to contain a near-optimal solution. As an example, for two highly similar sequences, as they share large regions of similarity, not a large number of gaps are expected in total. Because of that, the optimal path cannot visit the cells too far away from the diagonal. In case, the dynamic programming graph can be reduced to a smaller subgraph. Here we are going to introduce some heuristics approaches applied on dynamic programming in order to achieve fast sequence alignments.

3.3.1 Banded Dynamic Programming

Banded dynamic programming is originated from the idea that if two aligned sequences are similar, the number of insertions and deletions is below some upper bound. In the alignment graph, this situation corresponds to a diagonal band centered around the diagonal from the source to the target cell. The width of this diagonal band is determined by the tolerance of the number of insertions and deletions allowed. Obviously, the number of cell scores to be computed is proportional to this bandwidth as well as the sequence length. A narrow bandwidth offers a quick solution, but it may miss the higher scoring paths beyond the search band.

Combining banded dynamic programming solutions with other variations such as Hirschberg's algorithm is applicable and used in some early sequence alignment programs [16].

3.3.2 Bounded Dynamic Programming

Selecting a subset from the alignment grid using a constant band is simple but it has some serious drawbacks. An optimal alignment might follow a path off the banded coverage and there is no effective way of mitigating such a problem for a static method. Adapting the search space by including potentially optimal paths in the search space is capable of handling a more general class of alignment problems. Instead of limiting the search space using spatial bounds, alignment cell scores determine the boundaries. The potential search space is explored by enabling the children of cells with scores higher than an upper bound, and this methodology is generally called as bounded dynamic programming [17].

Assume we compute the score $S(i, j)$. For a given threshold τ, the children of $c(i, j)$ are included in the search space if $S(i, j) + P(i, j) > \tau$. Here, $P(i, j)$ is an estimation of the optimal path score from $c(i, j)$ to the ending corner. In the ideal case, if we had the exact values of $P(i, j)$, the search space would be greatly reduced which only consist of the near-optimal paths with scores greater than τ. Good approximations of the matrix P and the threshold value can provide decent reduction in the space and time complexity. The approximation of the remaining path score is a function of the distance of $c(i, j)$ to the lower-right corner. As an example, we can choose an upper approximation with the scenario that the path returns to the diagonal (i.e., gaps) and travels to the destination on the diagonal (i.e., matches). The corresponding score is

$$P(i,j) = |(|X| - i) - (|Y| - j)| \times \textit{gap penalty} + \min\{(|X| - i), (|Y| - j)\} \times \textit{match score}. \quad (14)$$

In Fig. 8, the alignment matrix of human and gorilla mitochondrial genomes are shown. The red line shows the paths with the optimal alignment, where the black region is the section selected by the thresholding described. Bounded dynamic programming in this example finds the exact solution with 60 % less number of cell computations (i.e., 2.5 times faster).

Determining the bounding threshold τ is difficult, and the complexity of the algorithm is sensitive on this threshold. A large threshold includes large number of cells in the search space, while a low threshold prunes large number of cells and it cannot return a

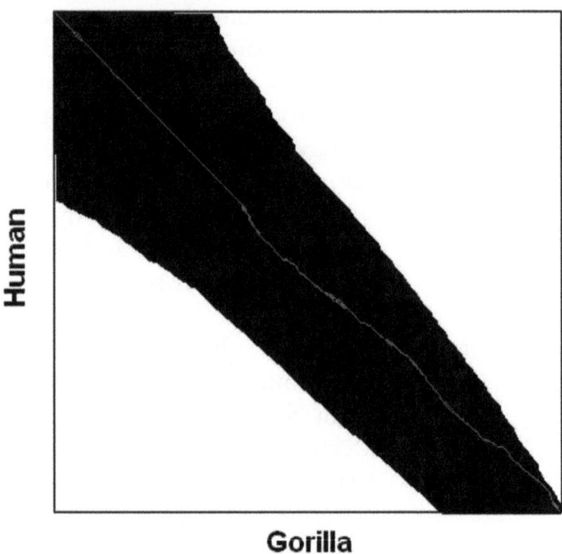

Fig. 8 Alignment matrix reduction by bounded dynamic programming (Color figure online)

solution. Iterative deepening procedures employ naive searching for threshold [18]. Iterations are repeated until the problem is solved with increasing τ starting from low values. Furthermore, regressing τ after a few iterations using the alignment depth reached at each step is known to be a more promising than the naive approach [19].

3.3.3 Seeding Dynamic Programming Using Conserved Fragments

According to the problem definition of sequence alignment, given M edges of the optimal alignment path, the problem can be split into $M + 1$ subproblems. Perhaps, the most popular modern heuristics is to estimate these split points by anchoring conserved fragments. A conserved fragment of length k indicates consecutive diagonal edges in the alignment grid. Dynamic programming is seeded by locating these diagonal sections as a preprocess. Employing the alignment procedures between the seed candidates restricts the search space substantially. For example in an ideal scenario, conserved regions are distributed uniformly throughout the sequences resulting in average distances of adjacent seeds ℓ. The overall complexity is around $\mathcal{O}(\ell^2 M)$ where ℓ is constant and M increases linearly with the sequence length. Of course, heuristics leading to this great complexity reduction comes with the cost of the difficulty in determining good seeds. To date, program series FASTA [20, 21] and BLAST [22, 23] have proposed the most accepted methods on seeded dynamic programming.

FASTA programs search for high-scoring ungapped segments of a fixed length. This is achieved by listing all 4–6 nucleotide (for DNA/RNA) and 1–2 amino acid (for protein) subsequences for both sequences in the pair. The matching subsequences are determined and located in the alignment grid. Finding matching pairs can be implemented efficiently using indexing techniques like hash-tables, so that it can be completed in linear time. Subsequences of selected sizes are likely to occur in several alignments other than the optimal ones. Therefore a filtering phase to find the significant hotspots is followed. It is assumed that if selected substring matchings are placed on the same diagonal, and they are close to each other this cluster is likely to be on an optimal path. Since the cells $c(i, j)$ located on the same diagonal line have the same $i - j$ value, they can easily be grouped together and the clusters within a close neighborhood are selected as the seeds. FASTA picks ten top scoring clusters. The best scoring cluster is selected as the center of an alignment band, and banded dynamic programming is employed to connect the seeds.

Using short subsequence matches as in the case of FASTA program is sensitive, but this approach usually generates high number of regions which are not a part of an optimal path. BLAST programs, on the other hand, use longer fragments (default 3 residues for proteins, 11 nucleotides for DNA/RNA). Statistical significance of each hit is determined by a statistical model using the random occurrence of a seed candidate, where the random occurrence is estimated from the background frequencies of the residues. The statistical modelling provides confidence scores for each alignment found, that is useful since the alignments are approximate solutions. The seeds are extended in each direction to find local alignments. A simple bounded dynamic program is recruited for the gapped extension. The algorithm searches for the optimal paths not dropping below a predetermined threshold score at some iteration.

Seeding operation can be improved to an adaptive procedure by using variable length subsequences. The hash-table data structures indexing the fixed-length subsequences as in the case of FASTA and BLAST class of programs are replaced by suffix-based structures such as suffix array and FM-index in variable length seeding [24]. The seed length is extended iteratively until a certain number of sequences below a threshold score are left. The dynamic programming procedure followed after the seeding phase is similar to BLAST programs.

4 Dynamic Programming and Multiple Sequence Alignment

Pairwise sequence alignment techniques can be generalized to multiple sequence alignment in a straightforward way due to the multiple alignment scoring scheme called the sum-of-pairs scoring. In sum-of-pairs, every pair of residues aligned is evaluated. According to this, the scores every residues pair (or gaps) in an alignment column are added together. Because of the linearity, the total score turns out to be

$$\mathrm{SP}(\mathcal{A}) = \sum_{i,j} s_{\mathcal{A}}(X_i, X_j), \quad 1 \le i < j \le M, \tag{15}$$

where $s_{\mathcal{A}}(X_i, X_j)$ is the pairwise score of the sequences X_i and X_j in an alignment $\mathcal{A}$, and M is the number of sequences to be aligned. Defining $c(k_1, k_2, \ldots, k_M)$ as a cell in a multidimensional grid, where each coordinate of the grid corresponds to a sequence residue, the alignment grid analogy can be extended to M dimensions. Similarly, a path from the smallest indexed corner to the ending corner has one-to-one correspondence with an alignment $\mathcal{A}$. The linear scoring scheme preserves the recursive scoring function definition (i.e., the scores of the paths can be calculated by adding the edge scores up linearly). Therefore, the multilayered network structure in Fig. 3 is preserved and the same dynamic programming model defined applies.

In case of multiple sequence dynamic programming, there are $\Pi_i^M |X_i|$ cells to be evaluated. This implies the computational complexity $\mathcal{O}(n^M)$ which is exponential with the number of sequences. The same requirement of space can be reduced to $\mathcal{O}(n^{M-1})$ by Hirschberg's algorithm. Unfortunately, exact solution of optimal alignment using dynamic programming is not practical considering the length and number of the biological sequences to be aligned in a task and the resources of modern computers. Heuristics based on bounded dynamic programming, divide-and-conquer methods, and seeded dynamic programming have been proposed in early days of computational biology. However, the success of these approximate solutions was reported to be limited to a few sequences. Starting from late 1980s, greedy methods such as progressive alignment gained popularity over the multidimensional dynamic programming solutions. We briefly mention the general methodology of extending the heuristics on multidimensional case and the use of pairwise alignment in popular methods.

4.1 Bounded Dynamic Programming in Multiple Alignment

Limiting the search space to potentially important cells of the multidimensional grid is the main approach to mitigate the problem of exponential complexity [25]. Bounded dynamic programming is a methodology to serve this purpose. The approximation of the potential function has better alternatives than Eq. 14 as pairwise alignment scores are available. Assume the score of a cell $c(k_1, k_2, \ldots, k_M)$ is computed, and we need to approximate the score of the optimal path from this cell to the target cell. The sum of pairwise alignments of the remaining sequence suffixes can be used as an approximation of $P(k_1, k_2, \ldots, k_M)$. The alignment scores of all sequence suffixes have to be recorded for this computation. This basically done by backward calculation of the optimal path scores from the target to the current coordinate pair.

$M(M+1)/2$ matrices holding $S_R(i, j)$ scores is called to sum up the pairwise scores for the coordinate pairs of $c(k_1, k_2, \ldots, k_M)$. This approximation can be extended from pair scores to score combinations of multiple alignment N-tuples ($N < M$). While estimating $P(k_1, k_2, \ldots, k_M)$ from multiple alignment scores is a better approximation, calculating multiple alignment combinations of N sequences might be costly. Nevertheless, using triplet scores is reported to be more efficient than the using pairwise alignments when the overall procedure is considered for a few sequences [26]. The threshold determining the search subset can be done by iterative deepening as in the case of pairwise algorithm, or a fast heuristics approximating the multiple alignment (e.g., progressive alignment) can be used [27].

Further reduction in space complexity is achieved by combining bounded dynamic programming with Hirschberg's insight. The corresponding method is usually referred to as the divide-and-conquer frontier search [26]. The frontier is the active region of the search space that is used in the computations. Since the grandparents of the cells are not used in the score calculation, a cell is not kept in the memory after the scores of its children are computed. Obviously, the memory to be used in the backtrace phase is lost. Hirschberg's algorithm keeps the ancestors of each active cell located in a selected median hyperplane of the grid. The ancestor list is updated at each cell computation. Finally, the ancestor of the target cell is determined to be the point where the optimal path crosses the hyperplane. The problem is then split into two subproblems as in the pairwise application, and the optimal path is found recursively.

4.2 Divide-and-Conquer Methods in Multiple Sequence Alignment

Divide-and-conquer methods have the similar subproblem splitting insight with Hirschberg's algorithm, but they differ in how the split points are found. The sequences are parsed with fast methods avoiding dynamic programming in multidimensional grid. Once a number of subproblems are defined, each problem is solved by a multiple sequence alignment algorithm of choice.

Some methods determine the binary split points directly using pairwise alignments [28]. Combining the forward and the backward alignment matrices S_F, and S_R the optimal split of the prefix of the suffix of a sequence be found as in pairwise Hirschberg's algorithm. An approximate cut point is found for each sequence by summing up the score matrices for each pairwise alignment associated. The overall maximum score determines an average split point. This procedure is repeated a few times and the corresponding subproblems are solved conventionally. The final multiple sequence alignment is simply the concatenation of all subsolutions.

4.3 Use of Seeds in Multiple Sequence Alignment

Seeding the dynamic program using consecutive regions of conserved subsequences is directly generalized to multiple sequence alignment problems. In the general form, fixed length of subsequences shared by all sequences are used as seed candidates and significant clusters are selected as seeds [29]. Bounded dynamic programming algorithms are employed to bridge the seeds. Greedy versions of seeding by utilizing the conserved subsequences of sequence pairs and assembling the alignment by combining pairs progressively [30] can also be accepted as in this class multiple alignment algorithms.

4.4 Dynamic Programming and Progressive Sequence Alignment

When the nature of biological sequence alignment problem is investigated, a tree-based alignment model appears to be realistic as similar sequences are expected to evolve from a close common ancestor. In such an approach a divergence from dynamic programming in multidimensional grids is appropriate because sum-of-pairs scoring scheme assigns equivalent importance to each sequence, disregarding their evolutionary distances. The popularity of tree-based alignment algorithms stems from this fact, as well as requiring much less computations.

Progressive alignment programs generally have two sequential steps. First, an evolutionary tree is estimated. Second, consensus sequences are generated and updated iteratively using pairwise alignments. Pairwise sequence alignment is the core procedure employed in progressive alignment.

While pairwise alignment is the major part of the second phase, it can also be employed in estimating the evolutionary tree. For example, CLUSTAL series of programs [31] use linear-space dynamic programming for building the similarity matrix to be used in building the guide tree. Dynamic programming schemes we have discussed are the progressive alignment algorithms employed in successful multiple alignment programs such as CLUSTALW, T-Coffee [32], MUSCLE [33], MAFFT [34], and GramAlign [35].

References

1. Waterman MS (1995) Introduction to computational biology. Chapman & Hall, London
2. Bellman R (1958) On a routing problem. Q Appl Math 16:87–90
3. Fredman ML, Tarjan RE (1987) Fibonacci heaps and their uses in improved network optimization algorithms. J Assoc Comput Mach 3:596–615
4. Needleman SB, Wunsch CD (1970) A general method applicable to the search for similarities in the amino acid sequences of two proteins. J Mol Biol 48:443–453
5. Waterman MS, Smith TF, Beyer WA (1976) Some biological sequence metrics. Adv Math 20:367–387
6. Waterman MS, Smith TF (1981) Identification of common molecular subsequences. J Mol Biol 147:195–197
7. Waterman MS (1981) Sequence alignments in the neighborhood of the optimum with general application to dynamic programming. Proc Natl Acad Sci USA 80:3123–3124
8. Gotoh O (1982) An improved algorithm for matching biological sequences. J Mol Biol 162:705–708

9. Hirschberg DS (1975) Linear space algorithm for computing longest common subsequences. Commun Assoc Comput Mach 18:341–343
10. Myers E, Miller W (1988) Optimal alignments in linear space. CABIOS 4:11–17
11. Chao KM, Pearson WR, Miller W (1992) Aligning two sequences within a specified diagonal band. Comput Appl Biosci 8(5):481–487
12. Chao KM, Hardison RC, Miller W (1993) Constrained sequence alignment. Comput Appl Biosci 55(3):503–524
13. Hirschberg DS (1997) Serial computations of Levenshtein distances. In: Pattern matching algorithms. Oxford University Press, New York, pp 123–141
14. Huang X, Hardison RC, Miller W (1990) A space-efficient algorithm for local similarities. Comput Appl Biosci 6(4):373–381
15. Huang XQ, Miller W (1991) A time-efficient, linear-space local similarity algorithm. Adv Appl Math 12(3):337–357
16. Chao KM, Hardison RC, Miller W (1994) Recent developments in linear-space alignment methods: a survey. J Comput Biol 1:271–291
17. Spouge JL (1989) Speeding up dynamic programming algorithms for finding optimal lattice paths. SIAM J Appl Math 49 (5):1552–1566
18. Korf RE (1985) Depth-first iterative-deepening: an optimal admissible tree search. Artif Intell 27(1):97–109
19. Ukkonen E (1985) Algorithms for approximate string matching. Inf Control 64:100–118
20. Lipman DJ, Pearson WR (1985) Rapid and sensitive protein similarity searches. Science 227(4693):1435–1441
21. Pearson WR, Lipman DJ (1988) Improved tools for biological sequence comparison. Proc Natl Acad Sci USA 85(8):2444–2448
22. Altschul S, Gish W, Miller W, Myers E, Lipman DJ (1990) Basic local alignment search tool. J Mol Biol 215(3):403–410
23. Altschul S, Madden TL, Schaffer AA, Zhang J, Zhang Z, Miller W, Lipman DJ (1997) Gapped blast and psi-blast: a new generation of protein database search programs. Nucleic Acids Res 25:3389–3402
24. Kielbasa SM, Wan R, Sato K, Horton P, Frith MC (2011) Adaptive seeds tame genomic sequence comparison. Genome Res 21(3):487–493
25. Hohwald H, Thayer I, Korf RE (2003) Comparing best-first search and dynamic programming for optimal multiple sequence alignment. In: Proceedings of the 18th international joint conference on artificial intelligence, IJCAI'03. Morgan Kaufmann Publishers, San Francisco, pp 1239–1245
26. Korf RE, Zhan W (2000) Divide-and-conquer frontier search applied to optimal sequence alignment. In: Proceedings of the 7th conference on artificial intelligence (AAAI-00) and of the 12th conference on innovative applications of artificial intelligence (IAAI-00). AAAI Press, Cambridge, MA, pp 910–916
27. Lipman DJ, Altschul SF, Kececioglu JD (1989) A tool for multiple sequence alignment. Proc Natl Acad Sci USA 86:4412–4415
28. Stoye J, Perrey SW, Dress AWM (1997) Improving the divide- and-conquer approach to sum-of-pairs multiple sequence alignment. Appl Math Lett 10(2):67–73
29. Sobel E, Martinez HM (1986) A multiple sequence alignment program. Nucleic Acids Res 14(1):363–374
30. Morgenstern B, Frech K, Dress A, Werner T (1998) Dialign: finding local similarities by multiple sequence alignment. Bioinformatics 14(3):290–294
31. Thompson JD, Higgins DG, Gibson TJ (1994) Clustal w: improving the sensitivity of progressive multiple sequence alignment through sequence weighting, position-specific gap penalties and weight matrix choice. Nucleic Acids Res 22:4673–4680
32. Notredame C, Higgins DG, Heringa J (2000) T-coffee: a novel method for fast and accurate multiple sequence alignment. J Mol Biol 302:205–217
33. Edgar RC (2004) Muscle: multiple sequence alignment with high accuracy and high throughput. Nucleic Acids Res 32 (5):1792–1797
34. Katoh K, Misawa K, Kuma K, Miyata T (2002) MAFFT: a novel method for rapid multiple sequence alignment based on fast Fourier transform. Nucleic Acids Res 30:3059–3066
35. Russell DJ, Otu HH, and Sayood K (2008) Grammar-based distance in progressive multiple sequence alignment. BMC Bioinformatics, 9:306

Chapter 2

Heuristic Alignment Methods

Osamu Gotoh

Abstract

Computation of multiple sequence alignment (MSA) is usually formulated as a combinatory optimization problem of an objective function. Solving the problem for virtually all sensible objective functions is known to be NP-complete implying that some heuristics must be adopted. Several general strategies have been proven effective to obtain accurate MSAs in reasonable computational costs. This chapter is devoted to a brief summary of most successful heuristic approaches.

Key words Multiple sequence alignment, Progressive method, Iterative refinement, Consistency transformation, Maximal expected accuracy, Anchor, Homology extension

1 Introduction

Pairwise sequence alignment can be solved exactly by the well-known dynamic programing algorithm. In theory, the same algorithm is applicable to alignment of $N > 2$ sequences of average length of L. However, the computational time grows in at least $O(2^N L^N)$, so that $N = 3$ is the practical limit by this brute-force approach. Although some elaborate algorithms have been devised [1, 2], they can hardly obtain the optimal solution when $N > 10$ even for short sequences. The vastly common heuristics employed for MSA are progressive methods. Progressive alignment is not only used to produce preliminary MSA to be refined by the succeeding iterative procedure but also used at the MSA construction phase after consistency transformation applied to residue pairs to be aligned. Rapid identification of almost certainly aligned regions (anchors) prior to the main MSA procedure often greatly accelerates overall computation. Anchoring is an inevitable step in large-scale genomic sequence alignment and also in non-progressive greedy MSA approaches. Thus, the progressive method, iterative refinement, consistency transformation, and anchoring form the major backbone of heuristic MSA algorithms. Additional information, such as inclusion of extra homologous sequences (homology

David J. Russell (ed.), *Multiple Sequence Alignment Methods*, Methods in Molecular Biology, vol. 1079,
DOI 10.1007/978-1-62703-646-7_2, © Springer Science+Business Media, LLC 2014

extension), predicted or observed higher-order structures, constraints on conserved linear motifs, may also improve the quality of MSA.

2 Methods

2.1 Pairwise Alignment

Alignment of two sequences, $\mathbf{a} = a_1...a_m$ and $\mathbf{b} = b_1...b_n$, taking account of the mutational events of substitutions, deletions, and insertions is the fundamental procedure in any biological sequence alignment problems. In most situations, deletions in one sequence cannot be distinguished from insertions in the other, and hence they are collectively called indels or gaps. Note that a gap means a contiguous run of blanks or nulls indicating the absence of a residue. Given an alignment of two sequences, we can easily calculate the alignment score $H(\mathbf{a}, \mathbf{b})$ by adding the substitution scores $S(a_i, b_j)$ for matched pairs $a_i \sim b_j$ (a matched pair in an alignment will be indicated by the '~' symbol hereafter) and the gap penalties $g(k)$ for individual indels of length k. Although many amino-acid substitution matrices $S(a, b)$ have been published [3], most of them follow the log-odds scoring scheme first proposed by Dayhoff et al. [4]. A few nucleotide substitution matrices reflecting biased G + C composition and unequal rates of transitions and transversions are also available [5, 6].

Of the exponentially many possible alignments, we want to find the one that has the maximal $H(\mathbf{a}, \mathbf{b})$ (*see* **Note 1**). In 1970, Needleman and Wunsch [7] first invented an efficient algorithm categorized in the dynamic programming paradigm. Needleman and Wunsch considered the case of semi-global alignment (*see* **Note 2**), and $g(k) = -v =$ constant with the computational complexity of $O(L^3 = m^2n)$. Later, Sellers [8] provided mathematically more rigorous formulation in the case of global alignment and "proportional gap penalty," $g(k) = -uk$ (u is a constant), with the computational complexity of $O(L^2 = mn)$. Waterman et al. [9] generalized the Sellers's formulation to any functional forms of $g(k)$, but the computational complexity went back to $O(L^3)$. Gotoh [10] showed that the Waterman et al.'s algorithm is solved in $O(L^2)$ for an "affine gap penalty," $g(k) = -(v + uk)$, where u and v are non-negative constants. The constant term v realizes the fact that creation of a new gap is more difficult than extension of an existing gap, and v and u are often called "gap open penalty" and "gap extension penalty," respectively (*see* **Note 3**). The Gotoh's algorithm is represented in a recurrent form as follows:

$$\begin{aligned} H_{i,j} &= Max\{H_{i-1,j-1} + S(a_i, b_j), E_{i,j}, F_{i,j}\} \\ E_{i,j} &= Max(H_{i-1,j} - v, E_{i-1,j}) - u \\ F_{i,j} &= Max(H_{i,j-1} - v, F_{i,j-1}) - u. \end{aligned} \tag{1}$$

The optimal alignment score is given by $H(\mathbf{a}, \mathbf{b}) = H_{m,n}$, and the associated alignment is obtained by a trace back procedure. Today, the affine gap penalty function is adopted in nearly all pairwise and multiple sequence programs. However, a slightly more general "piecewise linear gap penalty function" [11] may be preferred when existence of long gaps is expected, e.g., when genomic sequences are to be aligned. The so called "double affine gap penalty" corresponds to the simplest case of the number of pieces being two, for which the computational cost is only marginally (20 ~ 30 %) higher than that with a usual affine gap penalty function.

Although the above mentioned algorithms generally produce only one best alignment, optimal alignments are often degenerated, i.e., several alternative alignments have the same optimal score [11]. If we extend our attention to only slightly less optimal solutions, many optimal and near-optimal alignments may be found [12]. Instead of enumerating all these optimal/near-optimal alignments, however, we can obtain more informative statistical features associated with the all possible alignments of the two sequences by means of the so-called probabilistic alignment methods [13, 14]. While these initial studies attempted to mimic real evolutionary processes, Miyazawa [15] reached a related idea inspired by statistical physics; he considered that the optimal alignment mentioned above corresponds to the state of minimal energy, or minimal free energy at 0 K, whereas more realistic views might be obtained by minimizing the free energy at an ambient temperature, $T > 0$ K. To do so, the partial alignment scores shown in Eq. 1 are replaced by "partition functions" $Z_{i,j}^X$ ($X = H, E$, or F), which follow a set of recurrent relations analogous to Eq. 1:

$$\begin{aligned} Z_{i,j}^H &= Z_{i-1,j-1} z_{i,j}, \quad z_{i,j} = e^{S(a_i, b_j)/T} \\ Z_{i,j}^E &= \left(Z_{i-1,j} - Z_{i-1,j}^E\right) e^{-v/T} + Z_{i-1,j}^E e^{-u/T} \\ Z_{i,j}^F &= \left(Z_{i,j-1} - Z_{i,j-1}^F\right) e^{-v/T} + Z_{i,j-1}^F e^{-u/T} \\ Z_{i,j} &= Z_{i,j}^H + Z_{i,j}^E + Z_{i,j}^F \end{aligned} \tag{2}$$

To obtain interesting statistical features, e.g., $p_{i,j} = P(a_i \sim b_i) =$ the probability of a_i and b_j being aligned, one must calculate another recurrence for the "backward partition function" $\hat{Z}_{i,j}^X$ starting from the back end. The posterior probability $p_{i,j}$ is then obtained by:

$$p_{i,j} = Z_{i,j}^H \hat{Z}_{i,j}^H z_{i,j}^{-1} / Z(\mathbf{a}, \mathbf{b}), \tag{3}$$

where the factor $z_{i,j}^{-1}$ is introduced to compensate for the duplicate multiplication of $z_{i,j}$ in $Z_{i,j}^H \hat{Z}_{i,j}^H$, and $Z(\mathbf{a}, \mathbf{b}) = Z_{m,n}$ is the total partition function. Another formulation of probabilistic alignment was proposed by Durbin et al. [16] based on a generative statistical model called pair hidden Markov model (pair HMM). A pair HMM consists of three states, i.e., match, insert, and delete states. Match state emits a pair of residues and other states emit single reside at a certain probability, while transitions between states also occur probabilistically. Roughly speaking, the emission probability from a match state is proportional to $z_{i,j} = e^{S(a_i,b_j)/T}$ in the partition function model, the transition probability between match and insert/delete states to $e^{-v/T}$, and the duration probability of an insert or delete state to $e^{-u/T}$. However, these probabilities are not usually given from outside but learnt from a number of datasets of homologous sequences through the expectation-maximization algorithm [17].

If a pair HMM is decoded by the Viterbi algorithm [17] to produce the maximal likelihood path, i.e., to generate the alignment with the maximal probability, the resultant alignment is essentially the same as that obtained by the score-based algorithm mentioned above. To maximally utilize the advantage of pair HMM or other probabilistic alignment method, we may prefer another type of decoding known as "maximal expected accuracy" (MEA) [18]. To perform MEA decoding, we first calculate $p_{i,j}$ according to Eq. 3 or related one for every pair of $i \in [1, m]$ and $j \in [1, n]$, and then run another dynamic programming routine like Eq. 1 without gap penalties to obtain the path (alignment) that maximizes the sum of $p_{i,j}$ on the path (*see* **Note 4**). The MEA decoding or its variants are widely used in recent MSA programs as an important components of their architectures [19, 20].

2.2 Consistency and Consistency Transformation

The term "consistency" among pairwise alignments is used in the literature to refer to two related but different concepts. In this subsection, consistency in the narrower sense is discussed, whereas consistency in the broader sense will be discussed later under Subheading 2.5.

Consider three sequences **a**, **b**, and **c**, and pairwise alignments $\mathcal{A}$(**a**, **b**), $\mathcal{A}$(**b**, **c**), and $\mathcal{A}$(**c**, **a**) between them. In the narrower sense, the triplet (a_i, b_j, c_k) are said mutually consistent if $a_i \sim b_j$, $b_j \sim c_k$, and $c_k \sim a_i$ are present in the given pairwise alignments $\mathcal{A}$(**a**, **b**), $\mathcal{A}$(**b**, **c**), and $\mathcal{A}$(**c**, **a**), respectively. Strictly speaking according to the original definition [21], each element, a_i, b_j, or c_k, is not a residue but a node of the bipartite graph that represents the pairwise alignment (Fig. 1c), and a consistently aligned region is sandwiched by adjacent edges that connect such nodes (shadowed area in Fig. 1e). Alternatively, each residue may be assigned to the node of the alignment graph (Fig. 1b). An edge that connects aligned residues, e.g., $a_i \sim b_j$, is called "trace" [22]. The triplet of residues (a_i, b_j, c_k) and the corresponding traces may be defined as consistent in the same way as that defined above. As a consistently aligned region may include not only traces but also indels, the original

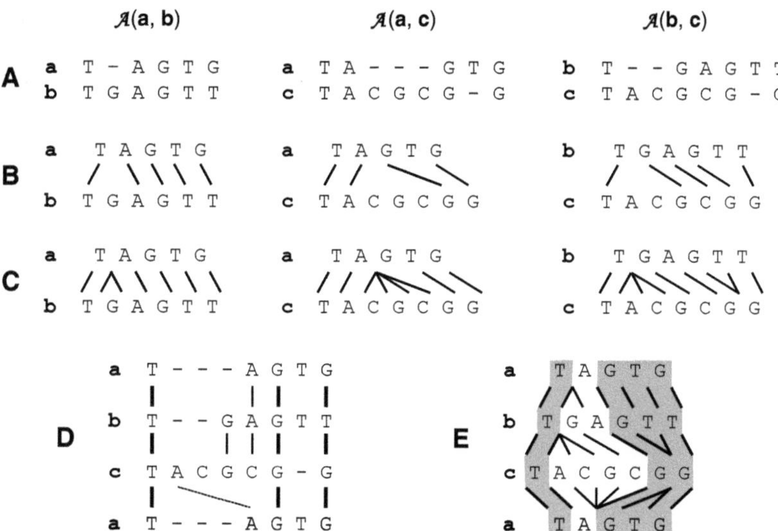

Fig. 1 Maximal consistency and maximal weight traces. (**A**) Pairwise alignments between three sequences a = TAGTG, b = TGAGTT, and c = TACGCGG. (**B**) Traces. (**C**) Bipartite graphs. (**D**) Maximal weight traces. (**E**) Consistently aligned regions (*shadowed area*). The consistent traces and edges are indicated by *bold lines* in (**D**) and (**E**), respectively. The *broken line* in (**D**) indicates a trace that is omitted in the final MSA

definition is slightly more general than the trace-based consistency. However, the latter is often used in consistency transformation, as described below.

The idea of consistency transformation was first proposed by Notredame et al. [23] to score an aligned residue pair. For the above example, a bonus (weight) is added to $S(a_i, b_j)$, if a_i and b_j are indirectly paired through c_k, i.e., if $a_i \sim c_k$, and $b_j \sim c_k$ are present in the pairwise alignments $\mathcal{A}(\mathbf{a}, \mathbf{c})$ and $\mathcal{A}(\mathbf{b}, \mathbf{c})$, respectively, irrespective of the presence of $a_i \sim b_j$ in $\mathcal{A}(\mathbf{a}, \mathbf{b})$. When we are aligning N sequences, the number of the "intermediate" sequences like **c** amounts to $N - 2$, with all of which the indirect pairing is examined. While this type of consistency-enhanced scoring system was first used as the objective function to be optimized by a stochastic algorithm [23], the same group soon came up with more efficient MSA program named T-Coffee that adopts a progressive method (see next subsection) [24]. Though named "consistency objective function," the scoring system of T-Coffee is more tightly related to the maximum weight trace problem studied by Kececioglu [25] than the concept of consistency discussed in the previous paragraph. T-Coffee assigns a reasonable but somewhat ad hoc weight to an indirectly aligned pair. Do et al. [19] elaborated theoretically more sound approach in their MSA program ProbCons, in which the posterior probability $p_{i,j}$ defined by Eq. 3 is used as the weight. Hence, their method is called probabilistic consistency transformation. Pairwise alignment between the N input sequences and rescoring of $S(a_i, b_j)$ may be repeated several times. ProbCons then uses a progressive method to construct the MSA, optionally followed by iterative refinement. Several protein MSA programs that follow the framework of ProbCons have been developed [26–28], which are among the most accurate MSA programs currently available. Moreover, genomic MSA program Pecan [29, 30] also uses basically the same strategy as ProbCons.

2.3 Progressive Method

The progressive method is not only the first practical MSA construction strategy [31] but also constitutes the core of a majority of contemporary MSA programs, besides "pure" progressive programs such as Aln3nn [32], Kalign2 [33], Prank [34], ClustalΩ [35], and ClustalW [36]. A progressive method consists of four steps, (1)–(4) as follows:

(1) Calculate a distance matrix from every pair of the N input sequences. In a standard way, $N(N-1)/2$ pairwise alignments are performed to count the numbers of matches, mismatch, and indels, which are then converted to the distance measures. This procedure is costly when N is large, as it takes $O(N^2L^2)$ computational time. Hence, more economical alignment-free methods [37] are often used to obtain the initial distance matrix. Among various alignment-free methods, Muth–Manber algorithm [38] that is tolerant to single mismatches/indels appears to be most sensitive. While Kalign2 [33] adopts this method, most other programs use simpler methods based on counts of common k-mers [39], sometimes in combination with reduced amino-acid alphabets [40, 41].

(2) Construct a guide tree by a hierarchical clustering algorithm. The UPGMA method [42] is most widely used. Unexpectedly, however, several reports suggest that single linkage clustering method produces consistently better alignment than the UPGMA method [43, 44]. Both UPGMA and single linkage trees can be constructed in $O(N^2)$ [41, 45]. The PartTree option of MAFFT [46] and the sequence-embedding technique [47] adopted by ClustalΩ [35] bypass a large part of step (1), enabling very rapid construction of a guide tree.

(3) A leaf of the guide tree corresponds to each input sequence, whereas an internal node corresponds to an MSA that is constructed by pairwise alignment of sequence(s) or MSA(s) corresponding to its children. Most MSA programs hold the MSA corresponding to an internal node in the form of profile [48] or "generalized profile" (see the next subsection), whereas several MSA programs use a directed acyclic graph (DAG) that can represent a set of alternative alignments rather than a unique MSA [49–51]. Generally speaking, the DAG-

based algorithms are better suited for alignment of close sequences, for which individual events of indels can potentially be traced back. In contrast, profile-based algorithms are better for distantly related sequences, as the resultant MSA reflects the core features that survive a saturating number of mutational events including indels.

(4) It is also a common practice to recalculate the distance matrix and the guide tree from the induced pairwise alignments after construction of the initial MSA [31, 40, 52, 53].

2.4 Iterative Refinement

The major drawback of the progressive method is well described by the paraphrase "once a gap, always a gap" [54], implying that if an erroneous gap is inserted at an early stage, it is propagated to the final result without any chance of correction. One method to circumvent this defect is to use the consistency transformation as already described in Subheading 2.2. Another effective approach relies on post process known as iterative refinement [55–58]. An iterative refinement consists of four steps:

(1) Construct an initial MSA by some method, e.g., by a progressive method.

(2) Horizontally divide the MSA into two groups. Remove the columns composed of nulls alone from each of the two groups.

(3) Realign the two groups by a pairwise sequence-to-group or group-to-group alignment method.

(4) Repeat (2) and (3) until no improvement in the alignment score is expected or by a predefined number of times.

Obviously, a precise definition of the objective function is indispensable for this procedure to work adequately. Most widely used objective function is the sum-of-pairs score [59] or slightly more general weighted sum-of-pairs score (WSP) [60] with affine gaps:

$$\begin{aligned}\mathrm{WSP}(A) &= \sum_{1\le p<q\le N} w_{p,q} H(\mathbf{a}_p, \mathbf{a}_q) \\ &= \sum_{1\le l\le L}\sum_{1\le p<q\le N} w_{p,q}\left[S(a_{p,l}, a_{q,l}) - v\cdot G(p,q,l)\right], \qquad (4)\end{aligned}$$

where $H(\mathbf{a}_p,\mathbf{a}_q)$ is the alignment score of a pair of rows in $\mathcal{A}$(matching nulls are ignored), $w_{p,q}$ is the weight given to the sequence pair $(\mathbf{a}_p,\mathbf{a}_q)$ ($w_{p,q} = 1$ in an unweighted case), and $G(p,q,l)$ is the Boolean variable indicating if a gap opens between $\mathbf{a}_p$ and $\mathbf{a}_q$ at the column position l. The heart of this method is the step (3), which should improve (at least not worsen) the alignment score compared with that before the division. This problem was first addressed by Gotoh [58] who devised a dynamic programming algorithm with a branch-and-bound extension that exactly optimizes group-to-group alignment under the sum-of-pairs scoring system. Gotoh [61] also succeeded in acceleration of the algorithm when N is large based on the "generalized profile" data structure, where "generalized" means that various lengths of internal gaps, as well as ordinary residues, are treated in the form of profile. The difficult point is to compute $G(p, q, l)$ exactly. In the worst case, this problem is proven to be NP-complete [62], but in practice not much more than $O(NL^2)$ operations are sufficient to reach the optimality [63, 64]. However, this type of exact method is still considerably slower than approximate methods that use only the ordinary profile vectors. Except for PRRN/PRRP [52, 63] and OPAL [43], currently popular MSA programs, such as ClustalW [36], MAFFT [40, 65] and MUSCLE [53], adopt a fast approximate method to evaluate gap open penalties. An intermediate approach that calculates $G(p, q, l)$ exactly but omits the branch-and-bound extension is employed in PRIME [66] to introduce double affine gap penalties in MSA. T-Coffee [24], ProbCons [19], and their derivatives are not bothered with the procedure to evaluate gap penalties in MSA, as gaps are not penalized at the progressive and refinement stages.

Note that many column scores other than the sum-of-pairs column score (the first term in the square brackets in Eq. 4) are compatible with progressive alignment and iterative refinement. In fact, MUSCLE [41] and HHsearch [67] (a component of ClustalΩ) use different implementations of column-specific log odds score. Two studies [68, 69] on relative performance of various column scores indicated that most of the scores they examined behave similarly to one another in alignment accuracy. However, further studies might be necessary on this point [70]. Meanwhile, Edgar [71] recently noticed that VTML amino acid substitution matrix [72] works best for accurate MSA construction by MUSCLE under the optimized gap penalties.

At the step (2), the existing MSA is horizontally divided into two groups. As there are $2^{N-1} - 1$ possible ways for division, trials of all of them [57, 58] are time consuming. Hirosawa et al. [73] examined various methods of partitioning, and reached the conclusion that tree-restricted partitioning (partitioning at each edge of the guide tree) is best in balance between accuracy and computational costs. Gotoh [74] also reached the same partitioning scheme from another point of view, i.e., this partitioning ideally fit calculation of WSP with the pair weights $w_{p,q}$ obtained by the three-way method, an approximation of the Rational II model of Altschul [60]. Weighting sequences or sequence pairs will be discussed in the last subsection in more detail. The tree-restricted partitioning is the default iterative refinement strategy of PRRN [52], MAFFT [65] and MUSCLE [53].

2.5 Anchoring

Fast homology search programs such as BLAST [75] and BLAT [76] generally adopts seed and extension strategies to find high-scoring segment pairs (HSPs). The concept of HSP is easily extended to $N > 2$ sequences. Such well-conserved local MSA segments can work as anchor points, by which the expensive MSA routines can be confined in the narrower range between neighboring anchors or sequence ends, leading to great reduction in both computational time and space.

Anchoring is a virtually indispensable strategy in genome sequence alignment. In one of the first attempts to align whole genomes, Delcher et al. [77] introduced the concept of maximal unique match (MUM), defined as a locally longest subsequence that occurs exactly once in the sequences under comparison. MUM anchoring is used in a local multiple genome alignment program, Mauve [78]. Instead of MUMs, several genomic MSA programs, such as MAG [79], GAME [80], and MISHIMA [81], use maximal exact matches (MEMs) that are locally longest common subsequences without the uniqueness condition of MUM. MUMs and MEMs can be efficiently detected by a suffix tree, suffix array, or simple *k*-mer table [82]. Simultaneous occurrence of MUMs or MEMs in multiple sequences is often too stringent a condition. To improve the sensitivity, MLAGAN [83] and MAVID [84] rely on the standard progressive/iterative strategy, where anchoring is used at the pairwise alignment phase, as implemented in GLASS [85], LAGAN [83], and AVID [86].

Anchoring also plays the essential role in segment-based MSA methods, such as DIALIGN [87] and SeqAn::T-Coffee [88]. A segment means a locally best matched alignment typically found by the Smith-Waterman-Gotoh algorithm [10, 89], and thus may contain internal mismatches and indels unlike MUM or MEM. Those segments are added in the order of their alignment scores to build up an MSA. The most critical issue in this process is to keep consistency in the broader sense between the new segment and those already incorporated in the MSA under construction. Here, the consistency in the broader sense implies a right positional order (partial order) between independently obtained segments. This type of consistency also plays the pivotal role in the greedy MSA construction algorithm called sequence annealing [20, 90] and related strategy adopted in PicXAA [91], in which an MSA column or residue pair, instead of a segment, is the building block of MSA. A directed acyclic graph is used to encode partially ordered building blocks and to facilitate efficient check of consistency among them.

The initial motivation of the study on consistency in the narrower sense [21] was to find anchor points to accelerate overall MSA computation. Instead of consistency, however, several contemporary MSA programs, such as PRRN/PRRP [52], RASCAL [92], MAFFT [65], and MUSCLE [41], rely on local alignment scores to heuristically identify anchor points in the iterative refinement phase, at an expense of slight loss of alignment quality [93].

2.6 Other Heuristic Methods

In the field of protein fold recognition, currently the most powerful tools are almost exclusively based on profile-profile or HMM-HMM (these terms are used nearly synonymously) comparison [67, 94–96]. Such profiles or HMMs are first constructed by using an iterative homology search program, typically PSI-BLAST [75]. This process is often called homology extension. In addition to sequence profiles, the most sensitive tools also incorporate predicted 1-D structural information, typically secondary structure [97] and solvent accessibility [98] propensities. Thus, it is natural to incorporate homology-extended sequence profiles and predicted 1-D structural information into the objective function of MSA to improve its quality. Earlier attempts are reviewed in [63]. PRALINE [99] and SPEM [100] are representatives of more systematic approaches. MAFFT [65] also has an option for homology extension, although no 1-D structural information appears to be considered. PCMA [101] is the first program that combines profile-profile comparison with consistency transformation. The same group has extended their approach to incorporate wider structure-related knowledge to yield MUMMALS [27] and PROMALS [102].

It is noteworthy that development of sequence pair weight $w_{p,q}$ shown in Eq. 4 [74] was coupled with homology extension. As a result of homology search, a various number of homologous sequences with various degrees of similarity to the query and to one another are obtained. The pair weights are designed to exert two effects: (1) correction for over/under representations of different groups of sequences, and (2) more favorable treatment of more closely related sequence pairs than distant pairs. The results of examinations of the influence of sequence weights or pair weights on MSA quality are somewhat controversial; Wheeler and Kececioglu [43] reported that several weighting schemes they examined, including unweighted case ($w_{p,q} = 1$), performed similarly in accuracy, whereas several other groups [28, 41, 52, 74] suggested positive effects of weighting. The fact that the experiments of [52, 74] were done after homology extension to the members in the structural alignment dataset might explain, at least in part, the cause of the discrepancy.

Incorporation of outer knowledge, such as linear motifs [103] and protein tertiary structures [104, 105], is promised to improve the accuracy of MSA. However, it is ambiguous where to limit the category of "sequence alignment," and these topics are not included in this chapter.

3 Notes

1. Although some algorithms are formulated to minimize distance rather than maximize similarity, these two approaches are proven to be mathematically equivalent [106].

2. An alignment is said global when internal and terminal gaps are not distinguished, whereas said semi-global when the terminal gaps are not penalized. Global and semi-global alignments are obtained with nearly the same dynamic programming algorithms except for the initial conditions and the choice of the traceback point [107].
3. Sometimes slightly different definition of gap open penalty is used: $g(k) = a + b(k - 1)$. It is easy to see that $a = -(u + v)$ and $b = -u$.
4. The original formulation [18] considers only aligned pairs ignoring gaps, which can lead to over prediction of aligned pairs. By taking into account of the probabilities of $P(a_i \sim$ '-') and $P($'-' $\sim b_j)$, a better balance between sensitivity and specificity may be achieved [20]. For more detailed discussion about the balance between sensitivity and specificity, *see* [108].

Acknowledgment

I thank Dr. Kentaro Tomii for instruction about protein fold recognition methods. This work was partly supported by Kakenhi (Grant-in-Aid for Scientific Research) B (grant number 22310124) from the Ministry of Education, Culture, Sports, Science and Technology of Japan.

References

1. Carrillo H, Lipman D (1988) The multiple sequence alignment problem in biology. SIAM J Appl Math 48:1073–1082
2. Gupta SK, Kececioglu JD, Schaffer AA (1995) Improving the practical space and time efficiency of the shortest-paths approach to sum-of-pairs multiple sequence alignment. J Comput Biol 2:459–472
3. Kawashima S, Pokarowski P, Pokarowska M, Kolinski A, Katayama T, Kanehisa M (2008) AAindex: amino acid index database, progress report 2008. Nucleic Acids Res 36: D202–D205
4. Dayhoff MO, Schwartz RM, Orcutt BC (1978) A model of evolutionary change in proteins. In: Dayhoff MO (ed) Atlas of protein sequence and structure, vol 5, Suppl. 3. National Biomedical Research Foundation, Silver Spring, MD, pp 345–352
5. Chiaromonte F, Yap VB, Miller W (2002) Scoring pairwise genomic sequence alignments. In: Altman RB, Dunker AK, Hunter L, Klein TED, Lauderdale K (eds) Pacific symposium on biocomputing. World Scientific, Singapore, pp 115–126
6. Frith MC, Hamada M, Horton P (2010) Parameters for accurate genome alignment. BMC Bioinformatics 11:80
7. Needleman SB, Wunsch CD (1970) A general method applicable to the search for similarities in the amino acid sequence of two proteins. J Mol Biol 48:443–453
8. Sellers PH (1974) On the theory and computation of evolutionary distances. SIAM J Appl Math 26:787–793
9. Waterman MS, Smith TF, Beyer WA (1976) Some biological sequence metrics. Adv Math 20:367–387
10. Gotoh O (1982) An improved algorithm for matching biological sequences. J Mol Biol 162:705–708
11. Gotoh O (1990) Optimal sequence alignment allowing for long gaps. Bull Math Biol 52:359–373
12. Waterman MS, Byers TH (1985) A dynamic-programming algorithm to find all solutions

in a neighborhood of the optimum. Math Biosci 77:179–188
13. Bishop MJ, Thompson EA (1986) Maximum likelihood alignment of DNA sequences. J Mol Biol 190:159–165
14. Thorne JL, Kishino H, Felsenstein J (1991) An evolutionary model for maximum likelihood alignment of DNA sequences. J Mol Evol 33:114–124
15. Miyazawa S (1995) A reliable sequence alignment method based on probabilities of residue correspondences. Protein Eng 8:999–1009
16. Durbin R, Eddy S, Krogh A, Mitchison G (1998) Biological sequence analysis: probabilistic models of proteins and nucleic acids. Cambridge University Press, Cambridge
17. Rabiner LR (1989) A tutorial on hidden Markov models and selected applications in speech recognition. Proc IEEE 77:257–286
18. Holmes I, Durbin R (1998) Dynamic programming alignment accuracy. J Comput Biol 5:493–504
19. Do CB, Mahabhashyam MS, Brudno M, Batzoglou S (2005) ProbCons: probabilistic consistency-based multiple sequence alignment. Genome Res 15:330–340
20. Bradley RK, Roberts A, Smoot M, Juvekar S, Do J, Dewey C, Holmes I, Pachter L (2009) Fast statistical alignment. PLoS Comput Biol 5:e1000392
21. Gotoh O (1990) Consistency of optimal sequence alignments. Bull Math Biol 52:509–525
22. Kruskal JB, Sankoff D (1983) An anthology of algorithms and concepts for sequence comparison. In: Sankoff D, Kruskal J (eds) Time warps, string edits, and macromolecules: the theory and practice of sequence comparison. Addison-Wesley, Reading, MA, pp 265–310
23. Notredame C, Holm L, Higgins DG (1998) COFFEE: an objective function for multiple sequence alignments. Bioinformatics 14:407–422
24. Notredame C, Higgins DG, Heringa J (2000) T-Coffee: a novel method for fast and accurate multiple sequence alignment. J Mol Biol 302:205–217
25. Kececioglu JD (1993) The maximum weight trace problem in multiple sequence alignment. Lect Notes Comput Sci 684:106–119
26. Roshan U, Livesay DR (2006) Probalign: multiple sequence alignment using partition function posterior probabilities. Bioinformatics 22:2715–2721
27. Pei J, Grishin NV (2006) MUMMALS: multiple sequence alignment improved by using hidden Markov models with local structural information. Nucleic Acids Res 34:4364–4374
28. Liu Y, Schmidt B, Maskell DL (2010) MSA-Probs: multiple sequence alignment based on pair hidden Markov models and partition function posterior probabilities. Bioinformatics 26:1958–1964
29. Paten B, Herrero J, Beal K, Birney E (2009) Sequence progressive alignment, a framework for practical large-scale probabilistic consistency alignment. Bioinformatics 25:295–301
30. Paten B, Herrero J, Beal K, Fitzgerald S, Birney E (2008) Enredo and Pecan: genome-wide mammalian consistency-based multiple alignment with paralogs. Genome Res 18:1814–1828
31. Hogeweg P, Hesper B (1984) The alignment of sets of sequences and the construction of phyletic trees: an integrated method. J Mol Evol 20:175–186
32. Kruspe M, Stadler PF (2007) Progressive multiple sequence alignments from triplets. BMC Bioinformatics 8:254
33. Lassmann T, Frings O, Sonnhammer EL (2009) Kalign2: high-performance multiple alignment of protein and nucleotide sequences allowing external features. Nucleic Acids Res 37:858–865
34. Loytynoja A, Goldman N (2005) An algorithm for progressive multiple alignment of sequences with insertions. Proc Natl Acad Sci U S A 102:10557–10562
35. Sievers F, Wilm A, Dineen D, Gibson TJ, Karplus K, Li W, Lopez R, McWilliam H, Remmert M, Soding J, Thompson JD, Higgins DG (2011) Fast, scalable generation of high-quality protein multiple sequence alignments using Clustal Omega. Mol Syst Biol 7:539
36. Thompson JD, Higgins DG, Gibson TJ (1994) CLUSTAL W: improving the sensitivity of progressive multiple sequence alignment through sequence weighting, position-specific gap penalties and weight matrix choice. Nucleic Acids Res 22:4673–4680
37. Blaisdell BE (1986) A measure of the similarity of sets of sequences not requiring sequence alignment. Proc Natl Acad Sci USA 83:5155–5159
38. Muth R, Manber U (1996) Approximate multiple string search. Lect Notes Comput Sci 1075:75–86

39. Jones DT, Taylor WR, Thornton JM (1992) The rapid generation of mutation data matrices from protein sequences. Comput Appl Biosci 8:275–282
40. Katoh K, Misawa K, Kuma K, Miyata T (2002) MAFFT: a novel method for rapid multiple sequence alignment based on fast Fourier transform. Nucleic Acids Res 30:3059–3066
41. Edgar RC (2004) MUSCLE: a multiple sequence alignment method with reduced time and space complexity. BMC Bioinformatics 5:113
42. Sneath PHA, Sokal RP (1973) Numerical taxonomy. Freeman, San Francisco, CA
43. Wheeler TJ, Kececioglu JD (2007) Multiple alignment by aligning alignments. Bioinformatics 23:i559–i568
44. Plyusnin I, Holm L (2012) Comprehensive comparison of graph based multiple protein sequence alignment strategies. BMC Bioinformatics 13:64
45. Gronau I, Moran S (2007) Optimal implementations of UPGMA and other common clustering algorithms. Inform Process Lett 104:205–210
46. Katoh K, Toh H (2007) PartTree: an algorithm to build an approximate tree from a large number of unaligned sequences. Bioinformatics 23:372–374
47. Blackshields G, Sievers F, Shi W, Wilm A, Higgins DG (2010) Sequence embedding for fast construction of guide trees for multiple sequence alignment. Algorithm Mol Bio 5:21
48. Gribskov M, McLachlan AD, Eisenberg D (1987) Profile analysis: detection of distantly related proteins. Proc Natl Acad Sci U S A 84:4355–4358
49. Hein J (1989) A new method that simultaneously aligns and reconstructs ancestral sequences for any number of homologous sequences, when the phylogeny is given. Mol Biol Evol 6:649–668
50. Lee C, Grasso C, Sharlow MF (2002) Multiple sequence alignment using partial order graphs. Bioinformatics 18:452–464
51. Loytynoja A, Vilella AJ, Goldman N (2012) Accurate extension of multiple sequence alignments using a phylogeny-aware graph algorithm. Bioinformatics 28:1684–1691
52. Gotoh O (1996) Significant improvement in accuracy of multiple protein sequence alignments by iterative refinement as assessed by reference to structural alignments. J Mol Biol 264:823–838
53. Edgar RC (2004) MUSCLE: multiple sequence alignment with high accuracy and high throughput. Nucleic Acids Res 32:1792–1797
54. Feng DF, Doolittle RF (1987) Progressive sequence alignment as a prerequisite to correct phylogenetic trees. J Mol Evol 25:351–360
55. Barton GJ, Sternberg MJ (1987) A strategy for the rapid multiple alignment of protein sequences. Confidence levels from tertiary structure comparisons. J Mol Biol 198:327–337
56. Subbiah S, Harrison SC (1989) A method for multiple sequence alignment with gaps. J Mol Biol 209:539–548
57. Berger MP, Munson PJ (1991) A novel randomized iterative strategy for aligning multiple protein sequences. Comput Appl Biosci 7:479–484
58. Gotoh O (1993) Optimal alignment between groups of sequences and its application to multiple sequence alignment. Comput Appl Biosci 9:361–370
59. Altschul SF (1989) Gap costs for multiple sequence alignment. J Theor Biol 138:297–309
60. Altschul SF, Carroll RJ, Lipman DJ (1989) Weights for data related by a tree. J Mol Biol 207:647–653
61. Gotoh O (1994) Further improvement in methods of group-to-group sequence alignment with generalized profile operations. Comput Appl Biosci 10:379–387
62. Ma B, Wang Z, Zhang K (2003) Alignment between two multiple alignments. Lect Notes Comput Sci 2676:254–265
63. Gotoh O (1999) Multiple sequence alignment: algorithms and applications. Adv Biophys 36:159–206
64. Kececioglu JD, Starrett D (2004) Aligning alignments exactly. In: Gusfield D, Bourne P, Istrail S, Pevzner P, Waterman M (eds) Proceedings of the 8th ACM conference on computational molecular biology (RECOMB). ACM Press, New York, pp 85–96
65. Katoh K, Kuma K, Toh H, Miyata T (2005) MAFFT version 5: improvement in accuracy of multiple sequence alignment. Nucleic Acids Res 33:511–518
66. Yamada S, Gotoh O, Yamana H (2006) Improvement in accuracy of multiple sequence alignment using novel group-to-group sequence alignment algorithm with piecewise linear gap cost. BMC Bioinformatics 7:524

67. Soding J (2005) Protein homology detection by HMM-HMM comparison. Bioinformatics 21:951–960
68. Edgar RC, Sjolander K (2004) A comparison of scoring functions for protein sequence profile alignment. Bioinformatics 20:1301–1308
69. Wang G, Dunbrack RL Jr (2004) Scoring profile-to-profile sequence alignments. Protein Sci 13:1612–1626
70. Altschul SF, Wootton JC, Zaslavsky E, Yu YK (2010) The construction and use of log-odds substitution scores for multiple sequence alignment. PLoS Comput Biol 6:e1000852
71. Edgar RC (2009) Optimizing substitution matrix choice and gap parameters for sequence alignment. BMC Bioinformatics 10:396
72. Muller T, Spang R, Vingron M (2002) Estimating amino acid substitution models: a comparison of Dayhoff's estimator, the resolvent approach and a maximum likelihood method. Mol Biol Evol 19:8–13
73. Hirosawa M, Totoki Y, Hoshida M, Ishikawa M (1995) Comprehensive study on iterative algorithms of multiple sequence alignment. Comput Appl Biosci 11:13–18
74. Gotoh O (1995) A weighting system and algorithm for aligning many phylogenetically related sequences. Comput Appl Biosci 11:543–551
75. Altschul SF, Madden TL, Schaffer AA, Zhang J, Zhang Z, Miller W, Lipman DJ (1997) Gapped BLAST and PSI-BLAST: a new generation of protein database search programs. Nucleic Acids Res 25:3389–3402
76. Kent WJ (2002) BLAT–the BLAST-like alignment tool. Genome Res 12:656–664
77. Delcher AL, Kasif S, Fleischmann RD, Peterson J, White O, Salzberg SL (1999) Alignment of whole genomes. Nucleic Acids Res 27:2369–2376
78. Darling AC, Mau B, Blattner FR, Perna NT (2004) Mauve: multiple alignment of conserved genomic sequence with rearrangements. Genome Res 14:1394–1403
79. Hohl M, Kurtz S, Ohlebusch E (2002) Efficient multiple genome alignment. Bioinformatics 18(Suppl 1):S312–S320
80. Choi JH, Cho HG, Kim S (2005) GAME: a simple and efficient whole genome alignment method using maximal exact match filtering. Comput Biol Chem 29:244–253
81. Kryukov K, Saitou N (2010) MISHIMA–a new method for high speed multiple alignment of nucleotide sequences of bacterial genome scale data. BMC Bioinformatics 11:142
82. Crochemore M, Hancart C, Lecroq T (2007) Algorithms on strings. Cambridge University Press, Cambridge
83. Brudno M, Do CB, Cooper GM, Kim MF, Davydov E, Green ED, Sidow A, Batzoglou S (2003) LAGAN and Multi-LAGAN: efficient tools for large-scale multiple alignment of genomic DNA. Genome Res 13:721–731
84. Bray N, Pachter L (2004) MAVID: constrained ancestral alignment of multiple sequences. Genome Res 14:693–699
85. Batzoglou S, Pachter L, Mesirov JP, Berger B, Lander ES (2000) Human and mouse gene structure: comparative analysis and application to exon prediction. Genome Res 10:950–958
86. Bray N, Dubchak I, Pachter L (2003) AVID: a global alignment program. Genome Res 13:97–102
87. Morgenstern B (1999) DIALIGN 2: improvement of the segment-to-segment approach to multiple sequence alignment. Bioinformatics 15:211–218
88. Rausch T, Emde AK, Weese D, Doring A, Notredame C, Reinert K (2008) Segment-based multiple sequence alignment. Bioinformatics 24:i187–i192
89. Smith TF, Waterman MS (1981) Identification of common molecular subsequences. J Mol Biol 147:195–197
90. Schwartz AS, Pachter L (2007) Multiple alignment by sequence annealing. Bioinformatics 23:e24–e29
91. Sahraeian SM, Yoon BJ (2010) PicXAA: greedy probabilistic construction of maximum expected accuracy alignment of multiple sequences. Nucleic Acids Res 38:4917–4928
92. Thompson JD, Thierry JC, Poch O (2003) RASCAL: rapid scanning and correction of multiple sequence alignments. Bioinformatics 19:1155–1161
93. Yamada S, Gotoh O, Yamana H (2009) Improvement in speed and accuracy of multiple sequence alignment program PRIME. Inform Media Tech 4:317–327
94. Sadreyev RI, Baker D, Grishin NV (2003) Profile-profile comparisons by COMPASS predict intricate homologies between protein families. Protein Sci 12:2262–2272
95. Tomii K, Akiyama Y (2004) FORTE: a profile-profile comparison tool for protein fold recognition. Bioinformatics 20:594–595
96. Soding J, Remmert M (2011) Protein sequence comparison and fold recognition: progress and good-practice benchmarking. Curr Opin Struct Biol 21:404–411

97. Jones DT (1999) Protein secondary structure prediction based on position-specific scoring matrices. J Mol Biol 292:195–202
98. Rost B, Sander C (1994) Conservation and prediction of solvent accessibility in protein families. Proteins 20:216–226
99. Simossis VA, Heringa J (2005) PRALINE: a multiple sequence alignment toolbox that integrates homology-extended and secondary structure information. Nucleic Acids Res 33: W289–W294
100. Zhou H, Zhou Y (2005) SPEM: improving multiple sequence alignment with sequence profiles and predicted secondary structures. Bioinformatics 21:3615–3621
101. Pei J, Sadreyev R, Grishin NV (2003) PCMA: fast and accurate multiple sequence alignment based on profile consistency. Bioinformatics 19:427–428
102. Pei J, Grishin NV (2007) PROMALS: towards accurate multiple sequence alignments of distantly related proteins. Bioinformatics 23:802–808
103. Papadopoulos JS, Agarwala R (2007) COBALT: constraint-based alignment tool for multiple protein sequences. Bioinformatics 23:1073–1079
104. O'Sullivan O, Suhre K, Abergel C, Higgins DG, Notredame C (2004) 3DCoffee: combining protein sequences and structures within multiple sequence alignments. J Mol Biol 340:385–395
105. Pei J, Kim BH, Grishin NV (2008) PROMALS3D: a tool for multiple protein sequence and structure alignments. Nucleic Acids Res 36:2295–2300
106. Smith TF, Waterman MS, Fitch WM (1981) Comparative biosequence metrics. J Mol Evol 18:38–46
107. Sellers PH (1980) The theory and computation of evolutionary distances: pattern recognition. J Algorithm 1:359–373
108. Hamada M, Asai K (2012) A classification of bioinformatics algorithms from the viewpoint of maximizing expected accuracy (MEA). J Comput Biol 19:532–549

Chapter 3

Objective Functions

Haluk Doğan and Hasan H. Otu

Abstract

Multiple sequence alignment involves alignment of more than two sequences and is an NP-complete problem. Therefore, heuristic algorithms that use different criteria to find an approximation to the optimal solution are employed. At the heart of these approaches lie the scoring and objective functions that a given algorithm uses to compare competing solutions in constructing a multiple sequence alignment. These objective functions are often motivated by the biological paradigms that govern functional similarities and evolutionary relations. Most existing approaches utilize a progressive process where the final alignment is constructed sequentially by adding new sequences into an existing multiple sequence alignment matrix, which is dynamically updated. In doing this, the core scoring function to assess accuracies of pairwise alignments generally remains the same, while the objective functions used in intermediary steps differ. Nevertheless, the overall assessment of the final multiple sequence alignment is generally calculated by an extension of pairwise scorings. In this chapter, we explore different scoring and objective functions used in calculating the accuracy and optimization of a multiple sequence alignment and provide utilization of these criteria in popularly used multiple sequence alignment algorithms.

Key word Scoring-functions, Scoring-matrices, Sum-of-pairs, Progressive-alignment, Consistency-based-scoring

1 Introduction

Multiple sequence alignment plays a major role in various areas of bioinformatics such as function and structure prediction, finding highly conserved regions, and resolving evolutionary relationship between a set of taxa. Alignment of biological sequences (DNA, RNA, or protein) reveals significant similarities, characterizes conserved regions, construes genetic diseases, and discovers functional motifs. Multiple sequence alignment, abbreviated from now on as MSA, is an extension of pairwise alignment where at least three sequences are aligned and the problem can be defined as follows:

Definition: *Let* S be a set of sequences $s_1, s_2, \ldots, s_n$ of lengths $l_1, l_2, \ldots, l_n$ defined in some alphabet Σ (e.g. $\Sigma_{\text{DNA}} = \{A, C, G, T\}$).

David J. Russell (ed.), *Multiple Sequence Alignment Methods*, Methods in Molecular Biology, vol. 1079,
DOI 10.1007/978-1-62703-646-7_3,

An MSA of S is defined as a set $S^{'}$ of sequences $s_1^{'}, s_2^{'}, \ldots, s_n^{'}$ whose lengths are m, defined in the same alphabet with an additional gap symbol. In other words, gaps are inserted in the sequences $s_1, s_2, \ldots, s_n$ so that their lengths become the same and a matrix representation where rows represent sequences and columns represent aligned symbols depicts the multiple alignment. The goal is to maximize a function $s : S^{'} \rightarrow \mathbb{R}$ that returns the score of probable alignments. Our aim is formally defined as follows:

$$a^{*} = \underset{a \in S'}{\operatorname{argmax}}\ s(a(S)),$$

where $a(S)$ denotes an alignment of a given set of sequences S.

Each column of an alignment must contain at least one symbol of Σ which implies that there cannot be any columns composed of only gap characters. Finding a certain optimal alignment is a computationally intensive task and dynamic programming methodology, which guarantees the optimal alignment is not directly applicable to MSA as in the case of pairwise alignment. The reason for this is the vast number of probable alignments, which is exponentially related to the number of given sequences and length of sequences. Indeed, assuming a simple scheme to calculate the score of each column is $O(n)$, the complexity of an MSA is $O(n2^nL^n)$ where L is $\max(|s_1|, |s_2|, \ldots, |s_n|)$, n is the number of sequences, and $|s_i|$ denotes the length of s_i [1]. Therefore, heuristic methods are employed to the MSA problem in order to perform faster; however, these methods do not guarantee the optimal alignment.

2 Scoring a Multiple Sequence Alignment

Cost of an alignment must be defined to distinguish the optimal multiple alignment from other possible alignments. Basically, we need an assessment, which is carried out using an objective function to score an MSA [2].

2.1 Sum of Pairs

Sum of Pairs (SP) [3] is one of the most popular objective function measures used in multiple sequence alignment. The SP score of an MSA involving n sequences is calculated by adding scores of all $\binom{n}{2}$ induced pairwise alignments. The SP score is defined as follows:

$$\mathrm{SP}(S') = \sum_{1 \leq i \leq j \leq n} \mathrm{PS}(R_i, R_j),$$

where n denotes the number of sequences, PS is a pairwise scoring function, and R denotes a row of an alignment.

Since pairwise scoring is column based, SP scoring can also be viewed as such, where each column is considered independently:

$$\begin{aligned}\mathrm{SP}(S') &= \sum_{1 \le i \le j \le n} \mathrm{PS}(R_i, R_j) \\ &= \sum_{1 \le i \le j \le n} \sum_{1 \le x \le y} d(R_i[x], R_j[x]) \\ &= \sum_{1 \le x \le y} \sum_{1 \le i \le j \le n} d(S'_{ix}, S'_{jx}) \\ &= \sum_{1 \le x \le y} \begin{pmatrix} S'_{1,c} \\ \vdots \\ S'_{n,c} \end{pmatrix}\end{aligned}$$

where $d : \Sigma_x, \Sigma_y \rightarrow \mathbb{R}$ is the scoring function, $R_i[x]$ (also denoted as S'_{ix}) is the xth symbol of the ith row in the MSA, and y is the row length. The scoring function d can also be considered as a matrix of predefined scores, where each cell represents the score of aligning the two corresponding symbols.

The following example illustrates the SP method in detail:

Example: Given the following sequences:

- S_1: ACCCGA
- S_2: ACTA
- S_3: TCCTA

and their alignment S':

$$S'(S_1, S_2, S_3) = \begin{cases} S_1 : ACCCGA \\ S_2 : AC--TA \\ S_3 : TCC-TA \end{cases}$$

The SP score of this alignment is:

$$\begin{aligned}\mathrm{SP}(S') = {} & [S(A,A) + S(A,T) + S(A,T)] + [S(C,C) + S(C,C) + S(C,C)] \\ & + [S(C,-) + S(C,C) + S(-,C)] + [S(C,-) + S(C,-) + S(-,-)] \\ & + [S(G,T) + S(G,T) + S(T,T)] + [S(A,A) + S(A,A) + S(A,A)].\end{aligned}$$

In practice, mismatch and gap penalty scores are negative values and scoring a match between two gaps is ignored. In each step of the alignment, the SP method calculates the scores of all pairs of residues for every column, which increases the MSA algorithm complexity by $O(n^2)$ where n denotes the number of sequences.

In aligning DNA/RNA sequences, the scoring schemes tend to be more egalitarian and independent of the symbols; however, protein sequence alignments require more sophisticated approaches as amino acids can be divided into various functional classes based on different similarity parameters. The two most popular score matrices used for aligning protein sequences are the PAM and BLOSUM matrices [4]. The motivating idea in developing scoring matrices for protein sequences is that the substitution of different amino acid pairs should be treated differently. For example, the substitution of a hydrophilic amino acid with another hydrophilic amino acid, which can be considered as a mutual substitution, should be punished less severely than substitution with a hydrophobic amino acid. In order to obtain a biological justification for the penalty of various substitutions, Dayhoff and colleagues worked on 34 protein superfamilies divided into 71 groups of homologous proteins [5]. Within each group, sequences were more than 85 % similar and the total number of changes was 1,572. Based on known evolutionary trees built for each group, a reversal on the tree represents substitution frequencies for different amino acid pairs.

2.2 Point Accepted Mutation

"*P*oint *A*ccepted *M*utation (PAM)" scoring matrices are calculated based on the substitution rates obtained by the aforementioned tree reversals done by Dayhoff et al. Entries in the scoring matrix represent the likelihood of replacing an amino acid X by an amino acid Y. PAM matrices are denoted by a rate, e.g., a 1-PAM matrix assumes the sequences to be aligned are 99 % identical, hence the accepted point mutation rate is 1 %. The score of a given substitution is the ratio of the frequency of this substitution to the expected mutation rate. This value is usually represented in the logarithmic scale and a higher level PAM matrix is calculated by successive multiplications of the 1-PAM matrix. For example, a 3-PAM matrix is the 1-PAM matrix taken to the power of three. However, as one residue may have mutated to another one and then reverted to the original residue, or a residue may have mutated more than once, an *X*-PAM matrix does not imply *X* % expected difference between the sequences to be aligned. For example, the 250-PAM matrix assumes a 20 % similarity, while the 80-PAM matrix assumes a 50 % similarity between the sequences to be aligned. In Table 1, we show the 250-PAM matrix, which is popularly used for aligning distant sequences.

2.3 Block Substitution Matrix

Block Substitution Matrix, shortly BLOSUM, is also designed for scoring protein alignments [6]. The idea is similar to that of PAM, but BLOSUM matrices use a larger amount of sequence data and consider local alignment blocks or highly conserved regions rather than independent residue alignments. BLOSUM matrices are calculated by processing sequences with different degrees of similarities. For example, the BLOSUM62 matrix is generated from sequences that are more than 62 % identical. BLOSUM matrix entries M_{ij} are calculated using:

$$M_{ij} = \frac{1}{\lambda} \log\left(\frac{p_{ij}}{q_i \times q_j}\right),$$

where p_{ij} is the probability of observing a substitution between amino acids i and j; q_i and q_j is the probability of observing i and j,

Table 1
The PAM250 scoring matrix

	A	R	N	D	C	Q	E	G	H	I	L	K	M	F	P	S	T	W	Y	V
A	2																			
R	−2	6																		
N	0	0	2																	
D	0	−1	2	4																
C	−2	−4	−4	−5	4															
Q	0	1	1	2	−5	4														
E	0	−1	1	3	−5	2	4													
G	1	−3	0	1	−3	−1	0	5												
H	−1	2	2	1	−3	3	1	−2	6											
I	−1	−2	−2	−2	−2	−2	−2	−3	−2	5										
L	−2	−3	−3	−4	−6	−2	−3	−4	−2	2	6									
K	−1	3	1	0	−5	1	0	−2	0	−2	−3	5								
M	−1	0	−2	−3	−5	−1	−2	−3	−2	2	4	0	6							
F	−4	−4	−4	−6	−4	−5	−5	−5	−2	1	2	−5	0	9						
P	1	0	−1	−1	−3	0	−1	−1	0	−2	−3	−1	−2	−5	6					
S	1	0	1	0	0	−1	0	1	−1	−1	−3	0	−2	−3	1	3				
T	1	−1	0	0	−2	−1	0	0	−1	0	−2	0	−1	−2	0	1	3			
W	−6	2	−4	−7	−8	−5	−7	−7	−3	−5	−2	−3	−4	0	−6	−2	−5	17		
Y	−3	−4	−2	−4	0	−4	−4	−5	0	−1	−1	−4	−2	7	−5	−3	−3	0	10	
V	0	−2	−2	−2	−2	−2	−2	−1	−2	4	2	−2	2	−1	−1	−1	0	−6	−2	4

Table 2
The BLOSUM62 scoring matrix

	A	R	N	D	C	Q	E	G	H	I	L	K	M	F	P	S	T	W	Y	V
A	4																			
R	−1	5																		
N	−2	0	6																	
D	−2	−2	1	6																
C	0	−3	−3	−3	9															
Q	−1	1	0	0	−3	5														
E	−1	0	0	2	−4	2	5													
G	0	−2	0	−1	−3	−2	−2	6												
H	−2	0	1	−1	−3	0	0	−2	8											
I	−1	−3	−3	−3	−1	−3	−3	−4	−3	4										
L	−1	−2	−3	−4	−1	−2	−3	−4	−3	2	4									
K	−1	2	0	−1	−3	1	1	−2	−1	−3	−2	5								
M	−1	−1	−2	−3	−1	0	−2	−3	−2	1	2	−1	5							
F	−2	−3	−3	−3	−2	−3	−3	−3	−1	0	0	−3	0	6						
P	−1	−2	−2	−1	−3	−1	−1	−2	−2	−3	−3	−1	−2	−4	7					
S	1	−1	1	0	−1	0	0	0	−1	−2	−2	0	−1	−2	−1	4				
T	0	−1	0	−1	−1	−1	−1	−2	−2	−1	−1	−1	−1	−2	−1	1	5			
W	−3	−3	−4	−4	−2	−2	−3	−2	−2	−3	−2	−3	−1	1	−4	−3	−2	11		
Y	y2	−2	−2	−3	−2	−1	−2	−3	2	−1	−1	−2	−1	3	−3	−2	−2	2	7	
V	0	−3	−3	−3	−1	−2	−2	−3	−3	3	1	−2	1	−1	−2	−2	0	−3	−1	4

respectively; and λ is the scaling factor. In Table 2, we show the entries of the BLOSUM62 scoring matrix.

2.4 Minimum Entropy

Given a column of an MSA, it might be reasonable to argue that the relative proportions of the symbols in this column should be related to the score of the column. The SP scoring scheme fails to cope with this approach. For instance, the following two columns have the same $A{:}C$ ratio, yet their SP scores can be quite different.

$$S_i' = \begin{matrix} A \\ C \\ C \\ C \end{matrix} \qquad S_j' = \begin{matrix} A \\ A \\ C \\ C \\ C \\ C \\ C \\ C \\ C \\ C \end{matrix}$$

$$\mathrm{SP}(S_i') = 4S(A,C) + \binom{4}{2} S(C,C) = 4S(A,C) + 6S(C,C)$$

$$\begin{aligned} \mathrm{SP}(S_j') &= S(A,A) + 16S(A,C) + \binom{8}{2} S(C,C) \\ &= S(A,A) + 16S(A,C) + 28S(C,C). \end{aligned}$$

This difference implies that the SP scoring scheme is not scalable with sequence sizes. An alternative approach calculates the Shannon's entropy for each column as the column's score. Shannon's entropy [7] is used to calculate the information content of a sequence of symbols.

Given a column S_i' in an MSA, the score for this column is directly related to Shannon's entropy but formally defined in a slightly different way as follows:

$$S(S_i') = -\sum_a c_{ia} \log_2 p_{ia},$$

where

- $c_{i\,a}$: number of times symbol a occurs in column i
- $p_{i\,a}$: probability of symbol a in column i

There are two extreme cases in this scoring scheme. When all symbols in the column are the same, then the entropy score is 0. On the other hand, the entropy score is maximum when all symbols in the column are equally distributed. A good alignment is the one that achieves low entropy scores for each column as the total score of the MSA is the sum of column entropy scores. The entropy-based scoring schemes are not affected by the number of sequences in the MSA as the entropy calculation involves relative frequencies of each symbol in the column.

2.5 NorMD

One of the drawbacks of SP scoring is the assumption that substitution probabilities are uniformly distributed and time-invariant. However, substitution probabilities may also depend on structural and functional properties of the proteins [8]. Normalized mean distance score (norMD), a column-based MSA scoring method, is proposed to overcome with this deficiency of SP scoring [9]. NorMD is formally defined as follows:

$$\text{NorMD} = \frac{\text{MD} - \text{GAPCOST}}{\text{MaxMD} \times \text{LQRID}},$$

where

- MD: mean distance.
- GAPCOST: affine gap cost.
- MaxMD: maximum obtainable MD score.
- LQRID (lower quartile range of the pairwise hash score): similarity measure of sequences based on a hash score which is obtained from dot plots of pairs of sequences.

The MD score is the negative exponential of the weighted pairwise distances between the sequences. The weights are inversely proportional to the percentage identities between pairs of sequences. The MaxMD value is included in the score as a normalization factor to eradicate the effects of high MD values of long sequences. Eventually, norMD is normalized into a value between 0 and 1. The advantage of the norMD scoring scheme is its independence from the number and length of the sequences. However, its major drawback is formidable hash computation during scoring.

3 Applications of Objective Functions

Implementation of MSA algorithms can be divided into five groups:

- Exact Methods: Dynamic Programming (DP) using an n-dimensional matrix
- Progressive Methods: Uses a guided tree to combine pairwise alignments to obtain the final multiple alignment (e.g. ClustalW, MUSCLE, GramAlign)
- Iterative Methods: First computes a sub-optimal solution and provides improvements via DP until solution converges (e.g. MAFFT)

- Consistency-Based Methods: Constructs a database of local and global alignments to find a final alignment (e.g. T-Coffee, DiAlign, ProbCons)
- Structure-Based Methods: Utilizes external knowledge such as protein structure (e.g. 3D-Coffee)

In heuristic implementation of MSA algorithms, the most widely used approach is the SP scoring scheme. The main idea is to progressively combine induced pairwise alignments to obtain the final MSA. Here, we briefly review a few such algorithms and indicate specific improvements the programs offer.

3.1 ClustalW

ClustalW is one of the earliest multiple sequence alignment programs and it is still widely used. It has three main steps. First, it starts by pairwise alignment of all pairs of sequences via global dynamic programming with a plausible scoring function. Second, it uses the pairwise alignment scores to build a phylogenetic tree employing the Neighbor-Joining algorithm. Finally, the sequences are aligned starting from leaves to the root and, as a result, the MSA of all sequences are obtained [10].

In the tree reversal process, highest scoring pairs are progressively combined. On the other hand, as new sequences are introduced to the MSA, initial alignment structures propagate. This introduces a twofold problem: the greediness of the progressive alignment approach calls for the possibility of staying at a local minimum as far as the overall MSA score is concerned and errors in the early alignments cannot be rectified. Therefore, CLUSTALW designates weights for sequences to overcome the aforementioned problems. If there exists an edge of length l and $n_i \in n$ that can be reached by traversing l, then the designated weight to n_i from this edge is $\frac{n}{l}$. This way, tree reversal is not solely dependent on the tree structure but also benefits from the distance between sequences based on pairwise alignment scores. One of the main improvements offered by CLUSTALW is the appropriate parameter value selection for scores involving gaps. As the protein core has less insertions and deletions, CLUSTALW considers short stretches of hydrophilic residues (e.g. 5 or more) as an indication of loops or random coil regions and reduces the gap opening penalty for these stretches. Besides, it increases the gap opening penalty for gaps that are less than eight residues apart based on the observation of alignments between sequences of known structures, where it is rare to find gaps within 8-residue segments [11]. The initial gap opening penalty and the extended gap penalty are defined as follows:

$$\mathrm{GOP} + \log(\min(N, M)) \times S(\bar{a}, b) \times \mathrm{ISF}$$

$$\mathrm{GEP} \times \left(1.0 + \left|\log\left(\frac{N}{M}\right)\right|\right)$$

where

- GOP: user defined gap opening penalty
- N, M: length of sequences
- $S(\bar{a}, b)$: average residue mismatch score
- ISF: percentage of identity scaling factor
- GEP: user-defined gap extension penalty

3.2 T-Coffee

Even though the idea proposed in ClustalW is simple, neat, and performs well, its main drawback is being too dependent on the initial global alignments. Errors in the early alignment phases are propagated and may lead to the exclusion of consistencies between close pairs and distant ones. T-Coffee, which stands for *T*ree based *C*onsistency *O*bjective *F*unction For Alignm*E*nt *E*valuation, is a progressive alignment approach as ClustalW but aims to overcome the aforementioned drawback [12]. T-Coffee starts with executing ClustalW for global alignment and Lalign (a local alignment algorithm [13]) for local alignment for all pairs of sequences and chooses the top scoring alignments. This collection of global and local alignments indicates two libraries and a weight is assigned to each pair of aligned residues. The two libraries are merged into a secondary library by assigning greater weight to pairs that match in both alignments and creating new entries for those pairs that do not match. For example, given the following two sequences:

$$S_1 : G\ A\ R\ F\ I\ E\ L\ D\quad T\ H\ E\quad L\ A\ S\ T\quad F\ A\ T\quad C\ A\ T$$
$$S_2 : G\ A\ R\ F\ I\ E\ L\ D\quad T\ H\ E\quad F\ A\ S\ T\quad C\ A\ T\quad -\ -\ -$$

There are 18 residues in the S_2 sequence, two of which are not matched. Hence, sequence identity is $100 \times (16/18) = 88$ which is the primary weight for this alignment. If this alignment also existed in the second library, which is built using local alignments, then the two alignments are merged into one and its new weight is $88 \times 2 = 176$ assuming the local alignment also has an 88 % identity. If in the local alignment of these after the primary library is constructed, T-Coffee alters the pairwise alignment weights by consulting a third sequence in order to improve the overall MSA at the cost of reducing pairwise alignment scores. The importance of incorporating a third sequence is illustrated as follows:

$$S_1 : -\ -\ -\ -\ -\ -\ -\ \ -\ -\ -\ S_{1,i}\ -\ -\ -\ -\ -\ -\ -\ -\ \ -$$
$$-\ -\ -\ -\ -\ -\ -$$
$$S_2 : -\ -\ -\ -\ -\ -\ S_{2,j}\ -\ -\ \ -\ -\ -\ -\ -\ -\ -\ -$$
$$S_{2,k}\ -\ -\ -\ -\ -\ -\ -$$

Let us assume $S_{1,\,i}$ aligns comparably well to both $S_{2,\,j}$ and $S_{2,\,k}$. Therefore, we are not sure which part of S_2 to align $S_{1,\,i}$ to.

However, if we know that $S_{1,i}$ aligns to $S_{3,l}$ in a third sequence S_3 and $S_{3,l}$ aligns well to $S_{2,k}$, then we can choose to align $S_{1,i}$ to $S_{2,k}$. For example in the given sequences S_1 and S_2, the "FASTCAT" substring of S2 can comparably be aligned to the "LASTFAT" and "FATCAT" substrings of S_2. The existence of a third sequence S_3 rectifies this ambiguity as follows:

$$S_1: G\ A\ R\ F\ I\ E\ L\ D\ \ T\ H\ E\ \ L\ A\ S\ T\ \ F\ A\ -\ T\ \ C\ A\ T$$
$$S_3: G\ A\ R\ F\ I\ E\ L\ D\ \ T\ H\ E\ \ V\ E\ R\ Y\ \ F\ A\ S\ T\ \ C\ A\ T$$
$$S_2: G\ A\ R\ F\ I\ E\ L\ D\ \ T\ H\ E\ \ -\ -\ -\ -\ \ F\ A\ S\ T\ \ C\ A\ T$$

Here, $w(S_1, S_3) = 77$ and $w(S_3, S_2) = 100$. The weight of the alignment S_1 and S_2 through S_3 is $w(S_1, S_2) = \min(w(S_1, S_3), w(S_3, S_2)) = 77$ so that we update the weight of the alignment S_1 and S_2 in the primary library with a new score $77 + 88 = 165$. Although this is lower than the optimum pairwise alignment of S_1 and S_2, we provide a better overall MSA.

Finally, T-Coffee produces its final MSA by using the traditional progressive alignment-based approaches on the modified pairwise scores in the secondary library. An appealing option of T-Coffee is that the program welcomes user-provided input sequences for the primary library. Moreover, the latest version of T-Coffee includes structural information for improved multiple protein alignments [14].

3.3 MAFFT

MAFFT, a high speed multiple sequence alignment program, implements Fast Fourier Transform (FFT) to identify homologous regions quickly after converting amino acid sequences into two feature vectors [15]. These feature vectors, which are composed of six components in total, represent volume and polarity of amino acid sequences [16]. The motivating idea in MAFFT is that highly correlated sequences may have homologous regions and sequence correlation is calculated by FFT of normalized volume and polarity vectors, $v(a)$ and $p(a)$, respectively

$$\hat{v}(a) = [v(a) - \bar{v}]/\sigma_v$$

$$\hat{p}(a) = [p(a) - \bar{p}]/\sigma_p.$$

Correlation between two sequences is then defined as:

$$c(k) = c_v(k) + c_p(k),$$

where

- $c_v(k) = \sum_{1\le n\le N, 1\le n+k\le M} \hat{v}_1(n)\hat{v}_2(n+k)$
- $c_p(k) = \sum_{1\le n\le N, 1\le n+k\le M} \hat{p}_1(n)\hat{p}_2(n+k)$
- N and M denote the length of sequences.

Consequently, in FFT form $c_v(k)$ is represented as:

$$c_v(k) \Leftrightarrow V_1^*(m) \cdot V_2(m)$$

where " * " denotes complex conjugation. MAFFT applies a sliding window approach with a 30 residue window size to find out homolog segments positions. DP is then used to align these segments optimally and gradually it joins these segments into a full alignment. As in most of the MSA programs, MAFFT uses guided trees and similarity matrices. Another proposed improvement of MAFFT is to use normalized similarity matrix and gap penalties so that all pairwise scores are positive and cost of multi-position gaps can be computed quickly [15]. The following formula is used to fill in the entries of the similarity matrix

$$\hat{M}_{ab} = [(M_{ab} - \text{average}_2)/(\text{average}_1 - \text{average}_2)] + S^a$$

- $\text{average}_1 = \sum_a f_a M_{a\ a}$
- $\text{average}_2 = \sum_{a,\ b} f_a f_b M_{a\ a}$
- a and b denote residues, f_a denotes frequency of symbol a, and S_a is a gap extension penalty.

The MAFFT algorithm is employed in two sequential steps. The first part of phase one, which is called FFT-NS-1 (*FFT* algorithm and the *N*ormalized *S*imilarity matrix), involves calculating pairwise distances, UPGMA tree construction, and progressive alignment by using the initial guide tree. In the second part of this first phase, FFT-NS-2, MAFFT improves on the distance matrix and the guide tree. In the second phase, consistency-based scoring is employed with iterative refinement. The modules G-INS-i constructs the global alignment library of pairwise alignments, L-INS-i uses local pairwise alignments with affine gaps to form the library, and E-INS-i uses local alignments with a generalized affine gap cost [17].

3.4 MUSCLE

MUSCLE (*MU*ltiple *S*equence *C*omparison by *L*og *E*xpectation) is an efficient progressive alignment method to align large numbers of nucleic acid and protein sequences accurately [18]. MUSCLE has two fundamental steps; progressive and iterative refinement. First, MUSCLE produces a temporary MSA by using k-mer distance measures and the UPGMA clustering method. Since MUSCLE starts alignment without any prior knowledge and the k-mer distance measure is used for unaligned sequences, in the next step MUSCLE opts to employ the Kimura distance [19]. Namely, the temporary MSA found in the first phase is used to assess a more accurate distance measure [18]. Subsequently, MUSCLE uses the

UPGMA to construct a guided tree and a progressive alignment is performed by considering only subtrees. Finally, in the refinement step, an edge from the previous guided tree is deleted. This splits the guided tree into two subtrees and the profile of the multiple alignment for each subtree is calculated. The new MSA produced by realigning the two profiles is compared with the previous alignment. If the SP score of the new MSA is improved, then alignment of the two profiles is kept for the next iteration. Refinement phase is repeated until convergence is achieved.

The term profile alignment resembles alignment of two alignments by matching up corresponding columns and gives scores based on composition of columns. Profile sum of pairs function and log-expectation scoring function of MUSCLE are defined as follows:

$$\mathrm{PSP}^{XY} = \sum_i \sum_j f_i^X f_j^Y S_{ij} = \sum_i \sum_j f_i^X f_j^Y \log\left(\frac{p_{ij}}{p_i p_j}\right)$$

$$\mathrm{LE}^{XY} = (1 - f_G^X)(1 - f_G^Y) \log \sum_i \sum_j f_i^X f_i^Y \left(\frac{p_{ij}}{p_i p_j}\right)$$

where f_G^X is frequency of gaps in profile column X.

4 Conclusions

Evolutionary and functional relationship between sequences can be inferred by assessing their alignments. However, finding the optimal MSA between a set of sequences is an NP-complete problem and assessing the results of a multiple sequence alignment is not an easy task as in the pairwise sequence alignment. Determining the performance and accuracy of an MSA relies on how diverged the given sequences are. Scoring functions give an insight to assess the accuracy of an alignment and highlight various criteria that a given algorithm tries to optimize. The MSA algorithms used in practice employ heuristic solutions; therefore, in most cases a manual inspection and intervention in parameter, data set, scoring/objective function, and algorithm selection would help enhance the overall performance. There exist MSA editors such as JalView [20] that can be used to inspect and amend alignment result. To best interpret the results of an MSA, an expert opinion that factors in biological basis in the findings constitutes a critical and valuable step.

5 Appendix Web Resources

Name	Web Link
ClustalW	http://www.ebi.ac.uk/Tools/msa/clustalw2/
T-Coffee	http://www.ebi.ac.uk/Tools/msa/tcoffee/
Kalign	http://www.ebi.ac.uk/Tools/msa/kalign/
MAFFT	http://www.ebi.ac.uk/Tools/msa/mafft/
MUSCLE	http://www.ebi.ac.uk/Tools/msa/muscle/

References

1. Setubal C, Meidanis J (1997) Introduction to computational molecular biology. PWS Publishing, Boston
2. Thompson JD, Poch O (2006) Multiple sequence alignment as a workbench for molecular systems biology. Curr Bioinform 1(1): 95–104
3. Lipman DJ, Altschul SF, Kececioglu JD (1989) A tool for multiple sequence alignment. Proc Natl Acad Sci USA 86(12):4412–4415
4. Lesk AM (2008) Introduction to bioinformatics. Oxford University Press, USA
5. Dayhoff MO, Schwartz RM, Orcutt BC (1978) A model of evolutionary change in proteins. In: Atlas of protein sequence and structure, vol. 5, Suppl 3. National Biomedical Research Foundation, Washington, pp 345–352
6. Henikoff S, Henikoff JG (1992) Amino acid substitution matrices from protein blocks. Proc Natl Acad Sci USA 89(22):10915–10919
7. Shannon CE, Weaver W, Blahut RE, Hajek B (1949) The mathematical theory of communication, vol 117. University of Illinois Press, Urbana
8. Teodorescu O, Galor T, Pillardy J, Elber R (2003) Enriching the sequence substitution matrix by structural information. Proteins Struct Funct Bioinform 54(1):41–48
9. Thompson JD, Plewniak F, Ripp R, Thierry J-C, Poch O et al (2001) Towards a reliable objective function for multiple sequence alignments. J Mol Biol 314(4):937
10. Thompson JD, Higgins DG, Gibson TJ (1994) Clustal w: improving the sensitivity of progressive multiple sequence alignment through sequence weighting, position-specific gap penalties and weight matrix choice. Nucleic Acids Res 22(22):4673–4680
11. Thompson JD (1995) Introducing variable gap penalties to sequence alignment in linear space. Comput Appl Biosci 11(2):181–186
12. Notredame C, Higgins DG, Heringa J et al (2000) T-coffee: a novel method for fast and accurate multiple sequence alignment. J Mol Biol 302(1):205–218
13. Huang X, Miller W (1991) Lalign-find the best local alignments between two sequences. Adv Appl Math 12:373–381
14. Armougom F, Moretti S, Poirot O, Audic S, Dumas P, Schaeli B, Keduas V, Notredame C (2006) Expresso: automatic incorporation of structural information in multiple sequence alignments using 3d-coffee. Nucleic Acids Res 34(Suppl 2):W604–W608
15. Katoh K, Toh H (2008) Recent developments in the mafft multiple sequence alignment program. Brief Bioinform 9(4):286–298
16. Katoh K, Misawa K, Kuma K-i, Miyata T (2002) Mafft: a novel method for rapid multiple sequence alignment based on fast fourier transform. Nucleic Acids Res 30(14): 3059–3066
17. Katoh K, Kuma K-i, Toh H, Miyata T (2005) Mafft version 5: improvement in accuracy of multiple sequence alignment. Nucleic Acids Res 33(2):511–518
18. Edgar RC (2004) Muscle: multiple sequence alignment with high accuracy and high throughput. Nucleic Acids Res 32(5): 1792–1797
19. Kimura M, Weiss GH (1964) The stepping stone model of population structure and the decrease of genetic correlation with distance. Genetics 49(4):561
20. Clamp M, Cuff J, Searle SM, Barton GJ (2004) The jalview java alignment editor. Bioinformatics 20(3):426–427

Chapter 4

Who Watches the Watchmen? An Appraisal of Benchmarks for Multiple Sequence Alignment

Stefano Iantorno, Kevin Gori, Nick Goldman, Manuel Gil, and Christophe Dessimoz

Abstract

Multiple sequence alignment (MSA) is a fundamental and ubiquitous technique in bioinformatics used to infer related residues among biological sequences. Thus alignment accuracy is crucial to a vast range of analyses, often in ways difficult to assess in those analyses. To compare the performance of different aligners and help detect systematic errors in alignments, a number of benchmarking strategies have been pursued. Here we present an overview of the main strategies—based on simulation, consistency, protein structure, and phylogeny—and discuss their different advantages and associated risks. We outline a set of desirable characteristics for effective benchmarking, and evaluate each strategy in light of them. We conclude that there is currently no universally applicable means of benchmarking MSA, and that developers and users of alignment tools should base their choice of benchmark depending on the context of application—with a keen awareness of the assumptions underlying each benchmarking strategy.

Key words Multiple sequence alignment, Benchmarking, Phylogenetic, Protein structure, Sequence evolution, Consistency, Homology

1 Introduction

Multiple sequence alignment (MSA) has become a common first step in the analysis of sequence data for downstream applications such as comparative genomics, functional analysis and phylogenetic reconstruction. Given their importance, MSA methods need to be objectively validated in order to ensure their output is both accurate and reproducible. Benchmarking is a crucial tool in the assessment of sequence alignment programs, as it allows their developers and users to compare the performance of different aligners objectively, identify strengths and weaknesses and help detect systematic errors in alignments. In recent years, there has been a growing

Stefano Iantorno and Kevin Gori contributed equally to this work.

David J. Russell (ed.), *Multiple Sequence Alignment Methods*, Methods in Molecular Biology, vol. 1079,
DOI 10.1007/978-1-62703-646-7_4, © Springer Science+Business Media, LLC 2014

appreciation of the importance of benchmarking measures and datasets to evaluate and critically examine the performance of different MSA software packages, as underscored by a number of recent articles addressing the subject [1–5].

At the same time, and despite these positive developments, the standard approach adopted by the great majority of scientists dealing with sequence alignment has remained reliance on aligners that have long been outperformed in benchmarks [6], or even manual and therefore inevitably subjective intervention in the alignment process [7]. It is unclear whether this is due to the simplicity of use and convenience of long-standing aligners ("historical inertia" [7]), reluctance to move away from customary practice, or unawareness or even distrust of newer, lesser-tested technologies. This trend is particularly worrying in light of the rapid spread of high-throughput technologies and the associated need for automation of analysis pipelines [8]. A reason for this state of affairs might lie upon the absence of a straightforward alignment benchmarking procedure and interpretation. In this chapter, we contribute to overcoming this problem by reviewing present alignment benchmarks, aiming to clarify their strengths and risks for MSA evaluation with a view towards having better (and better-trusted) benchmarks in the future. But before considering benchmarking strategies, we first need to review the alignment objectives we expect them to gauge.

1.1 What Should Sequence Aligners Strive for?

A conceptual complication lies in the fact that MSAs have multiple and potentially conflicting goals, depending on the biological question of interest [9]. Commonly, the residues aligned are those inferred to be related through homology, i.e., common ancestry. In other contexts, however, the emphasis might be more on functional or structural concordance among residues. A strictly evolutionary interpretation of homology in these cases could be counterproductive, as recognized also by Kemena and Notredame [1], since regions of the protein that carry out the same function or that occupy the same position in the three-dimensional conformation of the protein may have arisen independently by evolutionary convergence. For example, an alignment that pairs structurally analogous, but nonhomologous, residues would be informative and therefore "correct" to the structural biologist, although not so to the phylogeneticist. It should however be noted that functional and structural objectives are considerably less precise than the evolutionary objective: while common ancestry is an absolute, binary attribute, similarity in functional or structural role are context-dependent, continuous attributes, thus rendering any reduction to the aligned/unaligned dichotomy subjective at best, ill-defined at worst.

At the same time, the unambiguous nature of the evolutionary objective does not make it automatically easy to pursue (or, as we shall see below, ascertain). Indeed, the evolutionary history of biological sequences is mostly unknown and can only be inferred from present data under the (explicit or implicit) assumption of a model of sequence evolution.

In practice, most MSA methods muddle the distinction among homology-, structure-, or function-motivated alignment by employing strategies anchored in inconsistent objectives. Indeed, almost all well-established aligners assume and exploit evolutionary relationships among the sequences (e.g., by constructing the alignment using an explicitly phylogenetic guide tree and alignment scores derived from models of sequence evolution). Yet many use at the same time structural criteria in their parameters or heuristics, for example by training their parameters using structure-derived reference alignments [10, 11]. The complications of the strategies different aligners employ can however be divorced from the measurement of their success, and we wish to make no assumption that an aligner employing one strategy necessarily performs better when assessed according to criteria consistent with its internal methods. In the present context of alignment benchmarking, we therefore treat aligners as "black boxes" and refer the reader interested in the specifics of alignment methods to later chapters.

1.2 Aims and Desirable Properties of Alignment Benchmarks

As mentioned in the introduction, benchmarks provide ways of evaluating the performance of different MSA packages on standardized input. The output produced by the different programs is compared to the "correct" solution, the so-called gold standard, that is defined by the benchmark. The extent of similarity between the two then defines the quality of the aligner's performance.

Proper benchmarking is advantageous to both the user and the developer community: the former obtains standardized measures of performance that can be consulted in order to pick the most appropriate MSA tools to address a particular alignment problem, and the latter gains important insight into aspects of the software that need improvement, or new features to be implemented, thus promoting advancement of the field [2].

Which characteristics do benchmarks and the gold standard reference dataset need to satisfy in order to be useful to the user and developer community? Benchmarks can be critically examined by looking at their ability to yield performance measures that reflect the actual biological accuracy (whether defined in terms of shared evolutionary history or structural or functional similarity of the aligned sequence data) of the MSA method. This can most easily be done by defining a set of predetermined criteria for good benchmarking practice. We follow Aniba et al. [2] in their list of desirable properties of benchmarks, which states that a benchmark should be:

- *Relevant*, in that a benchmark should be reflective of actual MSA applications, i.e., tasks carried out by MSA in practice and not in an artificial or hypothetical setting.

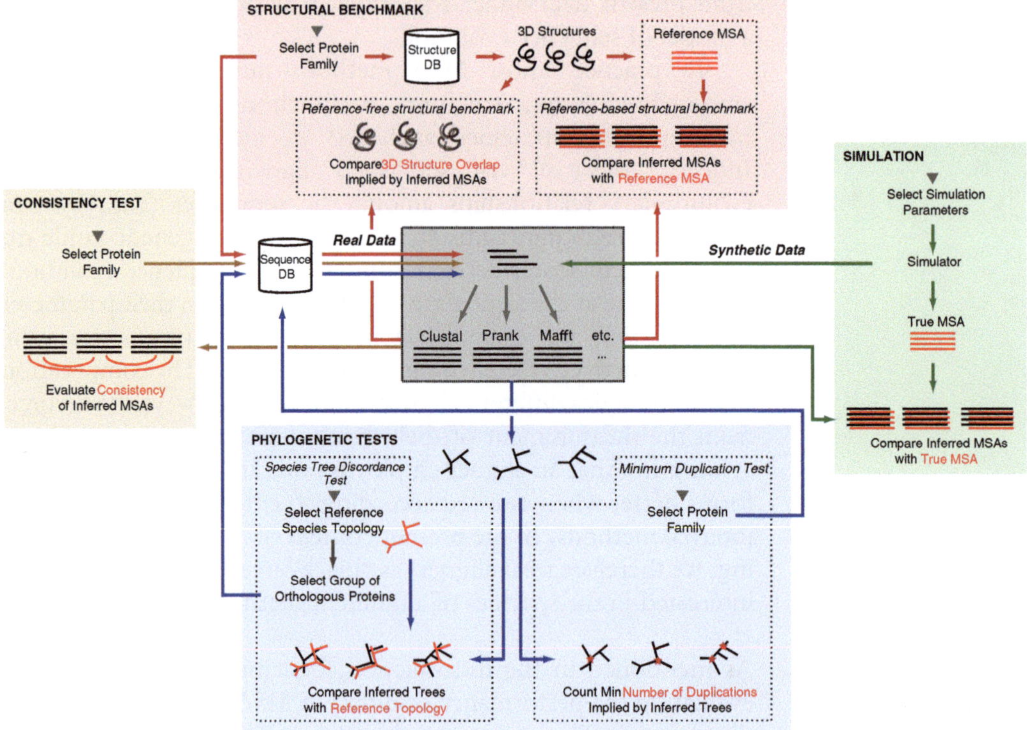

Fig. 1 Schematic of the four main MSA benchmarking strategies of this review: for each approach, the benchmarking process starts from the corresponding *downward-pointing arrow* and involves alignment by different MSA methods (*gray box* in center, illustrating example aligners that may be benchmarked)

- *Solvable*, in that it provides sufficient challenge to differentiate between poor and good performances, while remaining a tractable problem.
- *Scalable*, so that it can grow with the development of MSA programs and sequencing technologies.
- *Accessible*, in order to be widely used by developers and users.
- *Independent* from the methods used by programs under test, as benchmark datasets should avoid any overlap with the heuristics chosen for construction of MSA in order to constitute an objective reference.
- *Evolving*, to reduce the possibility of developers adapting their programs to a particular test set over time, thus artificially inflating their scores.

Although MSA methods employ different computational solutions to reconstruct sequence alignments, their performance needs to be assessed on the same benchmarks in order to be objectively evaluated and compared. In this chapter, we consider four broad MSA benchmarking strategies (Fig. 1):

1. Benchmarks based on simulated evolution of biological sequences, to create examples with known homology.

2. Benchmarks based on consistency among several alignment techniques.
3. Benchmarks based on the three-dimensional structure of the proteins encoded by sequence data.
4. Benchmarks based on knowledge of, or assumption about, the phylogeny of the aligned biological sequences.

In the remainder of this chapter, we analyze each of these benchmarking approaches to point out their pros and cons, and determine how well they satisfy the criteria defined above and summarized in Table 1.

2 Simulated Sequences

Given that a major objective of MSA is to identify residues that evolved from a common ancestor, i.e., to optimize for homology in the alignment, one approach to benchmarking involves generating families of artificial sequences by a process of simulated evolution along a known tree. Such simulation-based approaches adopt a probabilistic model of sequence evolution to describe nucleotide substitution, deletion, and insertion rates, while keeping track of "true" relationships of homology between individual residue positions. Since these are known, a "true" reference alignment and a test alignment based on the simulated sequence data, assembled by a particular MSA tool of choice, can be compared and measures of accuracy estimated (see below). There are many packages that will perform simulated sequence evolution, including Rose [12], DAWG [13], EvolveAGene3 [14], INDELible [15], PhyloSim [16], REvolver [17], and ALF [18].

To quantify the agreement between the reconstructed alignment and the true alignment (known from the simulation), two measures of accuracy are commonly employed: the sum-of-pairs (SP) and the true column (TC) scores [19]. The former is defined as the fraction of aligned residue pairs that are identical between the reconstructed and true alignment, averaged over all pairwise comparisons between individual sequences; the latter is defined as the fraction of correctly aligned columns that are reproduced in the reconstructed alignment. Given that the TC score considers whole columns in the alignment as comparable units, a single misaligned sequence can reduce the TC score to zero. For this reason, when considering numerous or divergent sequences, the finer-grained SP score is usually used. Yet even the SP score is not without problems. For instance, pairwise comparisons ignore correlations among sequences, meaning that closely related sequences contribute disproportionately more to the SP score than they do to the total phylogenetic information contained in the alignment; this can be misleading in phylogenetic applications. More generally, SP and TC

Table 1
The advantages and risks of the four approaches to MSA benchmarking. Examples are given of relevant software packages, benchmark databases and tests

Approach	Advantages	Risks	Examples	References
Simulation-based	Solvability: "true" homology is known	Relevance: simulated data might strongly differ from real biological data	Rose	[12]
	Evolving: different scenarios can be modelled	Independence: MSA parameters might resemble those used in simulation	DAWG	[13]
	Scalability: new data can be generated ad libitum		EvolveAGene3	[14]
			iSGv2.0	[48]
			INDELible	[15]
			PhyloSim	[16]
			ALF	[18]
Consistency-based	Scalability: not constrained to a particular reference set	Relevance: consistent MSA methods may be collectively biased	MUMSA	[26, 49]
	Accessibility: tests are easy and quick	Independence: similar scores might be used in MSA inference	HoT	[27]
Structure-based	Relevance: closely matches a major biological objective of MSA	Relevance: limited to structurally conserved regions; biological objective of MSA may vary	HOMSTRAD	[10, 30]
	Independence: empirical data is used as input	Scalability: only applicable to small subset of protein sequences	OXBench	[40]
			PREFAB	[33]
			SABMARK	[32]
			BAliBASE 3.0	[11, 31]
			STRIKE	[50]
Phylogeny-based	Relevance: closely matches a major biological objective of MSA	Relevance: biological objective of MSA may vary from phylogenetic reconstruction	Species-tree discordance test	[44]
	Independence: empirical data is used as input		Minimum duplication test	[44]
	Scalability: broad array of sequence data can be used as input			

are not proper metrics (they do not satisfy the conditions of symmetry or triangle inequality), which has motivated the recent development of better-founded alternatives [20].

Besides the advantage of knowing the true alignment, the fact that the parameters for simulated sequence evolution are user-defined directly translates into great flexibility to address specific questions or to investigate the effect of individual factors in isolation of others, which is particularly useful to gain insights into the behavior of complex alignment pipelines. For instance, Löytynoja and Goldman used simulated sequences to expose the systematic underrepresentation of the number of insertions by many aligners, which is especially true as sequence divergence and the number of sequences increases [21].

At the same time, the high level of flexibility afforded by simulation ties in with its biggest drawback: all observations drawn from simulated data depend on the assumptions and simplifications of the model used to generate these data. The vague notion of "realistic simulation" is often used to justify reliance on simulations capturing relevant aspects of real data, but simulations cannot straightforwardly, if at all, account for all evolutionary forces. The risk thus becomes the benchmarking of MSA programs in scenarios of little or no relevance to real biological data. For instance, Golubchik et al. investigated the performance of six aligners by simulating sequences in which gaps of constant size were placed in a staggered arrangement across all sequences [22]; although this scenario might be useful to emphasize a more general problem in aligning regions adjacent to gaps, its very artificial nature makes it a poor choice to gauge the extent of that problem on real data.

A further potential risk is the use of simulation settings more favorable to some packages than others [23]. For instance, the selected model of sequence evolution might resemble the underlying model of a particular aligner and thus provide it with an "unfair" advantage (i.e., presumably unrepresentative of typical situations) in the evaluation. Even when the evaluation is conducted in good faith, the high complexity of many MSA aligners—particularly in terms of implicit assumptions and heuristics—can make it challenging to design a fair simulation.

3 Consistency Among Different Alignment Methods

The key idea behind consistency-based benchmarks is that different good aligners should tend to agree on a common alignment (namely, the correct one) whereas poor aligners might make different kinds of mistakes, thus resulting in inconsistent alignments. Confusingly, this notion of consistency among aligners is different from that of consistency-based aligning, which is an alignment strategy that favors MSAs consistent with pairwise alignments [24, 25]. In the context of benchmarking, the relevant notion is the former—referred to by Lassmann and Sonnhammer as "inter-consistency," cf. "intra-consistency" for the latter [26].

Practically, benchmarking by consistency among aligners can be implemented using measures such as the overlap score [26], a symmetric variant of sum-of-pairs. From a set of input alignments, all paired aligned residues are determined over all sequences in every alignment. The overlap score for two alignments is calculated by counting the aligned pairs present in both alignments, and dividing by the average number of pairs in the alignments. Hence, two almost identical alignments have an overlap score close to one, while two very different alignments have an overlap score close to zero. Two additional scores based on this concept are the average overlap score, and the multiple overlap score. The average overlap score is simply the mean of the overlap scores measured over all pairs of input alignments, and represents the difficulty of the alignment problem. The multiple overlap score is a weighted sum of all pairs present in a single alignment, with the weight determined by the number of times each pair appears in the whole set of alignments. It is assumed that a high multiple overlap score, gained by an alignment with a high proportion of commonly observed pairs, corresponds to a good performance.

Another score that allows an internal control measure to estimate the consistency of different aligners is the heads-or-tails (HoT) score [27]. This consistency test is based on the assumption that biological sequences do not have a particular direction, and thus that alignments should be unaffected whether the input sequences are given in the original or reversed order. The agreement between the alignments obtained from the original and reversed sequences can be quantified with the overlap measures outlined above.

Both these consistency approaches—consistency among aligners and HoT score—are attractive because they assume no reference alignment or model of sequence of evolution, and thus can be readily and easily employed. Furthermore, high consistency is a necessary quality of a set of accurate aligners, thus making it desirable. The consistency criterion also appeals to the intuitive idea of "independent validation"—although most aligners have many aspects in common and are thus hardly "independent."

The biggest weakness of consistency is that it is no guarantee of correctness: methods can be *consistently wrong*. More subtly, consistency is sensitive to the choice of aligners in the set. This can be partly mitigated by including as many different alignments as possible [26]; nevertheless, it is easy to imagine cases where an accurate alignment, outnumbered by inaccurate, but similar, alignments, will be rated poorly. For instance, a new method solving a problem endemic to existing aligners will have low consistency scores.

Likewise, while low HoT scores can be indicative of considerable alignment uncertainty, the converse is not necessarily true. Hall reported that on simulated data at least, HoT scores tend to overestimate alignment accuracy [28]. That being said, considering the simplicity of HoT's scheme, the correlation Hall observed between HoT and simulation-based measures of alignment accuracy is strikingly high (depending on methods, Pearson ρ of 87–98 %). It remains to be seen whether this will remain the case over time—new aligners might be tempted to exploit HoT's idea in their inference algorithms or parameter optimization procedures, thus compromising its independence as a benchmarking criterion. For instance, a trivial way of "gaming" the HoT score is to align sequences with "centre-justification" (adding a gap character in the middle of sequences of even-numbered length). Such obviously flawed alignment procedure is nevertheless insensitive to joint sequence reversals, consistently obtaining a perfect HoT score.

4 Structural Benchmarks

Benchmarks have also been developed starting from protein structure data. Structural benchmarks are by far the most widely adopted type [2]. Most commonly these employ the superposition of known protein structures as an independent means of alignment, to which alignments derived from sequence analysis can then be compared using the sum-of-pairs and true-column measures discussed earlier.

Structural benchmarks are naturally highly relevant when sequence alignments are sought to identify structural concordance among amino-acid residues. Yet they are also relevant to an evolutionary interpretation of alignments. Indeed, the biological observation that forms the basis of using structure in the latter context is that homologous proteins often retain structural similarity even when sequence divergence is large [29]. Thus, at high levels of divergence, a greater degree of confidence may be placed on alignments based on structural conservation than on sequence similarity. If residues from different proteins can be shown to overlap in three-dimensional space, it is likely (though not certain) that they are homologous. An important advantage of structural benchmarks is that they provide a truly independent, empirically derived standard to test different alignment algorithms.

A number of structurally derived benchmark datasets exist. One of the oldest is HOMSTRAD [10, 30]. Although not originally intended for benchmarking, this dataset has been extensively used to rate the quality of alignments. The first purpose-built, large-scale structural benchmark was BAliBASE [11, 31], which was based on similarity of known protein structures. It is divided into a number of datasets, each suited to test a different alignment problem—for example, greater or lesser sequence diversity, the presence of large insertions or extensions or the presence of repeated elements. Each BAliBASE dataset was constructed by accessing information in structural databases, and alignments were verified by hand, at both the level of individual residues and of overall secondary structure. Other purpose-built structural benchmarks include SABMARK [32] and PREFAB [33], which differ from BAliBASE in that they are derived by automatic means, rather than by manual annotation of protein alignments. Reference sets also exist for RNA structures [34]. For further discussion of these datasets, we direct the reader to reviews by Aniba et al. [2], Edgar [3], Kim and Sinha [35], and Thompson et al. [4].

Regarding the desirable criterion of independence, although alignment algorithms incorporating structural aspects of sequence data do exist, such as Dynalign [36] and Foldalign [37]—for a more exhaustive discussion of RNA structural alignments, see Gardner et al. [34]—the parameters that go into constructing structure-based reference datasets are usually completely detached from the considerations that go into the development of MSA workflows.

Despite the degree of confidence structural alignment confers, it has been observed that sequence alignments used in BAliBASE and PREFAB are not always consistent with known annotations from external sources such as the CATH and SCOP databases, thus calling into question their biological accuracy [3]. Both manual and automated structural benchmark construction face considerable challenges. Manually curated structural benchmarks, while usually believed to generate more biologically accurate results than automated procedures, might also introduce subjective bias in the alignment. Automated procedures ensure reproducibility, but cannot avoid the existence of debatable parameter choices (e.g., the choice of the minimum spatial distance for two residues to be considered in the same fold) and potential systematic errors.

The nontrivial relationship between structural similarity of residues and alignment, however, is not the only cause of concern in structural benchmarks. Specifically, structure superpositions used for creating structural benchmarks are often not only based on experimentally derived structures, but also on primary sequence-based procedures such as BLASTP [38] and NORMD [39] which themselves employ amino acid substitution matrices and gap penalty scores, and thus make modelling assumptions about the sequences to be aligned [3]. If these parameters overlap with the parameters employed in MSA methods under test, then reference alignments obtained this way will be biased towards MSA-derived alignments that used those same parameters.

Problems arising from the use in benchmarking of reference alignments derived from structural comparisons can partially be overcome by the direct use of structural measures that are independent of any reference alignment. To evaluate the structure superposition implied by an MSA, Raghava et al. [40] adopted scores from a sequence-based multiple structure alignment algorithm [41]. Such structure similarity scores approximate the location of an amino acid in a test alignment by the location of its α-carbon (backbone carbon to which the amino acid side-chain attaches). Two aligned amino acid are then compared by the distance between their chains of α-carbon atoms, estimated by least squares over translations and rotations of their respective 3D protein structures (which are known a priori). A simple score is given by the root-mean-square deviations between superposed α-carbon atoms, whereas a more refined score also takes into account the orientation of these atoms [48].

Two final aspects of structural benchmarks further complicate their application in MSA assessment. The fact that reliable annotations exist only for structurally conserved sequences means that MSA of any region of the genome other than structured protein coding regions—be it intronic, regulatory, natively disordered, or simply poorly annotated—cannot be effectively assessed using existing structural benchmarks [4, 35]. This is particularly important given that only a very small fraction of genome sequences encode globular, folded protein domains, and that both structural benchmarks and MSA tools focus mainly on alignment of this very small portion of sequences. The current state of sequencing technologies also means that sequence data come with many artifacts due to sequencing errors, short read length, and/or poor gene prediction models [4, 8, 42, 43] which are only very recently starting to be accounted for in benchmarks [4].

Considering all these complications, it becomes apparent that the map between structure and alignment is neither straightforward nor unequivocal. And indeed, by annotating known domains in reference datasets (or estimating superfamilies when the domain was unavailable), and then comparing annotation agreement in the reference alignments by use of column scores, Edgar found inconsistencies in the assignment of aligned residues to specific secondary structure in both PREFAB and BAliBASE [3].

5 Phylogenetic Tests of Alignment

Our last type of benchmark is phylogenetic tests of alignment. Dessimoz and Gil [44] have recently introduced such tests, developing an MSA assessment pipeline that explicitly takes into consideration phylogenetic relationships within the input sequence data to evaluate the validity of alignment hypotheses generated by different MSA methods.

This approach to benchmarking involves deriving alignments of the test data from different MSA packages as the starting point for tree building. The principle of the tests is simple: the more accurate the resulting tree, the more accurate the underlying alignment is assumed to be. The quality of the tree is measured by its compliance with an auxiliary principle or model; auxiliary in the sense that the additional knowledge introduced be independent of sequence data. So far, two methods have been devised. In the first, referred to as the "species tree discordance test," test alignments are built from putative orthologous sequences, so that the resulting test trees can be expected to have the same topology as the underlying species tree. Each resulting tree is then compared to a reference species tree, comprising sufficiently divergent species that its branching order is deemed uncontroversial. The best performing aligners are taken to be those that most consistently generate alignments that yield test trees congruent with the species tree. Indeed, it can be expected that averaged over many hundreds or thousands of families, discordance due to non-orthology among the input sequences will affect the performance of all aligners equally, whereas discordance due to alignment error will vary among aligners.

The second method, termed the "minimum duplication" invokes a parsimony argument to interpret trees built from alignments of both orthologous and paralogous sequences, favoring trees which require fewer gene duplications to explain the data as more likely to reflect the true evolutionary history of the sequences.

One key advantage of phylogenetic benchmarks is that they provide a way of evaluating gap-rich and variable regions, regions for which structural benchmarks are often not applicable and simulation benchmarks lack realism [44]. In particular, the limited applicability of structural benchmarks to conserved protein core regions has quite possibly caused developers of alignment methods to focus their efforts on improving the performance of their tools on conserved regions at the expense of gap-rich or variable regions. Yet focusing on conserved regions can result in a loss of potentially informative data for multiple sequence alignment [21]. Adopting a simple tree inference method that looks only at presence or absence of gaps as a binary character within a maximum parsimony framework, Dessimoz and Gil reported that gap-only trees are sometimes even more accurate than nucleotide-based trees, thus highlighting the signal lost in neglecting variable or gap-rich regions [44].

At present, phylogeny-based benchmarks are the only ones that can be interpreted to be directly evaluating homology on real data. The premise of this interpretation is that more accurate trees on average necessarily ensue from a higher proportion of homologous positions in alignments on average, and therefore that the former is a good surrogate for the latter. Yet although we view the premise as highly plausible (and indeed fail to see how one could argue the opposite), there is no proof for it. If dismissed altogether, the interpretation has to be weakened so that these phylogeny tests only measure the effect of alignment on phylogenetic inference. In this case, phylogeny-based benchmarks are less meaningful even for other homology-based applications of alignments, such as detecting sites under positive selection [45].

6 Conclusions

Benchmarks for MSA applications have arisen in recent years as a crucial tool for bioinformaticians to keep a critical eye on existing software packages and reliably diagnose areas that need further development. The implementation of benchmarks to routinely assess the efficacy and accuracy of MSA methods has clearly provided important insights, and has pointed out to the developer community very serious shortcomings of existing methods that would not otherwise have been so apparent [4, 26, 44, 46]. Each benchmarking solution examined in this chapter—whether simulation-, consistency-, structure-, or phylogeny-based—entails risks of bias and error, but each is also useful in its own right when applied to a relevant problem. It is interesting to note that simulation benchmarks rank MSA methods differently from empirical benchmarks [21, 46, 47]. It is clear that no single benchmark can be uniformly used to test different MSA methods. Instead, due to both the computational and biological issues raised by the problem of sequence alignment optimization, a multiplicity of scenarios need to be modelled in benchmark datasets.

A telling symptom of the current state of affairs is the fact that subjective manual editing of sequence alignments remains widespread, reflecting perhaps an overall lack of confidence in the performance of automated multiple alignment strategies. The criteria used when editing sequence alignments "by eye" are vague and may introduce individual biases and aesthetic considerations into sequence alignment [9, 21].

In order to ensure reproducibility of experimental results, one of the most important goals of scientific practice, this trend needs to change. Context-specific, effective benchmarking with well-defined objectives represents a sensible way forward.

Acknowledgments

The authors thank Julie Thompson for helpful feedback on the manuscript. CD is supported by SNSF advanced researcher fellowship #136461. This article started as assignment for the graduate course "Reviews in Computational Biology" at the Cambridge Computational Biology Institute, University of Cambridge.

References

1. Kemena C, Notredame C (2009) Upcoming challenges for multiple sequence alignment methods in the high-throughput era. Bioinformatics 25(19):2455–2465
2. Aniba MR, Poch O, Thompson JD (2010) Issues in bioinformatics benchmarking: the case study of multiple sequence alignment. Nucleic Acids Res 38(21):7353–7363

3. Edgar RC (2010) Quality measures for protein alignment benchmarks. Nucleic Acids Res 38(7):2145–2153
4. Thompson JD, Linard B, Lecompte O, Poch O (2011) A comprehensive benchmark study of multiple sequence alignment methods: current challenges and future perspectives. PLoS One 6(3):e18093
5. Löytynoja A (2012) Alignment methods: strategies, challenges, benchmarking, and comparative overview. Methods Mol Biol 855:203–235
6. Thompson JD, Higgins DG, Gibson TJ (1994) CLUSTAL W: improving the sensitivity of progressive multiple sequence alignment through sequence weighting, position-specific gap penalties and weight matrix choice. Nucleic Acids Res 22(22):4673–4680
7. Morrison DA (2009) Why would phylogeneticists ignore computerized sequence alignment? Syst Biol 58(1):150–158
8. Mardis ER (2008) The impact of next-generation sequencing technology on genetics. Trends Genet 24(3):133–141. doi:10.1016/j.tig.2007.12.007
9. Anisimova M, Cannarozzi G, Liberles D (2010) Finding the balance between the mathematical and biological optima in multiple sequence alignment. Trends Evol Biol 2(1):e7
10. Stebbings LA, Mizuguchi K (2004) HOMSTRAD: recent developments of the Homologous Protein Structure Alignment Database. Nucleic Acids Res 32(Database issue): D203–D207
11. Thompson JD, Koehl P, Ripp R, Poch O (2005) BAliBASE 3.0: latest developments of the multiple sequence alignment benchmark. Proteins 61:127–136
12. Stoye J, Evers D, Meyer F (1998) Rose: generating sequence families. Bioinformatics 14(2):157–163
13. Cartwright RA (2005) DNA assembly with gaps (Dawg): simulating sequence evolution. Bioinformatics 21(Suppl 3):iii31–iii38
14. Hall BG (2008) Simulating DNA coding sequence evolution with EvolveAGene 3. Mol Biol Evol 25(4):688–695
15. Fletcher W, Yang Z (2009) INDELible: a flexible simulator of biological sequence evolution. Mol Biol Evol 26(8):1879–1888
16. Sipos B, Massingham T, Jordan GE, Goldman N (2011) PhyloSim – Monte Carlo simulation of sequence evolution in the R statistical computing environment. BMC Bioinformatics 12(1):104
17. Koestler T, Av H, Ebersberger I (2012) REvolver: modeling sequence evolution under domain constraints. Mol Biol Evol 29(9): 2133–2145
18. Dalquen DA, Anisimova M, Gonnet GH, Dessimoz C (2012) ALF-a simulation framework for genome evolution. Mol Biol Evol 29(4): 1115–1123
19. Thompson JD, Plewniak F, Poch O (1999) A comprehensive comparison of multiple sequence alignment programs. Nucleic Acids Res 27(13):2682–2690, gkc432 [pii]
20. Blackburne BP, Whelan S (2012) Measuring the distance between multiple sequence alignments. Bioinformatics 28(4):495–502. doi: 10.1093/bioinformatics/btr701
21. Löytynoja A, Goldman N (2008) Phylogeny-aware gap placement prevents errors in sequence alignment and evolutionary analysis. Science 320(5883):1632–1635. doi:10.1126/science.1158395
22. Golubchik T, Wise MJ, Easteal S, Jermiin LS (2007) Mind the gaps: evidence of bias in estimates of multiple sequence alignments. Mol Biol Evol 24(11):2433–2442
23. Huelsenbeck JP (1995) Performance of phylogenetic methods in simulation. Syst Biol 44(1): 17–48
24. Do CB, Mahabhashyam MS, Brudno M, Batzoglou S (2005) ProbCons: probabilistic consistency-based multiple sequence alignment. Genome Res 15(2):330–340. doi:10.1101/gr.2821705
25. Notredame C, Higgins DG, Heringa J (2000) T-Coffee: a novel method for fast and accurate multiple sequence alignment. J Mol Biol 302 (1):205–217. doi:10.1006/jmbi.2000.4042
26. Lassmann T, Sonnhammer ELL (2005) Automatic assessment of alignment quality. Nucleic Acids Res 33(22):7120–7128
27. Landan G, Graur D (2007) Heads or tails: a simple reliability check for multiple sequence alignments. Mol Biol Evol 24(6):1380–1383
28. Hall BG (2008) How well does the HoT score reflect sequence alignment accuracy? Mol Biol Evol 25(8):1576–1580
29. Chothia C, Lesk AM (1986) The relation between the divergence of sequence and structure in proteins. EMBO J 5(4):823
30. Mizuguchi K, Deane CM, Blundell TL, Overington JP (1998) HOMSTRAD: a database of protein structure alignments for homologous families. Protein Sci 7(11):2469–2471. doi:10.1002/pro.5560071126
31. Thompson JD, Plewniak F, Poch O (1999) BAliBASE: a benchmark alignment database for the evaluation of multiple alignment programs. Bioinformatics 15(1):87–88, btc017 [pii]

32. Van Walle I, Lasters I, Wyns L (2005) SABmark – a benchmark for sequence alignment that covers the entire known fold space. Bioinformatics 21(7):1267–1268. doi:10.1093/bioinformatics/bth493
33. Edgar RC (2004) MUSCLE: multiple sequence alignment with high accuracy and high throughput. Nucleic Acids Res 32(5): 1792–1797. doi:10.1093/nar/gkh340
34. Gardner P, Wilm A, Washietl S (2005) A benchmark of multiple sequence alignment programs upon structural RNAs. Nucleic Acids Res 33(8):2433–2439
35. Kim J, Sinha S (2010) Towards realistic benchmarks for multiple alignments of non-coding sequences. BMC Bioinformatics 11:54
36. Mathews DH (2005) Predicting a set of minimal free energy RNA secondary structures common to two sequences. Bioinformatics 21(10): 2246–2253. doi:10.1093/bioinformatics/bti349
37. Havgaard JH, Lyngso RB, Stormo GD, Gorodkin J (2005) Pairwise local structural alignment of RNA sequences with sequence similarity less than 40%. Bioinformatics 21(9): 1815–1824. doi:10.1093/bioinformatics/bti279
38. Altschul SF, Gish W, Miller W, Myers EW, Lipman DJ (1990) Basic local alignment search tool. J Mol Biol 215:403–410
39. Thompson JD, Fdr P, Ripp R, Thierry J-C, Poch O (2001) Towards a reliable objective function for multiple sequence alignments1. J Mol Biol 314(4):937–951. doi:10.1006/jmbi.2001.5187
40. Raghava GP, Searle SM, Audley PC, Barber JD, Barton GJ (2003) OXBench: a benchmark for evaluation of protein multiple sequence alignment accuracy. BMC Bioinformatics 4:47. doi:10.1186/1471-2105-4-47
41. Russell RB, Barton GJ (1992) Multiple protein sequence alignment from tertiary structure comparison: assignment of global and residue confidence levels. Proteins 14(2):309–323. doi:10.1002/prot.340140216
42. Pop M, Salzberg SL (2008) Bioinformatics challenges of new sequencing technology. Trends Genet 24(3):142–149. doi:10.1016/j.tig.2007.12.006
43. Berger SA, Stamatakis A (2011) Aligning short reads to reference alignments and trees. Bioinformatics 27(15):2068–2075. doi:10.1093/bioinformatics/btr320
44. Dessimoz C, Gil M (2010) Phylogenetic assessment of alignments reveals neglected tree signal in gaps. Genome Biol 11(4):R37
45. Jordan G, Goldman N (2011) The effects of alignment error and alignment filtering on the sitewise detection of positive selection. Mol Biol Evol 29:1125. doi:10.1093/molbev/msr272
46. Blackshields G, Wallace IM, Larkin M, Higgins DG (2006) Analysis and comparison of benchmarks for multiple sequence alignment. In Silico Biol 6(4):321–339
47. Lassmann T, Sonnhammer EL (2002) Quality assessment of multiple alignment programs. FEBS Lett 529(1):126–130, S0014579302031897 [pii]
48. Strope CL, Abel K, Scott SD, Moriyama EN (2009) Biological sequence simulation for testing complex evolutionary hypotheses: indel-Seq-Gen version 2.0. Mol Biol Evol 26(11): 2581–2593. doi:10.1093/molbev/msp174
49. Lassmann T, Sonnhammer EL (2006) Kalign, Kalignvu and Mumsa: web servers for multiple sequence alignment. Nucleic Acids Res 34 (Web Server issue):W596–W599. doi: 10.1093/nar/gkl191
50. Kemena C, Taly JF, Kleinjung J, Notredame C (2011) STRIKE: evaluation of protein MSAs using a single 3D structure. Bioinformatics 27 (24):3385–3391. doi:10.1093/bioinformatics/btr587

Chapter 5

BLAST and FASTA Similarity Searching for Multiple Sequence Alignment

William R. Pearson

Abstract

BLAST, FASTA, and other similarity searching programs seek to identify homologous proteins and DNA sequences based on excess sequence similarity. If two sequences share much more similarity than expected by chance, the simplest explanation for the excess similarity is common ancestry—homology. The most effective similarity searches compare protein sequences, rather than DNA sequences, for sequences that encode proteins, and use expectation values, rather than percent identity, to infer homology. The BLAST and FASTA packages of sequence comparison programs provide programs for comparing protein and DNA sequences to protein databases (the most sensitive searches). Protein and translated-DNA comparisons to protein databases routinely allow evolutionary look back times from 1 to 2 billion years; DNA:DNA searches are 5–10-fold less sensitive. BLAST and FASTA can be run on popular web sites, but can also be downloaded and installed on local computers. With local installation, target databases can be customized for the sequence data being characterized. With today's very large protein databases, search sensitivity can also be improved by searching smaller comprehensive databases, for example, a complete protein set from an evolutionarily neighboring model organism. By default, BLAST and FASTA use scoring strategies target for distant evolutionary relationships; for comparisons involving short domains or queries, or searches that seek relatively close homologs (e.g. mouse–human), shallower scoring matrices will be more effective. Both BLAST and FASTA provide very accurate statistical estimates, which can be used to reliably identify protein sequences that diverged more than 2 billion years ago.

Key word BLAST, FASTA, Homology, Similarity, Expectation value, Scoring matrices

1 Introduction

Identification of homologous sequences is an essential first step before Multiple Sequence Alignment. If multiply aligned sequences are not homologous, their alignment has no biological meaning. Unfortunately, Multiple Sequence Alignments do not provide the measures of statistical significance that are required to infer homology. The selection of a set of sequences for multiple alignment presumes that they are homologous; in this chapter we will discuss the inference of homology from sequence similarity searches using the popular programs BLAST and FASTA.

David J. Russell (ed.), *Multiple Sequence Alignment Methods*, Methods in Molecular Biology, vol. 1079,
DOI 10.1007/978-1-62703-646-7_5, © Springer Science+Business Media, LLC 2014

Homology has become a ubiquitous term in genome analysis and computational biology, but the inference and implications of homology—descent from a common ancestor—can be confusing. Two protein or DNA sequences are either homologous or they are not, but our ability to infer homology depends on context: the particular sequences and programs used, the library selected, and the statistical threshold chosen.

We infer that two sequences are "homologous" from excess similarity. When two sequences share more similarity than would be expected by chance, the most parsimonious explanation is that the sequences diverged from a common ancestor. Thus, this simple and widely accepted understanding of significance and homology has a statistical foundation; we cannot infer homology without some estimate of how often a similarity score might occur by chance. The distribution of chance scores depends on the search context; searches against large databases will produce higher scores on average, simply because there are more opportunities to produce a high score by chance. Thus, a similarity score that is clearly significant and provides strong evidence for homology in a search of the human protein set (about 40,000 sequences) might not be significant in the context of a search of 20,000,000 sequences, the current size of the largest protein databases. Context dependence is one of the several unsettling properties of homology inference; a statistically significant similarity score *can* be used to infer homology, but a nonsignificant score *cannot* be used to infer non-homology.

Likewise, our ability to infer homology from similarity searches depends on the query sequence used for the search. The significance/nonsignificance problem is frequently encountered in diverse protein families, where many members of a family share significant similarity to one member of the family, but others do not. A sequence from a highly populated part of a protein family tree, e.g., human protein kinases, can easily detect thousands of homologs with very significant scores, but a protein kinase from slime mold may find only a few dozen clear homologs. Strategies like `psiblast` that build models of protein families can reduce these differences by capturing a much larger fraction of the members of a family, but most diverse protein families will have members that are hard to identify from sequence alone.

A Multiple Sequence Alignment can still make sense when not every sequence shares significant similarity with every other sequence in the multiple alignment, as long as some combination of significant similarities can connect the members of the family. In diverse families, a sequence A may share significant similarity with family members B, C, and D, but not with E, F, and G. In this case, if B, C, or D shares significant similarity with E, F, and G, then one can infer that they all belong to the same family, and that a Multiple Sequence Alignment makes biological sense.

However, the transitive inference of homology between sequences that do not share significant similarity requires that the *same regions* (domains) of the sequences align. If protein B contains domains B1 and B2, and the B1 domain shares significant similarity with A, C, and D, and B2 with E, F, and G, then there is no reason to believe that A, C, and D are homologous to E, F, and G. However, if all seven full-length proteins A–D, rather than the individual domains, were included in a Multiple Sequence Alignment, many programs would align the unrelated residues, as many Multiple Sequence Alignment programs assume that the sequences being aligned are globally homologous.

To ensure that Multiple Sequence Alignments are built from homologous sequences, BLAST, FASTA, and other pairwise similarity searching programs have two central goals: (1) to identify sequences sharing excess similarity; (2) to ensure that statistical estimates are accurate. In this chapter, we describe programs and search strategies to perform sensitive searches that reliably identify homologs.

2 Using BLAST and FASTA

2.1 Selecting a Search Program

The most effective similarity searches perform sequence comparisons at the protein sequence level, either by comparing a protein sequence to a protein sequence database with `blastp`, `fasta`, or `ssearch`, or by using `blastx` or `fastx` (Table 1) to translate DNA query sequences "on-the-fly" and compare them to a protein sequence database at the protein level. Protein sequence comparison is 5–10-fold more sensitive than DNA:DNA comparison, and protein sequence databases are considerably less redundant than DNA sequence databases. Together, this means that protein or

Table 1
BLAST and FASTA programs

BLAST program	FASTA[a] program	Query	Library	Comments
`blastp`	`fasta`	Protein	Protein	Fast, sensitive protein comparison [6, 8, 23, 24]
`blastn`	`fasta`	DNA	DNA	*Only* for non-protein coding sequences. `blastn -task blastn` is required for sensitive searches
`blastx`	`fastx/ fasty`	DNA	Protein	Performs 6-frame translation of query with frameshifts [25]
`tblastn`	`tfastx/ tfasty`	Protein	DNA	

[a]The names of the FASTA programs are typically followed by a major version number, e.g., `fasta36` or `ssearch36`. These numbers are not shown

Table 2
BLAST-specific programs

BLAST program	Query	Library	Comments
`psiblast`	Protein	Protein	Highly sensitive iterative protein similarity searches [8]
`rpsblast`	Protein	PSSMs	Protein searches against a PSSM library of conserved domains [8]
`tblastx`	DNA	DNA	More sensitive DNA:DNA comparison at the translated DNA level [6, 8]
`megablast`	DNA	DNA	(`blastn -task megablast`) high-speed DNA matching for Near-identical sequences (e.g. ESTs vs genomes). The default for `blastn`

translated DNA searches return 10–50-times as many homologs as DNA:DNA searches, with much more reliable statistical estimates. Both the BLAST[1] and FASTA software packages offer different programs for different similarity searching programs, depending on the source of the query sequences. Table 1 summarizes the programs in the BLAST and FASTA packages that have very similar functions; Table 2 lists programs found only in the BLAST package; Table 3 lists FASTA-specific programs.

Both packages offer heuristic strategies for rapid sequence comparison (`blast`, `fasta`). For more sensitive searches the BLAST package provides `psiblast` and `rpsblast`, which use models of protein families encoded as Position Specific Scoring Matrices. Almost all the BLAST package programs use some version of the heuristic approach to calculate *local* similarity scores implemented by the BLAST algorithm. BLAST package typically requires specially formatted databases, created with `makeblastdb`, except when only two sequences are compared.[2]

In addition to heuristic methods, the FASTA package offers a variety of optimal search algorithms, including the local Smith –Waterman algorithm [2] implemented in `ssearch`, as well as optimal global alignment algorithms (`ggsearch` and `glsearch`). An implementation of the `psisearch` strategy that uses the optimal `ssearch` program is available at the European Bioinformatics Institute (www.ebi.ac.uk/Tools/sss, [3]). FASTA also provides the `lalign` program, which uses the `sim` [4] implementation

[1]The programs listed in Tables 1 and 2 are part of the NCBI BLAST+ distribution [1]. An earlier version of the BLAST distribution used the `blastall` program with the `-p blastp` option to specify the specific program.

[2]Previous versions of BLAST provided `bl2seq` to compare two sequences in FASTA format. The BLAST programs now provide the "Blast2Sequences" mode by using `-subject` option, rather than the `-db` option.

Table 3
FASTA-specific programs

FASTA[a] program	
`ssearch`	Compare a protein sequence to a protein sequence database or a DNA sequence to a DNA sequence database using the Smith–Waterman [2] algorithm
`ggsearch/glsearch`	Compare a protein sequence to a protein sequence database or a DNA sequence to a DNA sequence database using global:global or global:local alignment
`lalign`	Compare two protein sequences or DNA sequences reporting all significant non-overlapping local alignments using the Waterman–Eggert algorithm [5] as implemented by Huang [26] (`sim4`)
`fasts/m/ tfasts/m`	Compare a set of short un-ordered [S] or ordered [M] peptides or oligonucleotides to a protein/translated DNA or DNA database [27]
`fastf/ tfastf`	Compare a set of short "mixed peptide" sequences to a protein or translated DNA [28]

[a] The names of the FASTA programs are typically followed by a major version number, e.g., `fasta36` or `ssearch36`. These numbers are not shown

of the Waterman–Eggert algorithm for non-overlapping local alignments [5] to identify internally repeated domains.

The *local* alignment strategies used by BLAST and FASTA are ideal for searches for shared homologous domains, or with partial query sequences, because they identify the best local alignment between two sequences, ignoring the unrelated sequence context. *Global* sequence alignment programs require alignments to extend from the beginning to the end of the sequences and can be more effective in capturing conserved domain morphology over an entire protein. The FASTA package also offers two optimal global alignment programs. `ggsearch` computes an alignment score that is global for both the query sequence and library sequence; it is particularly useful for functional inference, since it requires all the domains in the homologous protein to be present. `glsearch` calculates an alignment that is global in the query sequence (e.g., a full-length domain) but can be local in the library sequence. Because of its requirement for global similarity, `ggsearch` only aligns library sequences that are between 75 and 133 % the length of query; likewise `glsearch` only aligns library sequences that are more than $\geq$ 75 % the length of the query.

The FASTA package also includes several programs designed to align unordered short peptides (`fasts`) or ordered sets of non-contiguous oligonucleotides (`fastm`). `fasts` is particularly useful for aligning the peptides produced by Mass Spectrometry proteomic sequencing.

2.2 BLAST and FASTA Differences

Although they share similar goals and strategies, BLAST and FASTA differ in several respects: (1) BLAST and FASTA use a different strategies for estimating statistical significance (though the resulting estimates are very similar); (2) FASTA supports more database formats and output alignment options; (3) there are cosmetic differences on how the query, database, scoring parameters, and options are specified.

Statistical differences—The BLAST program introduced rapid heuristic similarity searching based on statistical thresholds [6]. With the development of "Karlin–Altschul" local similarity score statistics [7], it became possible to set thresholds based on statistical parameters; only sequence alignments that could produce "significant" scores were examined, minimizing alignment computations on unrelated sequences. For the original BLAST, which focused on combining ungapped alignments (HSPs), the statistical parameters could be calculated analytically, but the introduction of gapped-BLAST [8] required that the parameters be estimated for standard scoring matrices and gap penalties by simulating unrelated sequences. As a result, the BLAST programs offer a fixed set of scoring matrices and gap penalties.

FASTA uses a different approach, which calculates an approximate similarity score for every sequence in the database. FASTA assumes that it has calculated thousands of unrelated similarity scores in every database search and uses these scores to estimate the required statistical parameters (if only a few sequences are compared, unrelated sequences are produced by shuffling the library sequences). As a result of this assumption, FASTA includes an option to shuffle every sequence in the library if the library is not "representative" (`-z 11`). Since the FASTA programs estimate statistical parameters in every search, the programs provide much more flexibility in scoring matrix and gap-penalty schoice. About a dozen scoring matrices are built-in to the FASTA programs; other scoring matrices can be provided from files. The FASTA programs also allow arbitrary gap penalties.

The most common cause of misleading statistical significance estimates (*E*()- or expect-values) is low-complexity regions in proteins. Older versions of `blastp` used the `seg` program [9] to identify and mask-out low-complexity regions. The current version of BLAST uses a more sophisticated strategy by default [10]. The FASTA programs can search sequence databases that are "soft-masked" by indicating low-complexity regions with lowercase amino acids by using the `-S` option. The `pseg` program, available from the National Center for Biotechnology Information (NCBI) (ftp://ftp.ncbi.nlm.nih.gov/pub/seg/pseg), or the `segmasker` program, part of the BLAST distribution, can be used to soft-mask entire sequence databases with lowercase characters for low complexity.

Table 4
FASTA sequence file formats

0	FASTA (>SEQID - comment/sequence)
1	Uncompressed Genbank (LOCUS/DEFINITION/ORIGIN)
3	EMBL/Uniprot (ID/DE/SQ)
6	GCG (version 8.0) Unix Protein and DNA (compressed)
7	FASTQ (sequence only, quality ignored)
10	Library subset list
12	NCBI Blast (`makeblastdb` format
16	MySQL (requires special compilation)
17	Postgres (requires special compilation)

Because `fasta` estimates statistical parameters from the unrelated sequences in the database, `fasta` DNA:DNA expectation values are probably more accurate than `blastn` values, but both expectation values are far less reliable than protein:protein and translated-DNA:protein estimates. For protein:protein and translated-DNA:protein searches, expectation ($E()$) values $< 10^{-3}$ provide strong evidence for homology (one non-homolog in 1,000 searches). For DNA:DNA searches, $E()$-values $> 10^{-10}$ are suspect.

Library formats—The BLAST programs can either compare two sequences in `FASTA` format, or search databases formatted with the `makeblastdb` command, which converts a `FASTA` or `ASN.1` format file into a set of indexed binary sequence files that can be searched very efficiently. As noted above, current versions of comprehensive protein and DNA databases that are pre-formatted for BLAST searches are available from the NCBI (ftp.ncbi.nlm.nih.gov/blast/db), and subsets of these databases can be constructed using the `gilist` option. But restricting searches to files that can be downloaded from the NCBI restricts searches to the NCBI *ecosystem*; some bioinformatics resources, such as the Pfam database and GeneOntology links are more easily accessed using UniProtKB/Swiss-Prot proteins and accessions, so investigators running local copies of the BLAST programs will need to run `makeblastdb`. Likewise, `makeblastdb` is required when searching local sequence collections.

The FASTA program can read query and library databases in popular formats, including `FASTA`, `makeblastdb` (BLAST), and `FASTQ`, formats (Table 4). In addition, the FASTA programs can read databases comprised of multiple files in different sequence formats. In addition to conventional "flat-file" (`FASTA`, GenBank, EMBL/Uniprot) and binary `makeblastdb`/BLAST formats, the

FASTA programs can read sequences from MySQL and PostgreSQL databases. FASTA can also read library subsets; format 10 libraries work like BLASTP `-gilist` searches, but also allow a more general strategy for identifying sequence subset identifiers. Protein databases can use lower-case residues to indicate low-complexity residues, which are ignored when the `-S` option is used. While `makeblastdb` can be used to produce FASTA format 12 databases, it is rarely necessary. FASTA can search most widely used sequence formats.

Command line differences—BLAST and FASTA both offer a diverse set of command line options that modify the behavior and output of the programs. Both BLAST and FASTA provide a list of popular options with the `-h` option, and a more comprehensive list with the `-help` option, e.g., `blastp -help`. Popular BLAST options are outlined in Table 5; popular FASTA options are listed in Table 6. The BLAST programs use command line options (`-db`, `-query`, `-matrix`) to provide all the information the program needs, including the name of the query file, database, scoring matrix, etc. FASTA uses command line options (`-s matrix`, `-f gap-open`, `-g gap-extend`) to modify default search parameters but expects the `query.file` and `library.file` to be specified after all program options. Thus, the scoring matrix `-s BP62` option below:

```
fastx -s BP62 query.file library.file
```

must precede the `query.file` and `library.file` arguments.

2.3 Where to Search?

Searching—Web-based or local?—Widely used searching programs like BLAST and FASTA can be run either through web interfaces, on local computers, or in cloud computing environments like Amazon Web Services. The BLAST programs were developed at the NCBI and are tightly integrated into the NCBI's web site (blast.ncbi.nlm.nih.gov/Blast.cgi). All the programs in the FASTA package are available at the European Bioinformatics Institute (EMBL-EBI) web site (www.ebi.ac.uk/Tools/sss); the EMBL-EBI also provides the BLAST programs. Similarity searching on the web is convenient; investigators can be confident that they are using a current version of the search program to search comprehensive and up-to-date databases. Interactive web access is often the quickest way to build a comprehensive set of sequences from a protein family for Multiple Sequence Alignment. For more time-consuming analyses (e.g., characterization of the thousands of sequences from a finished microbial genome), both the NCBI and EMBL-EBI offer programmatic access to their web sites, so that a computer script or program can launch large numbers of searches and collect the results.

For large-scale analyses, for example from millions of metagenomics sequence reads, the similarity searching programs will typically be run on local computers or a local computer cluster, or

Table 5
BLAST command-line options

BLASTP/N/X options	
-query file_in	Input file name
-query_loc	[start-stop] Location on the query sequence
-task	blastp -task blastp blastp-short; blastn -task megablast blastn blastn-short
-db	BLAST database name
-gilist	Restrict search of database to list of GI's
-out	Output file name
-evalue	Expectation value (E) threshold for saving hits; Default = '10'
-word_size	Word size (≥ 2) for wordfinder algorithm
-frame_shift_penalty	(blastx/tblastn) frameshift cost (not allowed by default)
-gapopen	Cost to open a gap
-gapextend	Cost to extend a gap
-matrix	Scoring matrix name (normally BLOSUM62), BLOSUM45, BLOSUM80, PAM70, PAM30
-ungapped	Perform ungapped alignment only
-comp_based_stats	Use composition-based statistics for blastp / blastx/ tblastn (on by default)
-outfmt	Alignment view options: 0 = pairwise [default]
-num_descriptions	Number one-line descriptions for database entries [500]
-num_alignments	Number alignments [250]
-html	Produce HTML output
BLAST-2-Sequences options	
-subject	Subject sequence(s) to search
-subject_loc	Location on the subject sequence (Format: start-stop)
Query filtering options	
-seg	Filter query sequence with SEG ([no], yes, "window locut hicut"
-soft_masking	Apply filtering locations as soft masks [false]
-lcase_masking	Use lower case filtering in query and subject sequences
-db_soft_mask	Filtering algorithm ID to apply to the BLAST database as soft masking
Restrict search options	
-gilist	Restrict search of database to list of GI's
-negative_gilist	Restrict search of database to everything except the listed GIs
Statistical options	
-dbsize	Effective length of the database
-searchsp	Effective length of the search space
Miscellaneous options	
-num_threads	Number of threads to use in the BLAST search [1]

in a cloud computing environment [11]. Running a local copy of the BLAST or FASTA programs provides the researcher with some control over the time required for the analysis, ensures that the searches are reproducible (the version of the program and reference database will remain constant), and allows searches to be performed against the most appropriate database for the research question being addressed. Moreover, web implementations of the search programs may impose output constraints that can be removed in local implementations. The NCBI BLAST programs can be downloaded from ftp://ftp.ncbi.nlm.nih.gov/blast/executables/LATEST. The FASTA programs can be downloaded from http://faculty.virginia.edu/wrpearson/fasta or ftp://ftp.ebi.ac.uk/pub/software/unix/fasta.

Sequence databases—The biggest challenge in running the programs locally is keeping the sequence databases current. While the programs change slowly, protein and DNA sequence databases change daily or weekly. Even when the download process is scripted and runs automatically, the downloading and reformatting process is time consuming and can fail unexpectedly. The scripting expertise required to keep sequence comparison programs and databases up-to-date may be better used to build scripted interfaces to the NCBI and EMBL-EBI web resources.

Comprehensive protein and DNA sequence databases can be downloaded from the NCBI (ftp://ftp.ncbi.nlm.nih.gov/blast/db/) and the EMBL-EBI (ftp://ftp.ebi.ac.uk/pub/databases/). Selection of the appropriate database for similarity searching is discussed below (Subheading 4.1), but the most sensitive and efficient searches are performed against protein databases, which are relatively compact (< 20 GB for the largest protein sets). DNA sequence datasets are many orders of magnitude larger and highly redundant; except in rare cases searches should be performed against protein sets, or selected DNA subsets should be found.

Similarity searching on the "Cloud"—Recently, comprehensive sets of bioinformatics programs, including BLAST and FASTA, have been packaged as instances for the Amazon Web Services cloud computing environment [11]. This packaging makes it easier to cheaply set up the computing infrastructure necessary for a large-scale analysis project, as programs are collected from diverse sources, installed and tested. The Cloud BioLinux environment also provides access to many model organism genomes, but the focus seems to be on DNA read mapping; few protein sequence databases are available within the Amazon Web Services environment. Using the Bio-Linux environment is more convenient than downloading and installing dozens of bioinformatics programs, but access to current protein sequence databases is much easier using the Web search interfaces at the NCBI and EMBL-EBI.

Table 6
FASTA command-line options

FASTA option		BLAST option
-b:	high scores reported (limited by -E by default);	-num_descriptions
-d:	number of alignments shown (limited by -E by default)	-num_alignments
-e:	expand_script to extend hits	
-E:	[10,1] E()-value,E()-repeat threshold	-evalue
-f:	[-10] gap-open penalty	-gapopen
-F:	[0] min E()-value displayed	
-g:	[-2] gap-extension penalty	-gapextend
-h	help—show options, arguments	-h
-m:	[0] output/alignment format;	-outfmt
-M:	filter on library sequence length	
-O:	write results to file	-out
-p/-n	protein/nucleotide query	blastp/blastn
-r:	[+5/-4] +match/-mismatch for DNA/RNA	-reward / -penalty
-s:	[BL50] Scoring matrix: (protein) BL50, BL62, PAM250, OPT5, VT160, VT120, BL80; VT80, VT40, VT20, VT10; scoring matrix file name;	-matrix
-s	?BL50 adjusts matrix for short queries;	
-S	filter lowercase (seg) residues	-lcase_masking
-T:	max threads/workers	-num_threads
-V:	annotation characters (phospho-sites, variation) in query/library for alignments	
-z:	[1] statistics estimation method: 1–6-regression, MLE, etc.; 11–16-estimates from shuffled library sequences; 21–26-E2()-stats from shuffled high-scoring sequences;	-comp_based_stats
-Z:	[library entries] database size for E()-value	-dbsize

Summary—Both BLAST and FASTA provide a comprehensive set of protein:protein, translated-DNA:protein, and DNA:DNA sequence similarity searching programs. The BLAST package extends the heuristic BLAST approach [6, 8] in two directions: `psiblast` for more sensitive iterated protein sequence comparisons, and `blastn -megablast`, for rapid mapping of DNA sequences against genomes. Recent improvements in the BLAST programs have focused on improved statistical estimates, particularly for protein query sequences with biased amino-acid composition [10, 12, 13].

The FASTA package has grown as well; in addition to heuristic strategies, it now offers accelerated optimal algorithms for Smith–Waterman local protein alignment, global–global and global–local alignment, and specialized algorithms for short sequences. For general-purpose protein and translated DNA sequence local similarity searches, the programs in Table 1 will give very similar results; both provide statistical significance estimates as expectation values (*E*()-values), and both provide comparable scaled "bit" scores for comparing results over different searches and database sizes.

Most performance differences between BLAST and FASTA reflect the different scoring matrices, gap, extend, and frameshift penalties used by default. The BLAST family of proteins typically use the BLOSUM62 [14] matrix with a gap-open penalty of −11 and an gap-extension penalty of −1 (a cost of −12 for one residue gap); the FASTA programs use BLOSUM50 with lower effective gap penalties. The FASTA parameters allow higher sensitivity for very distantly related sequences but require longer alignments.[3] By default the `fastx` and `fasty` programs allow frameshifts in alignments, just as they allow gaps; `blastx` can allow frameshifts with the `-frame_shift_penalty` option.

3 Inferring Homology: Interpreting Results

3.1 Use Expect or Bit Scores, Not Percent Identity, to Infer Homology

BLAST and FASTA provide a variety of similarity measurements from which one can infer homology. BLAST provides a `bit` score, the `E-value` or `Expect`, the percent identity, percent positives, and the alignment length. The FASTA programs provide a `bit` score, *E*()-value, percent identity, and percent similarity.[4] In addition, the FASTA programs provide a variety of "raw" similarity scores that reflect the various stages of the heuristic FASTA algorithm (e.g., `init1`, `initn`, `opt`), or the single optimal "raw" score (`s-w`) for `ssearch`. The `bit` score and `Expect/E()` values of BLAST and FASTA are comparable and describe the number of times the alignment score would be expected by chance. Thus, of all the different scores provided by BLAST and FASTA, the `Expect/E-value` is the *one* score that unambiguously reports the statistical significance of the match.

---

[3]The FASTA programs provide a variable scoring matrix option that shifts the scoring matrix for shorter query sequences. The BLAST programs provide the `-task blastp-short` or `-task blastn-short` for short protein:protein and DNA:DNA searches.

[4]BLAST's percent positive counts aligned residues with a score > 0; FASTA's fraction similar includes aligned residues with scores ≥ 0.

Expectation values—The `E-value` of an alignment reflects both the alignment score and the size of the database from which the alignment was identified. Thus, FASTA *E*()-values are reported in the context of a number; `E(100000) = 1E-6` indicates that this alignment would occur one time in a million searches of a database of size 100,000.[5] For example, an alignment of two 400 residue proteins with a 40-bit alignment score would have an *E*(4,000)-value $\leq mn2^{-\text{bits}}D = 400 \times 400 \times 2^{-40} \times 4,000 \leq 0.0006$ (m, n are the lengths of the query and library sequence; D is the database size); the same 40-bit alignment found in the RefSeq database ($\sim$ 13 million entries) would have *E*(13,000,000) $\leq$ 1.9, a value 3,000 times greater (and no longer statistically significant, Subheading 4.1).

The *E*()-value or `Expect` reports the statistical significance of an alignment score in the context of a database search. *E*()-values between 0.001 and 0.01 are widely used as a threshold for inferring homology (`psiblast` uses 0.005 as its default for including a new sequence into the PSSM profile). An *E*()-value of 0.001 implies that the alignment score would happen only once in 1,000 searches by chance. However, in metagenomics and other large-scale analyses, millions of similarity searches may be run, so an alignment with an *E*()-value of 0.001 would be expected to occur $0.001 \times 1,000,000 = 1,000$-times by chance. Thus, for large-scale analyses, much more conservative statistical thresholds are often applied; 10^{-6} to 10^{-10} or even lower. While very strict (10^{-10}) thresholds for large-scale searches can dramatically reduce the number of false-positive assignments, such conservative significance thresholds increase the number of false negatives. Many distantly related clear homologs will have *E*()-values between 10^{-10} and 10^{-3}.

Bit scores—Because an *E*()-value is database-size (search-size) dependent, many investigators record bit scores, rather than *E*()-values, in large-scale analyses. The formula for converting bit scores to *E*()-values is shown above, but, as a rule of thumb, alignments with scores < 40-bits are never statistically significant; scores between 40 and 50 bits, are only significant in relatively small databases; and scores > 50-bits will be significant in databases as large as 10,000,000 entries. A one bit bit-score change corresponds to a twofold change in statistical significance, so a 10-bit increase in score improves the statistical significance $\sim$ 1,000-fold. At the NCBI web site BLAST summary, alignments with scores from 50 to 80 bits are plotted as green bars; these alignments will be statistically significant in almost any database context (likewise, alignments < 40-bits are plotted in black; they are never significant).

[5]The BLAST programs use a slightly different formulation of the Expect value; rather than using the number of entries in the database, BLAST uses the combined length of all the sequences in the database. For average length proteins, the result of the two calculations is identical.

Percent-identity—While *E*( )-values provide the most direct estimate of statistical significance, and bit scores provide a database-independent measure of alignment strength, investigators often use percent identity to describe the likelihood that two sequences are homologous. In general, if two sequences are 30 % identical across their entire length, they can reliably be inferred to be homologous. This "rule of thumb" correctly identifies homologs, but it misses large numbers of clearly homologous proteins.

Many alignments with *E*( )-values $< 10^{-6}$ and bit scores greater than 60 will be less than 30 % identical. For example, in a comparison of *E. coli* protein to human proteins, there are 10,417 human: *E. coli* homologs with *E*() $< 10^{-6}$, but only 36 % of these are $\geq$ 30 % identical. Percent identity is far less sensitive than expectation values and bit scores because it cannot distinguish between common and rare identities, and it does not count conservative amino-acid replacements. Percent identities can give a useful measure of evolutionary distance (for example, on average mammalian orthologs are about 80 % identical), but the 30 % identity threshold excludes large numbers of homologs that are readily identified with *E*( )-values and bit scores.

3.2 Confirming Statistical Significance

Ideally, if *E*()-values are accurate, then one can have confidence that sequences sharing a similarity score expected one time in 1,000 by chance are almost certainly homologous. Unfortunately, low-complexity regions, biased amino-acid composition, and unusual sequence lengths, can violate statistical assumptions about protein sequences, resulting in low-expectation values for unrelated sequences. When the FASTP program was introduced in 1985 [15], it included a program that estimated the statistical significance of a similarity score by shuffling one of the two aligned sequences, and recording the number of standard deviations separating original unshuffled alignment score from the mean of the shuffled sequence alignment scores. The guidelines for inferring homology in that paper did not account for database size, but the shuffling strategy is still available in the FASTA programs to evaluate the statistical significance of an alignment score. When any of the FASTA programs are used to compare two sequences, the statistical significance of the unshuffled alignment is estimated by shuffling the second sequence and applying the appropriate statistical distribution. The `-k shuffle-count` command line option sets the number of shuffles performed (`-k 250` by default). The `-v shuffle-window-size` performs local window shuffles; `-v 20` produces each shuffled sequence by shuffling residues 1–20, 21–40, etc. `-v` window-shuffled sequences preserve local composition biases in the shuffled proteins, e.g., transmembrane domain regions.

In pairwise comparisons involving a protein and translated-DNA sequence, e.g., `fastx` or `tfastx`, `fastx` will provide more accurate shuffled statistical estimates. All the FASTA programs shuffle the second (library) sequence to produce statistical estimates, and shuffles of protein sequences, which are produced by `fastx`, more accurately reflect the distribution of unrelated sequence scores. Randomly shuffled DNA sequences are a less accurate model of unrelated DNA sequences.

By default, the expectation values provided by the FASTA programs when only two sequences are compared, and by the BLAST programs in `bl2seq` mode (e.g., `blastp -query seq1 -target seq2`) are based on a database size of one sequence, rather than the size of the database that was initially searched to identify the candidate homolog. Since the *E*()-value is the product of the pairwise alignment score probability and the database size ($E = p \times D$), the two-sequence expectation values will be 10,000–10,000,000 times more significant than those calculated in the original search, depending on the original database size. For the FASTA programs, the expectation value can be adjusted with the `-Z dbentry` option, e.g., `ssearch -Z 500000 seq1 seq2` would increase the expectation value 500,000-fold, to reflect the fact that `seq2` was originally found in a search of UniProtKB/Swiss-Prot (which contains about 500,000 entries). Without this correction, an alignment found in a search of the `refseq_protein` database (13 million entries) with an expectation of 10 would have an apparent expectation of $10/13,000,000 = 7E-7$, which is incorrect, because it ignores the size of the database where the "similarity" was originally found. The expectation values produced by `blastp` in two-sequence mode must be corrected for the original database size as well.

In addition to the *automatic* shuffled statistical estimates that are produced whenever two sequences are compared by the FASTA programs, `fasta`, `fastx`, and `ssearch` can display two expectation values using the `-z 21` command line option. When `-z 21` is used, two expectation values are reported: (1) the standard expect value calculated from the distribution of similarity scores calculated in the search and (2) a second *E2*() value calculated by shuffling the high scoring sequences found in the initial search. For average composition proteins, the *E*() and *E2*() values will be very similar, but for biased composition proteins, the *E2*() value will be more conservative. The *E2*() value can be helpful in translated searches, where out-of-frame translations can produce biased composition low-complexity regions.

The BLAST programs do not explicitly provide statistical estimates based on shuffled sequences, but it is possible to confirm the accuracy of BLAST statistical estimates by looking for the highest scoring (lowest `expect`) unrelated sequence in the list of high-scoring sequences. If the statistical estimates are accurate, the highest scoring unrelated sequence should have an `expect` value ~ 1. Since lack of significant similarity cannot be used to infer non-homology (unrelatedness); additional analyses must be done to identify the highest scoring unrelated sequence. One strategy is to perform a "reverse" search with the candidate non-homolog, particularly if the query or library (target) sequence come from large protein families. If there are no sequences with significant alignment scores shared by the initial query sequence in the first search and the candidate non-homolog library sequence in the second search, it is much less likely that they are homologous. Alternatively, if the query and candidate non-homolog do not share any domains annotated by Pfam [16] or other domain database, they are probably not homologous.

3.3 Establishing Homology Boundaries

BLAST and FASTA (and SSEARCH) calculate *local* sequence alignments; the boundaries of the alignments are calculated to maximize the similarity score. If the alignment were longer or shorter, the similarity score would be worse. In contrast, global similarity scores require that the alignment extend to the ends of the aligned sequences. Local similarity scores will always be positive; global scores, even for proteins that contain homologous domains, can be positive or negative. Local alignment scores have been universally adapted for similarity searching for several reasons: (1) the statistical theory for local similarity scores is well understood; (2) local similarity scores can identify locally homologous domains in different protein contexts; (3) local scores work well for partial sequences; and (4) local sequences can be used to identify homologous exons in long stretches of chromosomal DNA.

While statistically significant *local* sequence similarity can be used to reliably infer homology, the overall homology of two aligned proteins or DNA sequences does not guarantee that every aligned residue-pair reflects homology, particularly at the ends of the alignment. For *local* sequence alignments, the boundaries of the alignment, i.e., whether it stops at residue n or residue $n + 5$, depends strongly on the scoring matrix. As discussed below (Subheading 4.2), an evolutionarily "deep" or sensitive scoring matrix (BLOSUM62 or BLOSUM50) will produce longer alignments than "shallow" scoring matrices (VT20, PAM30), even between unrelated sequences. (Unrelated or random alignments will not have statistically significant scores, but they will be longer with "deep" matrices.) Because they depend on the scoring matrix, alignment boundaries between two homologous domains flanked by non-homologous regions do not always stop at the end of the homology; the homologous alignment can be "over-extended" into non-homologous sequence [17].

Homologous over-extension was first recognized in genomic DNA sequence alignment [18]; more recently it was shown to be the major cause of `psiblast` Position Specific Scoring Matrix contamination [17], which can dramatically reduce the selectivity of `psiblast` searches [3, 17]. Versions of `psiblast` and

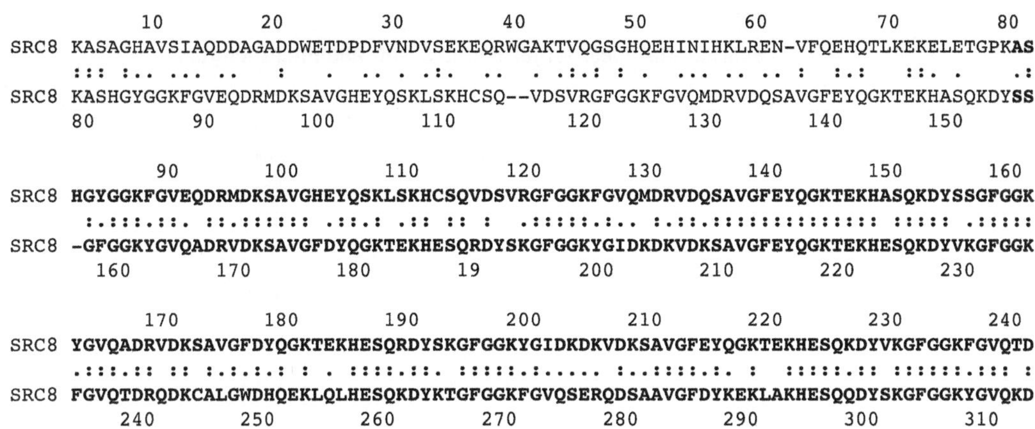

Fig. 1 Homologous over-extension—self-alignment of cortactin (SRC8_HUMAN) using the lalign program using BLOSUM62, −11/−1. Five significant alignments are produced in addition to the 100 % identity alignment. Shown is the second most significant alignment, with *E*() < 10^{-29}, which aligns cortactin domains 1–5 with 3–7

psisearch that dramatically reduce over-extension are available at the EMBL-EBI [3]. In proteins, alignment of relatively closely related domains (> 50 % identity) using a matrix like BLOSUM62 or BLOSUM50, which target alignments with 25–30 % identity, can cause the alignment to extend well beyond the homologous domain boundaries into non-homologous sequences (Fig. 1).

Homologous over-extension is most easily recognized when the average percent identity of the aligned sequences differs dramatically from one part of the alignment to another. For example, Fig. 1 shows a local lalign self-alignment of cortactin (SRC8_HUMAN) using BLOSUM62, −11/−1, between cortactin domains 1 and 5 and homologous cortactin domains 3–7. Across the homologous regions (residue 80− on the top and 155− on the bottom, highlighted in bold), the domains average 60 % identity. But the first 80 residues of the alignment shown are non-homologous and average less than 15 % identity.

When the SRC8_HUMAN is aligned with itself using VT80, a scoring matrix appropriate for sequences that are 50 % identical, the non-homologous over-extension part of the alignment disappears; the alignment begins at residue 80 in the first sequence and residue 155 in the second. The protein scoring matrices available with the FASTA programs make it possible to target the scoring matrix to the evolutionary distance of the aligned sequences using the target identity information in Table 7. Using the appropriate matrix can dramatically reduce homologous over-extension.

For very distantly related sequences, it is very difficult to identify and correct homologous over-extension, and homologous "under-extension"—an alignment that covers the most similar parts of two

Table 7
Scoring matrices, target identity, and alignment lengths

Scoring Matrix	Target % ident.	bits/ pos.	50-bit align len.
VT10	91.1	3.45	14
VT20	83.2	2.92	17
VT40	69.8	2.27	22
PAM30[a]	53.3	1.47	34
VT80	50.2	1.39	36
PAM70[a]	41.7	0.966	52
BLOSUM80[a]	41.7	1.04	48
VT120	39.3	1.06	47
BLOSUM62[a]	28.6	0.439	114
BLOSUM50	25.0	0.216	231
VT160	24.4	0.288	174

[a] Using default BLASTP gap penalties

very distantly related domains but does not extend over the full length of the homology—is equally likely to occur. Homologous under-extension can sometimes be recognized by identifying intermediate distance homologs, just as transitive similarity can be used to recognize distant homologs. If domain A aligns to domain B over 200 amino-acids, using the appropriate scoring matrix, and domain B aligns to domain C for 200 amino-acids, then it makes sense to include all 200 amino-acids of all three proteins in a Multiple Sequence Alignment, even if domain A only aligns to 100 amino-acids of domain C.

Summary—BLAST and FASTA produce accurate sequence alignment expectation values; expectation values < 0.001 can be used to reliably infer homology in single searches; lower (more stringent) thresholds are required when multiple searches are performed. Expectation values capture the effect of database size; larger databases produce larger (worse) expectation values for the same alignment score. For this reason, the bit score can be used to roughly characterize the significance of an alignment independent of algorithm or scoring parameters. Alignments scoring greater than 50 bits are almost always significant; 40–50-bit alignment scores are significant when small databases are searched; < 40 bits are never significant. The significance of very surprising, but weakly significant, alignments can be confirmed using shuffled sequence statistical estimates. Homology boundaries can be improved by matching the scoring matrix to the evolutionary distance of the homologous domains.

4 Improving Search Performance

4.1 Selecting the Database to Search

For most similarity searches, the choice of program (`blastp` vs `fasta` or `blastx` vs `fastx`) is far less important than the choice of database to be searched. As emphasized earlier, the most important choice an investigator can make is to search a protein sequence database, using either `blastp/fasta` for protein:protein comparisons, or `blastx/fastx` for translated-DNA queries against a protein sequence database. Protein sequence searches have dramatic advantages: (1) they provide an evolutionarily look-back time that is 5–10-times greater than DNA:DNA comparison; (2) the statistical estimates from protein/translated-DNA:protein alignments are many orders of magnitude more accurate than DNA:DNA statistical estimates; (3) even the largest protein databases are hundreds of times smaller than DNA datasets, and comprehensive protein sequence searches can be done against a few hundred thousand to a few million sequences. Twenty years ago, there were clear homologs in DNA sequence databases that had not yet been entered into the protein databases. This is no longer true; protein sequences are rapidly imported from genomic DNA sequencing projects, and the protein databases are so comprehensive that there are very few proteins to be found. Protein sequence databases should be searched first.

4.1.1 Statistics of Similarity Scores: Searches of Smaller Databases are More Sensitive

Protein databases have become very large in the post-genome decade. As this is written in Fall, 2012, comprehensive databases like NCBI's `nr` and `refseq_protein` databases have 13- (`refseq`) to 20-million sequences; the UniProtKB/TrEMBL database (uniprot.org) contains more than 25-million protein sequences. Much of the increase in the most comprehensive `nr`, `refseq`, and UniProtKB/TrEMBL databases over the past 5 years has been driven by genome sequencing projects. As a result, these databases have become very redundant,[6] reducing search sensitivity.

Homology can be inferred from statistically significant similarity; protein sequence alignment scores expected less than once in 1,000 searches ($E() < 10^{-3}$) are most easily explained by inferring common ancestry. But the use of a statistical criteria for inferring homology means that a similarity score that is significant in some contexts may not be significant in others.

[6]There are more than 2.5 million *E. coli* protein sequences from 200+ genomes available from the NCBI protein databases.

The change in expectation value from $E(4{,}000) \leq 0.0006$ to $E(13{,}000{,}000) \leq 1.9$ calculated in the previous section does not mean the sequences are no longer homologous; it simply means that their common ancestry cannot be distinguished from the 40-bit alignment scores that would be produced by chance because of the large size of the database. Thus, pairwise `blastp`, `fasta`, and `ssearch` searches should be performed against the smallest comprehensive databases that are likely to contain a homolog. For sequences from vertebrates, the human protein set (30,000–40,000 entries) is likely to contain homologs for all the sequences that can be detected. Likewise, searches against taxonomic subsets of sequence databases will improve sensitivity and dramatically reduce the computation required.

Protein sequence databases differ not only in their size and redundancy, but they also differ in their annotation quantity and quality. The Swiss-Prot [19] subdivision of the UniProtKB Knowledgebase provides a rich set of annotations and links to other biological databases. UniProtKB/Swiss-Prot entries typically provide links to popular protein domain databases, homologous structures, E.C. numbers for enzymes, and information on functionally critical residues and sequence variation. Both the NCBI and EMBL-EBI web sites provide searches against the Swiss-Prot database, which currently contains about 500,000 entries.

The NCBI's `refseq_protein` database can also provide rich links to other biological resources. Unlike UniProtKB/Swiss-Prot entries, each `refseq_protein` sequence is linked to a `refseq_mrna` entry; this allows Multiple Sequence Alignments with `refseq_proteins` to be converted to DNA-sequence multiple alignments, which can be used for DNA-based and codon-based evolutionary analyses. `refseq_protein` entries are also linked to the NCBI Entrez-Gene resource, which provides links to variation, clinical, and expression databases. Significant alignments in searches against taxonomic subsets of `refseq_protein` yield rich genetic information. Searches against full `refseq_protein` are less sensitive, because the database is almost as large as `nr`, the largest protein database offered at the NCBI.

Unfortunately, by default both the NCBI and EMBL-EBI web sites offer their largest protein databases (`nr` at the NCBI, `UniProtKB` at the EMBL-EBI) for searches. At the NCBI, the `refseq_protein` database is far more informative, and the NCBI offers organism-specific search pages that improve statistical significant 100-fold or more. At the EMBL-EBI, the UniProtKB/Swiss-Prot database provides the most richly functionally annotated protein sequence dataset available; the EMBL-EBI also offers a comprehensive set of organism-specific sequence sets.

Searching subsets of databases—Comprehensive protein sequence databases are very large (more than ten-million sequences) and can be very redundant (the largest protein families in `refseq_-protein` have more than 30,000 members). Thus, searching representative portions of sequence databases, or focussing on taxonomically close sequences, can be far more efficient than searching complete datasets. Both the BLAST and FASTA programs provide options for searching database subsets. The BLAST programs accept the `-gilist` option, which specifies the specific `gi` numbers to be searched in a much larger database. This option requires that the database being searched contains `gi` numbers (all the databases available from the NCBI do, but databases from the EMBL-EBI and UniProt do not). The `-gilist` option provides a powerful tool for searching sequences from selected representative organisms, as the NCBI Entrez site makes it very easy to download a list of sequences from an organism, or the results of an Entrez query. For example, searching Entrez protein sequences with the query:

srcdb_refseq[prop] AND txid9606[orgn]

provides a list of 35,615 `refseq` protein sequences from human (`txid9606`). The sequences can be downloaded in FASTA format from the search result page, but it is much easer to simply download the `GI List` to a file, and then use `blastp` or `blastx` with the `-gilist` option. The `gilist` option is available after any NCBI/Entrez query into protein and DNA databases; one can just as easily for all proteins with the term "glutathione transferase" in their name.

The FASTA programs provide multiple options for searching subsets of larger sequence databases. Unlike BLAST, which only searches one database format (BLAST `makeblastdb` format), the FASTA programs can search sequences in BLAST, FASTA, and several other less widely used formats. In addition, FASTA can search protein sequences stored in a MySQL or PostgreSQL relational database, or subsets of sequence databases defined by lists of GI numbers or accession identifiers. With the FASTA programs, the format of the sequence database is specified as part of its name; by default a FASTA format database is searched (Table 4). The FASTA version of the `gilist` option (format 10, library subset list) works both with NCBI sequence databases (which have a `gi` number) and with EMBL-EBI and Uniprot databases, which use sequence identifiers or accessions. The FASTA library subset list can use either numbers or strings to identify library sequence subsets.

The FASTA programs can also search portions of a sequence database by using `SELECT` statements on a MySQL or PostgreSQL database. Format 16 and 17 files provide SQL select statements for getting the complete set of sequences, getting individual sequence entries, and getting entry annotations.

While both BLAST and FASTA provide options for searching a subset of a sequence database, FASTA provides the additional ability to search a subset of a database, but then use the significant library "hits" to the query sequence to align to additional sequences by projecting the smaller database onto a larger, more comprehensive (and possibly redundant) database. The `-e expand.sh` script option specifies a script that can return an additional set of sequences to be aligned, based on the sequences that were found in the initial search. For example, if the initial search returns the sequence `GSTM1_HUMAN` from a search of human UniProt proteins, and a search of Swiss-Prot with human proteins identifies homologs from other vertebrates, then the scores and alignments are shown not only with the original `GSTM1_HUMAN` but also with the other sequences, which were not present in the initial search, but were linked and returned by the `-e expand.sh` expansion script. This allows searches to be performed against small, representative datasets, but return results as if the additional sequences were included. The strategy can also be used to align mRNAs (`fastx -e expand.sh`) against all known isoforms of a gene, after initially searching only the canonical form of the protein.

4.1.2 psiblast Works Best with Large Databases

Pairwise sequence similarity programs like `blastp` and `fasta` can become less sensitive as database size increases, because larger databases produce more high alignment scores by chance. Iterative programs, like `psiblast`, can take advantage of the diversity in large comprehensive database searches to dramatically improve search sensitivity. Thus, while smaller databases can make `blastp/blastx` and `fasta/fastx/ ssearch` more effective, `psiblast` performs best when used against larger databases, like `refseq_protein`. If there are only a small number of very distant homologs to the query, then smaller databases will be more effective. But if there are many homologs that lack useful annotations, `psiblast` can sometimes build a sensitive PSSM that can find a well-annotated homolog (however, very distantly related sequences are less likely to share a function).

4.2 Changing Scoring Matrices and Gap Penalties

The BLAST and FASTA programs are optimized for identifying distantly related sequences with full-length protein and gene-length DNA sequences. Most investigators searching for homologs to build a Multiple Sequence Alignment will do best by using the default search parameters provided by `blastp` (BLOSUM62 scoring matrix, −11/−1 for gap-open and gap-extend penalties) or `fasta/ssearch` (BLOSUM50, −10/−2). `blastp` and `fasta/ssearch` search parameters have been extensively evaluated over a very wide range of evolutionary distances and query sets; changing the parameters almost always reduces sensitivity.

The scoring matrix and gap penalties should be changed (1) for searches with partial-length (short) query sequences and (2) when maximum sensitivity is not the first priority. Scoring matrices and gap penalties have a implicit target evolutionary distance; the BLOSUM62 −11/−1 parameters used by `blastp` target alignments that are about 30 % identical; the `fasta/ssearch` scoring parameters (BLOSUM50, −12/−2) target 25 % identical alignments (Table 7). But the scoring parameters that work best for distant relationships require long alignments [20]; to provide 50 bits of statistical significance, BLOSUM62 −11/−1 must align > 100 amino-acids, and BLOSUM50 −12/−2 must align > 200 amino-acid residues (Table 7). BLOSUM62 and BLOSUM50 work well for full-length proteins or protein domains that are longer than 100–200 residues, but searches with short- (< 150 nt) or even medium- (300–400 nt) read lengths produce translated protein sequences in between 50 and 133 amino-acids and require shallower scoring matrices.

Shallow matrices for short sequences—Shallower scoring matrices allow short query sequences to produce significant similarity scores (bit scores). For searches with shorter query sequences, `blastp` provides the `-task blastp-short` option for query sequences shorter than 30 amino-acids, which shifts the scoring matrix to `-matrix PAM30`. `blastn` provides the `-task blastn-short` option, which strengthens the mismatch penalty from +1/−2 to +1/−3 and shifts the target percent identity from ~ 90 % to more than 99 %. `blastx` does not have a `-task blastx-short` option, but `-matrix PAM30` has a similar effect, and dramatically improves the expectation values in searches with query sequences shorter than 100 nt. Investigators performing large-scale `blastx` searches with datasets that include shorter DNA queries should sort their query sequences by length, and then use `-matrix PAM30` for query sequences shorter than about 120 nt, `-matrix PAM70` for queries from 120 to 300 nt long, and the default `-matrix BLOSUM62` for queries longer than 300 nt.

The FASTA programs provide a more finely graded set of scoring matrices, and the programs can automatically adjust the scoring matrix based on the length of the query sequence. FASTA scoring matrices are set using the `-s matrix-name` option, where `matrix-name` can be one of the sixteen matrices; scoring matrices include BLOSUM50, BLOSUM62 [14] and VT10 . . .VT200 [21]. In addition, the FASTA programs can use scoring matrix values provided in a file, so any scoring matrix can be used. To accommodate searches with different query lengths, the FASTA programs offer a *variable* scoring matrix option; the `-s ?BP62` option indicates that the BLOSUM62 matrix, with the −11/−1 gap penalties used by `blastp` be used for long queries, but the "?" indicates that the scoring matrix should be adjusted to ensure that the query can produce a 40-bit score against an average length protein sequence. When a short sequence is encountered, the program selects the VT series matrix with sufficient entropy (bits per position) to produce a significant score. The variable scoring matrix option ensures that an appropriate scoring matrix for the query sequence length will be used automatically.

Short queries produce short alignments that require shallower scoring matrices, but short alignments are also produced by short domains or short exons. Shallower scoring matrices can be used to annotate closely related genomes. For example, comparison of Drosophila gene models from *D. pseudoobscura* to *D. melanogaster* proteins (two organisms that diverged about 25 Mya and whose average ortholog identity is about 80 %) using BLOSUM62 would require a 80 amino-acid, or 240 nt exon, while VT20, the matrix appropriate for their evolutionary distance, could detect exons as short as 75 nt. With `blastx`, PAM30 is the shallowest matrix available, but it can be combined with `blastx -ungapped` to produce higher sensitivity with short sequences.

Shallow matrices for close relationships—`blastp`, `blastx`, `fasta`, `fastx`, and `ssearch` are very sensitive sequence comparison programs; they readily detect evolutionary relationships between sequences that diverged more than 2 billion years ago. But today, particularly when annotating a newly sequenced genome from an organism that is only a few hundred million years from a model organism, significant alignments against distant homologs can be very distracting. For example, if one wants to know how many class-mu glutathione transferases are present in the Chicken genome, it is not helpful to see a report that also includes much more distantly related family members from nematodes, flies, and bacteria.

Scoring matrices can be used to set the evolutionary look-back time of `blastp/blastx` or `fasta/fastx/ ssearch`. As shown in Table 7, the `VT20` matrix performs best for sequences that are about 80 % identical, the average evolutionary distance between primates and rodents. Using a scoring matrix to set evolutionary look-back time is much more reliable and efficient than trying to extract the closely-related sequences from a long list that includes both closely and distantly related sequences. For example, comparison of `GSTM1_HUMAN` with chicken `refseq` proteins using `fasta` with the BLOSUM62 matrix and $-11/-1$ gap penalties produces eight significant alignments, ranging from 24 to 66 % identity (the same results are found with `blastp`). Using the `VT80` matrix, which is appropriate for sequences sharing 50 % identity or more (Table 7; human–chicken orthologs share about 60 % identity, on average), reduces the number of significant homologs found from 8 to 3; the high scoring homolog ($E() < 10^{-120}$) is a 66 % identical orthologous class-mu chicken sequence, while the other two significant alignments align over fewer than 50 amino-acids.[7] Thus, by using a scoring matrix that is appropriate for the average protein identity between human and chicken, it is much easier to identify orthologous sequences—sequences that differ because of the mammal:reptile speciation event. Closely related (orthologous) sequences are more reliably identified using a scoring matrix that reflects the evolutionary distance of the organisms being compared; in general, searches between organisms that have diverged over the past 400 million years should certainly use "shallower" (less evolutionarily distant) scoring matrices.

Changing gap penalties—Like the default scoring matrices, the gap penalties used by default by BLAST and FASTA are designed to find distant evolutionary relationships. The BLAST program provides a limited range of gap penalties for the scoring matrices it supports and most alternative penalties are more stringent. Increasing the gap penalty, like choosing a shallower scoring matrix, will improve the statistical significance of shorter, or more closely related, alignments. The FASTA programs do not limit the gap-penalty choices, but scoring matrices have matched default gap penalties that are appropriate for the matrix target percent identity [22]. In general, the FASTA defaults should not be reduced (made less stringent); increasing the penalties can sometimes improve significance for shorter alignments. While strong arguments can be made for adjusting scoring matrices to match short domain/exon lengths and short evolutionary distances, current theory does not provide a rationale for changing the default gap penalties for `blastp/blastx` and `fasta/fastx/ssearch` similarity searches.

Summary—Protein sequence databases provide the most sensitive searches for homologs; but modern protein databases are so large that it is often more efficient, both statistically and computationally, to search smaller, representative database, such as complete protein sets from model organisms. The FASTA programs offer an option to align against a larger set of sequences selected by "expanding" the original set of hits. In general, BLAST and FASTA search parameters are set to find very distant relationships for long sequences (but `blastn` uses the much less sensitive rapid `-megablast` option by default). Searches with short queries, for short domains, or over relatively short evolutionary distances (< 500 My), should use shallower scoring matrices.

[7] Similar results are found with `blastp` with the PAM70 matrix, though the less stringent gap penalties used by `blastp` produce longer alignments.

5 Summary

Multiple Sequence Alignment requires homologous sequences; the BLAST and FASTA programs calculate both alignment scores and accurate estimates of their statistical significance that can be reliably used to infer homology. Protein alignment scores with $E() < 0.001$ in a single search reliably reflect homology—sequences that have descended from a common ancestor. Searches for homologs are far more sensitive at the protein level. Protein sequences change more slowly than DNA sequences, providing greater evolutionary look back time; protein alignments have more accurate statistical estimates; and protein databases are dramatically smaller than DNA databases. Today, most sequences are determined as DNA, but `blastx` and `fastx` can automatically translate those sequences and compare them to protein databases. Search sensitivity can be increased by searching smaller, representative databases, and using scoring matrices that are targeted to the length and evolutionary distance of the sequences of interest. Because protein sequence databases have become so diverse, it is rare that a query sequence does not find homologs; the most common reason for failing to find homologs is a query sequence that is largely low-complexity or strongly biased amino-acid composition. It is routine to find homologs between human and bacterial proteins that last shared a common ancestor more than 2.5 billion years ago. Sequence comparison has improved dramatically since it became generally available more than 25 years ago; sequence databases are far more comprehensive, and statistical estimates are far more reliable. It has become much easier to identify homologs, which can provide more data for Multiple Sequence Alignments.

References

1. Camacho C, Coulouris G, Avagyan V, Ma N, Papadopoulos J, Bealer K, Madden TL (2009) Blast+: architecture and applications. BMC Bioinformatics 10:421
2. Smith TF, Waterman MS (1981) Identification of common molecular subsequences. J Mol Biol 147:195–197
3. Li W, McWilliam H, Goujon M, Cowley A, Lopez R, Pearson WR (2012) PSI-Search: iterative HOE-reduced profile ssearch searching. Bioinformatics 28:1650–1651
4. Huang X, Hardison RC, Miller W (1990) A space-efficient algorithm for local similarities. Comput Appl Biosci 6:373–381
5. Waterman MS, Eggert M (1987) A new algorithm for best subsequences alignment with application to tRNA–rRNA comparisons. J Mol Biol 197:723–728
6. Altschul SF, Gish W, Miller W, Myers EW, Lipman DJ (1990) A basic local alignment search tool. J Mol Biol 215:403–410
7. Karlin S, Altschul SF (1990) Methods for assessing the statistical significance of molecular sequence features by using general scoring schemes. Proc Natl Acad Sci USA 87:2264–2268
8. Altschul SF, Madden TL, Schaffer AA, Zhang J, Zhang Z, Miller W, Lipman DJ (1997) Gapped BLAST and PSI-BLAST: a new generation of protein database search programs. Nucleic Acids Res 25:3389–3402
9. Wootton JC, Federhen S (1993) Statistics of local complexity in amino acid sequences and sequence databases. Comput Chem 17:149–163

10. Yu Y, Wootton JC, Altschul SF (2003) The compositional adjustment of amino acid substitution matrices. Proc Natl Acad Sci USA 100:15688–15693
11. Krampis K, Booth T, Chapman B, Tiwari B, Bicak M, Field D, Nelson KE (2012) Cloud biolinux: pre-configured and on-demand bioinformatics computing for the genomics community. BMC Bioinformatics 13:42
12. Schaffer AA, Aravind L, Madden TL, Shavirin S, Spouge JL, Wolf YI, Koonin EV, Altschul SF (2001) Improving the accuracy of PSI-BLAST protein database searches with composition-based statistics and other refinements. Nucleic Acids Res 29:2994–3005
13. Altschul SF, Wootton JC, Gertz EM, Agarwala R, Morgulis A, Schaffer AA, Yu Y (2005) Protein database searches using compositionally adjusted substitution matrices. FEBS J 272:5101–5109
14. Henikoff S, Henikoff JG (1992) Amino acid substitution matrices from protein blocks. Proc Natl Acad Sci USA 89:10915–10919
15. Lipman DJ, Pearson WR (1985) Rapid and sensitive protein similarity searches. Science 227:1435–1441
16. Bateman A, Birney E, Cerruti L, Durbin R, Etwiller L, Eddy SR, Griffiths-Jones S, Howe KL, Marshall M, Sonnhammer ELL (2002) The pfam protein families database. Nucleic Acids Res 30:276–280
17. Gonzalez MW, Pearson WR (2010) Homologous over-extension: a challenge for iterative similarity searches. Nucleic Acids Res 38:2177–2189
18. Zhang Z, Berman P, Miller W (1998) Alignments without low-scoring regions. J Comput Biol 5:197–210
19. UniProt Consortium (2011) Ongoing and future developments at the universal protein resource. Nucleic Acids Res 39:D214–D219
20. Altschul SF (1991) Amino acid substitution matrices from an information theoretic perspective. J Mol Biol 219:555–565
21. Mueller T, Spang R, Vingron M (2002) Estimating amino acid substitution models: a comparison of dayhoff's estimator, the resolvent approach and a maximum likelihood method. Mol Biol Evol 19:8–13
22. Reese JT, Pearson WR (2002) Empirical determination of effective gap penalties for sequence comparison. Bioinformatics 18:1500–1507
23. Pearson WR, Lipman DJ (1988) Improved tools for biological sequence comparison. Proc Natl Acad Sci USA 85:2444–2448
24. Pearson WR (1996) Effective protein sequence comparison. Methods Enzymol 266:227–258
25. Pearson WR, Wood TC, Zhang Z, Miller W (1997) Comparison of DNA sequences with protein sequences. Genomics 46:24–36
26. Huang X, Miller W (1991) A time-efficient, linear-space local similarity algorithm. Adv Appl Math 12:337–357
27. Mackey AJ, Haystead TAJ, Pearson WR (2002) Getting more from less: algorithms for rapid protein identification with multiple short peptide sequences. Mol Cell Proteomics 1:139–147
28. Damer CK, Partridge J, Pearson WR, Haystead TAJ (1998) Rapid identification of protein phosphatase 1-binding proteins by mixed peptide sequencing and data base searching. Characterization of a novel holoenzymic form of protein phosphatase 1. J Biol Chem 273:24396–24405

Part II

Alignment Techniques

Chapter 6

Clustal Omega, Accurate Alignment of Very Large Numbers of Sequences

Fabian Sievers and Desmond G. Higgins

Abstract

Clustal Omega is a completely rewritten and revised version of the widely used Clustal series of programs for multiple sequence alignment. It can deal with very large numbers (many tens of thousands) of DNA/RNA or protein sequences due to its use of the mBED algorithm for calculating guide trees. This algorithm allows very large alignment problems to be tackled very quickly, even on personal computers. The accuracy of the program has been considerably improved over earlier Clustal programs, through the use of the HHalign method for aligning profile hidden Markov models. The program currently is used from the command line or can be run on line.

Key words Multiple sequence alignment, Progressive alignment, Protein sequences, Clustal

1 Introduction

Clustal Omega [1] is a package for performing fast and accurate multiple sequence alignments (MSAs) of potentially large numbers of protein or DNA/RNA sequences. It is the latest version of the popular and widely used Clustal MSA package [2, 3]. Clustal Omega retains the basic progressive alignment MSA approach of the older ClustalX and ClustalW implementations, where the order of alignments is determined by a so called guide-tree, which in turn is constructed from pairwise distances amongst the sequences. The main improvements over ClustalW2 are (1) use of the mBed algorithm for creating guide trees of any size [4] and (2) a very accurate profile–profile aligner, based on the HHalign package [5].

As a first step a traditional progressive aligner calculates all $N(N-1)/2$ pairwise distances amongst all N input sequences. This may be computationally too demanding for much more than 10,000 sequences. The mBed algorithm, as implemented in Clustal Omega, reduces the time and memory complexity for guide tree calculation from $O(N^2)$ to $O(N(\log(N))^2)$. This is achieved by calculating the pairwise distances of all N sequences with respect

David J. Russell (ed.), *Multiple Sequence Alignment Methods*, Methods in Molecular Biology, vol. 1079, DOI 10.1007/978-1-62703-646-7_6, © Springer Science+Business Media, LLC 2014

to $(\log(N))^2$ randomly chosen seed sequences only. The fast pairwise distance calculation routines, based on a ktuple alignment algorithm, have been retained from the previous Clustal programs. The pairwise distances are then clustered, using a bisecting k-means algorithm [6]. Groups of sequences are bisected until a certain threshold for the cluster size is reached. In the current version this threshold is hard-wired to 100. Guide-tree construction within the clusters and amongst the clusters makes use of the tree building routines in Muscle [7]. This dendrogram is referred to as a guide-tree to emphasize that it is only used to guide the progressive alignment—it is not a reliable guide to the phylogeny of the sequences. Guide-tree construction will be skipped if only two sequences are to be aligned or if an externally constructed guide-tree is inputted.

In the profile–profile alignment phase sequences are aligned in larger and larger groups, according to the branching order in the guide-tree. At each stage of this final step, two alignments are aligned. Initially these are single sequences, but they grow with the addition of new sequences as one traverses the guide-tree. The alignment of residues and the positioning of gaps during each profile–profile alignment are fixed and cannot be undone at a later profile–profile alignment higher up in the tree. The main algorithmic change over ClustalW2 is a new profile–profile engine, based on the HHalign software [5]. HHalign is entirely based on Hidden-Markov Models (HMMs). Sequences and intermediary profiles are converted into HMMs, which are aligned in turn. It is also possible to input a HMM in addition to the unaligned sequences, and to use this external HMM during the profile–profile alignment stage. This is referred to External Profile Alignment (EPA). There are two HMM alignment algorithms: the accurate and memory-hungry Maximum Accuracy (MAC) algorithm and the faster, less accurate and more memory efficient Viterbi algorithm. The MAC algorithm is the default, and Viterbi is activated automatically only if the system resources are exhausted.

Sequence input to Clustal Omega is handled by the Squid routines (http://selab.janelia.org/software.html), and permissible input formats are a2m (fasta/vienna), clustal, msf, phylip, selex and stockholm. Output can be in the same formats.

The maximum number of sequences and lengths that can be aligned will depend on the machine being used. The number of sequences primarily affects the distance matrix calculation. Storing an mBed matrix for $N = 10{,}000$ sequences takes up approximately 14 MB of memory. A full distance matrix would take up almost 400 MB. Both alternatives are clearly feasible on a modern desktop computer. For $N = 100{,}000$ the mBed matrix will take up 220 MB, while the full distance matrix will require about 40 GB which may require a higher end machine. The length of the individual input sequences also contributes to the memory consumption of the tree-building phase via the ktuple alignment distance calculation amongst pairs of sequences. The length of the final alignment, and therefore by extension the lengths of the input sequences, however, mostly impacts on the profile–profile alignment phase. For every profile–profile alignment the MAC algorithm constructs six $L_1 \times L_2$ matrices of double variables, where L_1 and L_2 are the lengths of the two profiles to be aligned. An alignment of two profiles, each 100 residues in length will therefore require $8 \times 6 \times 100 \times 100 = 480{,}000$ bytes. The maximum alignment length for a machine with 2 GB would therefore be two profiles of $\sqrt{(2\text{GB}/(6 \times 8))} = 6,688$ positions in length each, or equivalent (*see* **Note 1**). The number of sequences affects the resource requirements of the profile–profile alignment stage only indirectly, in that it influences how much the lengths of the intermediary profiles grow from the lengths of the individual sequences. This growth is difficult to predict and depends, amongst other factors, on the similarity of the sequences.

The time required for the profile–profile alignment stage is a function of the number of sequences N, the lengths of the intermediary profiles L and the shape of the guide-tree. An MSA of N sequences requires $N - 1$ profile–profile alignments; increasing the number of sequences increases the number of profile–profile alignments linearly and therefore the alignment time will also grow in a linear fashion, at least in simple cases. Increasing the lengths of the input sequences clearly will increase the lengths of the intermediate profiles. Building up the HMM matrices requires a multiple of $L_1 \times L_2$ operations, so increasing the lengths of the sequences will increase the matrix construction times in a quadratic fashion. The guide-tree topology affects the profile–profile alignment times in a subtle way. Roughly speaking, alignments generated using a balanced tree will require less time than using an imbalanced (chained) tree. For example, on a single core of a 64 bit 3.0 GHz machine with 4 GB of RAM it takes just over 5 min to construct the tree and align 50,000 zinc-finger sequences of average length 23 residues, It takes 25 min for 20,000 and 68 min for 50,000 sdr sequences of average length 163 residues. It takes 106 min for 20,000 p450 sequences of average length 331 amino acids.

The current implementation of Clustal Omega is command-line driven. There is as of yet no GUI and no interactive menu but it is hoped to have one in place during 2013. A list of all permissible command-line arguments is available by typing -h (--help) on the command-line. There is an exhaustive help file explaining all command-line arguments and their usage in detail. The help file also contains many examples, elucidating the use of all individual command-line arguments and a range of typical combinations of command-line arguments.

2 Materials

Clustal Omega is available on-line for interactive usage. Two sites offering Clustal Omega are:

http://www.ebi.ac.uk/Tools/msa/clustalo/ (*see* Fig. 1).

http://mobyle.pasteur.fr/cgi-bin/portal.py#forms::clustalO-multialign.

Clustal Omega source code and executables can be downloaded from http://www.clustal.org/omega/.

Executables are provided for Linux (32/64 bit), Mac (64 bit), Windows (32 bit) and FreeBSD (32/64 bit).

To compile Clustal Omega from source one has to un-tar the distribution and cd into the un-tar-ed directory, configure and make. For example, if the tar-ball is called Clustal-Omega-1.0.3 then a typical installation might require (do not type the "$", this is

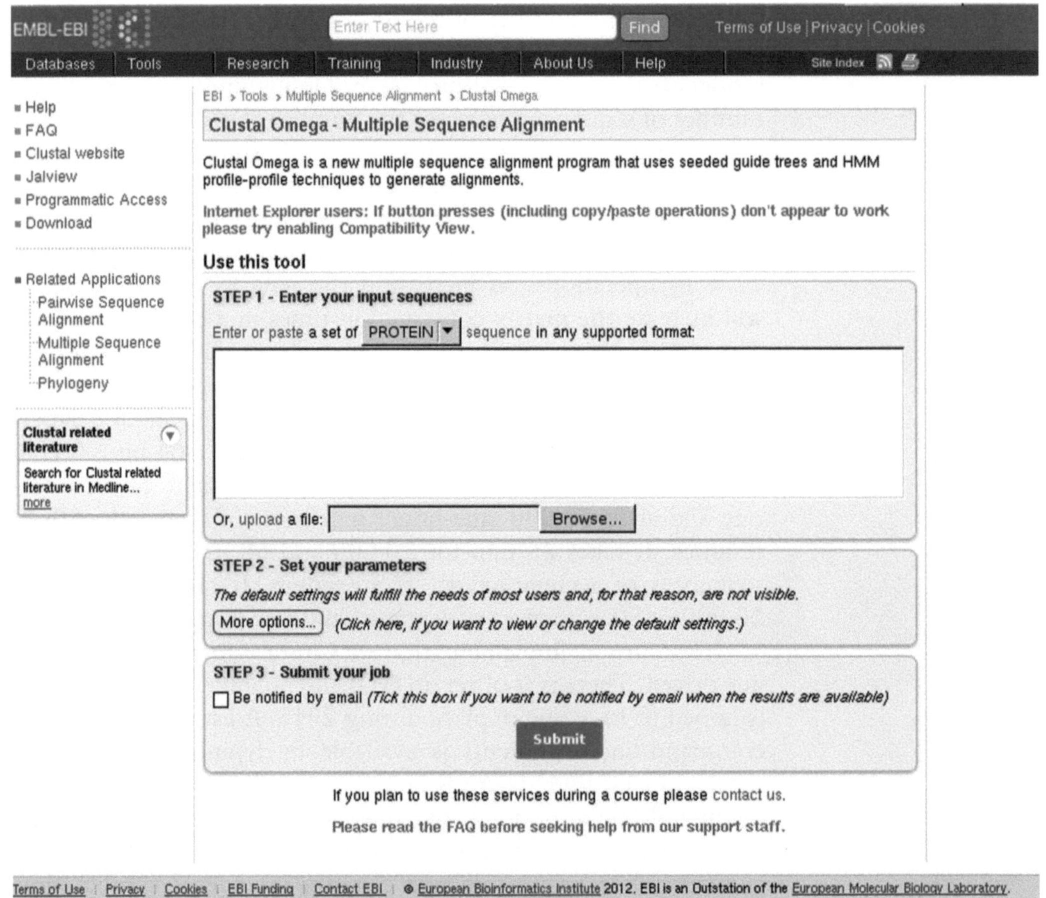

Fig. 1 Screen shot of the Clustal Omega Web page on the EBI Web site: http:/www.ebi.ac.uk/Tools/msa/clustalo/

the command-line prompt, different operating systems and/or shells may have different command-line prompts, like ">" or "%"):

```
$ tar -xvf Clustal-Omega-1.0.3
$ cd Clustal-Omega-1.0.3
$ ./configure
$ make
$ make install
```

This should configure, build, and install the Clustal Omega package (*see* **Notes 2–4**). The last step may require root/sudo privileges. However, make will still compile a Clustal Omega executable, which can be moved to any location in the user's file tree and can then be invoked by specifying the full path-name, for example:

/home/clustal-user/path/to/where/clustal/is/located/clustalo

3 Methods

For now, Clustal Omega can only be run from the command-line. To obtain a brief list of all available Clustal Omega command-line flags type:

```
$ clustalo -h
```

and hit `<return>`.

Parts of the Clustal Omega code have been multi-threaded, using OpenMP. By default, Clustal Omega will attempt to use all available threads. To limit the number of threads one can specify --threads.

Despite Clustal Omega being very fast for many purposes, some alignments may take a long time. In this case it may be reassuring to track the real-time progress of the alignment. This can be done by specifying the -v flag. This will print to screen what phase the calculation is in (distance matrix calculation, k-means clustering, guide-tree construction, progressive alignment). Repeating the -v flag a second time increases the level of verbosity, giving a more detailed progress report. Triple -v is the highest level of verbosity, giving details about distances, tree building and intermediate alignments. This level is only useful for the smallest alignments.

3.1 Basic Multiple Sequence Alignment

The most basic use of Clustal Omega involves aligning a number of unaligned sequences that are all contained in a single file. For example, if the file globin.fa contains more than two unaligned sequences in fasta format, then:

```
$ clustalo -i globin.fa
```

will read in the file, align the sequences and output the alignment to screen (default) in the default (fasta) format.

```
$ clustalo -i globin.fa -o globin.sto --outfmt=st
```

If the file globin.sto does not exist, then Clustal Omega reads the sequence file globin.fa, aligns the sequences and prints the result to globin.sto in Stockholm format. If the file globin.sto does exist already, then Clustal Omega terminates the alignment process before reading globin.fa.

```
$ clustalo -i globin.fa -o globin.aln --outfmt=clu
--force
```

Clustal Omega reads the sequence file globin.fa, aligns the sequences and prints the result to globin.aln in Clustal format, overwriting the file globin.aln, if it already exists.

```
$ clustalo -i globin.fa --guidetree-out=globin.dnd
--force
```

Clustal Omega reads the sequence file globin.fa, aligns the sequences, prints the result to screen in fasta/a2m format (default) and the guide-tree to globin.dnd, overwriting this file if it already exists (*see* **Notes 5** and **6**).

```
$ clustalo -i globin.fa --guidetree-in=globin.dnd
```

Clustal Omega reads the files globin.fa and globin.dnd, skipping distance calculation and guide-tree creation, using instead the guide-tree specified in globin.dnd. The alignment is outputted to screen in fasta format.

3.2 External Profile Alignment (EPA)

As mentioned in Subheading 1 Clustal Omega is a progressive aligner. This means that residues that are aligned and gaps that are positioned during an early stage of the alignment process remain fixed throughout the rest of the process and cannot be changed. An alignment of two residues that appears to be advantageous at an early stage may indeed turn out to be suboptimal in the later alignments. External Profile Alignment (EPA) is a way to provide the alignment process with a certain degree of "foresight." If the final alignment can be anticipated, then this prior knowledge can be encoded as a HMM. This may be, because the user knows they are aligning, for example, globins. Precomputed HMMs for globins are available from repositories such as Pfam (http://pfam.sanger.ac.uk/). Alternatively, the user could have produced a manually curated high-quality alignment of sequences that are homologous to the input set. This alignment can then be converted into a HMM using, for example, HMMER [8].

Clustal Omega accepts these external profiles-HMMs as input, accompanying the unaligned sequences. During the alignment stage sequences and profiles are first aligned to the external profile, and pseudo-counts from the HMM are transferred to the internal HMM used to align the sequences progressively. The desired effect of this is to "nudge" particular residues and gaps towards the position where they are expected to end up in the final alignment.

Individual sequences and small profiles are most vulnerable to misalignment. On the other hand large profiles have already built up more pseudo-counts and are more likely to resemble the final alignment. Clustal Omega therefore up-weights the pseudo-count transfer for single sequences and intermediate alignments with small numbers of sequences and reduces the transfer to larger intermediate alignments. Pseudo-count transfer to alignments larger than, say, 10 is negligible. Using EPA increases the profile–profile alignment time approximately threefold. Firstly, each of the two profiles is aligned to the external HMM, and finally the pre-aligned profiles have to be aligned themselves. As of yet Clustal Omega only accepts one external profile-HMM. This is not a problem for a HMM that extends over the entire length of the final alignment. However, if the unaligned sequences extend over multiple domains, and HMMs are only known for the individual domains, then only one of these HMMs can be submitted. This is a current limitation, which will hopefully be rectified in a future version of Clustal Omega.

To use a HMM in conjunction with unaligned sequences, first determine the appropriate HMM. For example, a search of the Pfam Web site for "globin" finds PF00042 which is the Pfam family for globins. Go to "Curation & model" of the PF00042 page and download the HMM called PF00042.hmm. Then type:

```
$ clustalo -i globin.fa --hmm-in=PF00042.hmm
```

3.3 Iteration

A useful means of refining an alignment is by "iterating" the alignment process. The guide-tree used for performing the initial alignment is based on pairwise distances between unaligned sequences. This may not be a reliable distance measure and the guide-tree derived from these distances may not be ideal. A better distance measure between sequences is one based on a full multiple alignment [9]. In Clustal Omega these distances are calculated from the initial alignment and are used to calculate a new, hopefully better, guide-tree. Any subsequent guide-tree refinement will again use the full alignment distances between sequences. These distances are expected to become more accurate as the alignments they are based upon become more accurate, leading in turn to better guide-trees and by extension to better alignments.

EPA required an externally computed HMM. This can be used to create a simple iteration scheme. In a first step unaligned sequences are aligned without any external profile. This produces an alignment which can be internally converted into a HMM and used in a second round of aligning in the same way as an EPA. Both of these steps: the initial unassisted alignment and the second alignment using a HMM and a new guide-tree derived from the first alignment, can be performed with one invocation of Clustal Omega:

```
$ clustalo -i globin.fa --iter=1
```

This will perform an initial alignment. It will then derive new distances between sequences from this alignment and construct a new guide-tree. It will also convert the initial alignment into a HMM and use this HMM in a second round of profile–profile alignment. After this second round the final alignment is written to the screen. Guide-tree iteration and HMM iteration can be decoupled in the following way:

```
$ clustalo -i globin.fa --iter=5 --max-guidetree-iterations=1
```

This performs an initial alignment. In a second round (first iteration) one reconstruction of the guide-tree is performed in tandem with one profile–profile alignment using a HMM derived from the initial alignment. Four subsequent refinement rounds will use HMMs derived from the previous alignments but will not recalculate the guide-tree. Conversely, one can restrict the number of HMM iterations, while repeatedly refining the guide-tree, by setting --max-hmm-iterations to a value less than the one specified by --iter. However, this variant is probably less useful as it does not use any HMM information at the last alignment step (*see* **Note 7**).

3.4 Profile Alignment

When reading in aligned sequences, Clustal Omega makes use of the alignment information (full alignment distances for guide-tree construction and HMM information for EPA) and then "dealigns" the sequences (removes all gaps) before realigning them. If the --dealign flag is specified, then the sequences are dealigned without making use of the alignment information. Sometimes this is not desirable. For example, one might have a high-quality, hand-curated alignment to which some unaligned sequences are to be added, while keeping the curated alignment fixed. Alternatively, one might want to align two profiles. In these cases one has to use the Clustal Omega --profile flag. To align two profiles use (*see* **Note 8**):

```
$ clustalo --profile1=globin1.aln --profile2=globin2.aln
```

If more than one unaligned sequences are to be added to an existing profile use:

```
$ clustalo --profile1=globin1.aln -i moreGlobins.fa
```

Clustal Omega extracts HMM information from the profile and uses this HMM as an external profile for EPA of the alignment of the unaligned sequences. Once all the unaligned sequences have been aligned this new profile is aligned to the previously existing profile.

4 Notes

1. By default, Clustal Omega uses the MAC algorithm to align profiles. This algorithm requires a certain amount of memory. If this amount of memory is not available, then Clustal Omega switches to the Viterbi algorithm. However, it is not straightforward to establish the amount of RAM available on different machines running different operating systems. Clustal Omega therefore assumes that it will have 2 GB = 2,048 MB of memory available for the MAC alignment. Should one be fortunate enough to have more than 2 GB of RAM then this can be communicated by setting the --MAC-RAM flag to the appropriate size (in MB). For example, on a machine with 4 GB of RAM one might specify:

```
$ clustalo -i globin.fa --MAC-RAM=4096
```

   where 4096 corresponds to 4 GB (=4 × 1,024 MB). In reality one might want to reduce the --MAC-RAM value, to allow for memory usage other than the MAC alignment.

2. The "configure" shell script attempts to guess correct values for various system-dependent variables used during compilation. It uses those values to create a "Makefile" in each directory of the package. It may also create one or more ".h" files containing system-dependent definitions. Finally, it creates a shell script "config.status" that one can run in the future to recreate the current configuration, and a file "config.log" containing compiler output (useful mainly for debugging "configure").

3. Clustal Omega needs argtable2 (http://argtable.sourceforge.net/). If argtable2 is installed in a nonstandard directory one may have to point configure to its installation directory. For example, if on a Mac with argtable installed via MacPorts then one should use the following command line:

```
$./configureCFLAGS='-I/opt/local/include' LDF
LAGS='-L/opt/local/lib'
```

4. Clustal Omega will automatically support multi-threading if the compiler supports OpenMP. For some reason automake's OpenMP detection for Apple's gcc is broken. OpenMP detection can be forced by calling configure as follows:

```
$./configure OPENMP_CFLAGS='-fopenmp' CFLAGS='-
DHAVE_OPENMP'
```

5. Distance matrix output can be initiated by specifying --distmat-out. However, this is not possible in default (mBed) mode because mBed does not calculate a full distance matrix. Therefore one must specify full distance matrix calculation by setting --full. For example

```
$ clustalo -i globin.fa --distmat-out=globin.mat --full
```

Distance matrix output in conjunction with iteration requires the --full-iter flag as well as the --full flag. If this flag is not specified then Clustal Omega will write out the first distance matrix, based on k-tuple distances and will then perform the preliminary alignment. During the first iteration it will then calculate new distances, based on the full multiple alignment, using the mBed algorithm. Since distance matrix output is not possible in mBed mode the preliminary matrix, based on k-tuples, will not be overwritten. Full alignment distances are used though for constructing an iterated guide-tree.

6. Guide-tree construction is based on pairwise distances. In this context fragments are particularly problematic. Fragments may be very short indeed and may therefore align perfectly with all sequences, leading to zero distances of all sequences with respect to the fragment. By transitivity, this insinuates that sequences that are in fact not close to each other appear very close indeed by proxy. This in turn will lead to a very bad guide-tree. This guide-tree will be extremely unbalanced (chained), which will lead to overly long execution times. It will also arrange the sequences in a more or less random order in this chained tree, leading to suboptimal alignments.
7. A Clustal Omega iteration can be broken up into two distinct steps. Performing:

```
$ clustalo -i globin.fa -o globin-0.out
$ clustalo -i globin-0.out -o globin-1.out
```

is equivalent to

```
$ clustalo -i globin.fa --iter=1 -o globin-1.out
```

The first invocation produces an alignment called globin-0.out. During the second invocation Clustal Omega reads in globin-0.out and detects that this is a valid alignment. It does so by ascertaining that all input sequences have the same length and that at least one input sequence contains at least one gap. The alignment information is used to build a guide-tree from the full alignment distances (rather than from the k-tuple distances that were used during the first invocation), as well as to produce a HMM, which is used during the second profile–profile alignment stage. This approach may be desirable for certain reasons. Firstly, it retains the intermediate alignment, which is lost using the --iter flag. Secondly, it allows one to use (and refine) existing alignments which may have been produced by aligners other than Clustal Omega. For example, for moderate numbers of sequences Kalign [10] is a faster alignment program than Clustal Omega, while still producing alignments of reasonable quality.

```
$ kalign-2.04 -i globin.fa -o globin-0.out -q -f
fasta
$ clustalo -i globin-0.out -o globin-1.out
```

This uses kalign to create a rough but high-speed initial alignment, which is then refined using Clustal Omega.
It is always advisable to ensure whether input sequences are actually aligned or not. In certain pipelines, unaligned sequences are arranged in such a way that sequences are padded at the end with gaps, such that all sequences have the same length. This is interpreted by Clustal Omega as a valid alignment, while in fact it is not. While the guide-tree that is derived from such an input is useless at best, the HMM information that is derived from this arrangement establishes the present, nonsensical, alignment. In this case one could either remove all gaps from the input by hand or specify the --dealign flag.

8. To align a single sequence to an existing profile use the profile–profile syntax:

```
$ clustalo --profile1=globin1.aln --profile2=
singleSequence.fa
```

When adding multiple sequences to a profile Clustal Omega first aligns all the unaligned sequences, taking regard of the HMM information derived from the profile, and then aligns the newly formed profile to the already existing profile. If the profile/sequences mode were to be used for adding a single sequence, then Clustal Omega would complain because there is only one sequence during the first round of alignments, which cannot be aligned against any other sequence.

Conversely, to add unaligned sequences one-by-one to an existing profile (rather than first aligning all the unaligned sequences and then aligning the new and the old profiles) one will have to distribute the unaligned sequences amongst multiple files and align the single sequences to the profile, overwriting the existing profile with the newly formed profile. One possible (bash) implementation to do this might be:

```
while read label; do
read seq;
echo -e $label"\n"$seq>in.vie;
clustalo--p1=globin-0.aln--p2=in.vie-o globin-
0.aln--force;
done < unaligned.vie
```

where unaligned.vie is the file that contains the unaligned sequence in Vienna format. Vienna format is the same as Fasta format but where all the residue information is in one (long) line. globin-0.aln is the file that originally contains the existing profile. At every stage it is overwritten with the alignment comprising of the previous profile and one extra added sequences. It is advisable to keep a copy of the original alignment. When performing this procedure the order in which unaligned sequences are added to the profile can impact the final alignment.

References

1. Sievers F, Wilm A, Dineen D et al (2011) Fast, scalable generation of high-quality protein multiple sequence alignments using Clustal Omega. Mol Syst Biol 7:539. doi:10.1038/msb.2011.75
2. Higgins DG, Sharp PM (1988) CLUSTAL: a package for performing multiple sequence alignment on a microcomputer. Gene 73(1): 237–244
3. Larkin MA, Blackshields G, Brown NP et al (2007) Clustal W and Clustal X version 2.0. Bioinformatics 23:2947–2948
4. Blackshields G, Sievers F, Shi W et al (2010) Sequence embedding for fast construction of guide trees for multiple sequence alignment. Algorithms Mol Biol 5:21
5. Söding J (2005) Protein homology detection by HMM-HMM comparison. Bioinformatics 21:951–960
6. Arthur D, Vassilvitskii S (2007) k-means++: the advantages of careful seeding. Proceedings of the eighteenth annual ACM-SIAM symposium on discrete algorithms, Philadelphia, PA, pp 1027–1035
7. Edgar RC (2004) MUSCLE: multiple sequence alignment with high accuracy and high throughput. Nucleic Acids Res 32:1792–1797
8. Finn RD, Clements J, Eddy SR (2011) HMMER web server: interactive sequence similarity searching. Nucleic Acids Res 39 (Suppl 2):W29–W37
9. Kimura M (1985) The neutral theory of molecular evolution. Cambridge University Press, Cambridge
10. Lassmann T, Sonnhammer ELL (2005) Kalign – an accurate and fast multiple sequence alignment algorithm. BMC Bioinformatics 6:298

Chapter 7

T-Coffee: Tree-Based Consistency Objective Function for Alignment Evaluation

Cedrik Magis, Jean-François Taly, Giovanni Bussotti, Jia-Ming Chang, Paolo Di Tommaso, Ionas Erb, José Espinosa-Carrasco, and Cedric Notredame

Abstract

T-Coffee, for Tree-based consistency objective function for alignment evaluation, is a versatile multiple sequence alignment (MSA) method suitable for aligning virtually any type of biological sequences. T-Coffee provides more than a simple sequence aligner; rather it is a framework in which alternative alignment methods and/or extra information (i.e., structural, evolutionary, or experimental information) can be combined to reach more accurate and more meaningful MSAs. T-Coffee can be used either by running input data via the Web server (http://tcoffee.crg.cat/apps/tcoffee/index.html) or by downloading the T-Coffee package. Here, we present how the package can be used in its command line mode to carry out the most common tasks and multiply align proteins, DNA, and RNA sequences. This chapter particularly emphasizes on the description of T-Coffee special flavors also called "modes," designed to address particular biological problems.

Key words MSA, 3D structure, Protein sequences, Transmembrane protein, Homolog sequences, DNA/RNA sequences, Promoter alignment, RNA secondary structure

1 Introduction

Multiple sequence alignment (MSA) is one of the most widely used bioinformatic methods in biology for the simultaneous comparison of evolutionarily related sequences [1, 2]. In an MSA, the relationship between all residues of the considered sequences is explicitly described, thus making it possible to identify highly conserved positions, or positions whose variability has a functional significance. These MSA models are rarely used for their own sake and their computation is usually an intermediate step towards more sophisticated applications: phylogenetic reconstruction, profile estimation (often referred to as hidden Markov models, HMM), structural predictions, promoter analysis, active site identification, and RNA secondary structure prediction. Building the most

David J. Russell (ed.), *Multiple Sequence Alignment Methods*, Methods in Molecular Biology, vol. 1079,
DOI 10.1007/978-1-62703-646-7_7,

accurate MSAs is thus essential for these downstream analysis and biological applications to be correct. However, computing an accurate alignment is not a trivial task, the problem of accuracy performing multiple comparisons being complex from a computational and a biological point of view.

From a computational point of view, estimating a correct multiple alignment has been shown to be a nondeterministic polynomial (NP)-complete problem, even when using very simple measurements such as sequence identity [3]. In practice it means that rather than estimating an optimal alignment, one has to use approximate model delivered by heuristic algorithm. The design of such algorithms has been and is still an intense focus of interest [4] and it might be argued that a majority of the available packages can be described as alternative heuristics designed for optimizing similar objective functions. Amongst all currently available algorithm, most of the new-generation aligners include a consistency-based component similar to the one originally described in T-Coffee (Tree-based consistency objective function for alignment evaluation) [5]. Consistency-based aligners such as T-Coffee, despite being slower than other algorithms, have also been shown to be much more accurate. T-Coffee combines a consistency-based evaluation with fast standard assembly algorithm such as the progressive alignment method used in ClustalW [6]. This combination does not only yield alignments with a higher accuracy, but it also results in a framework where methods, sequences, and structures can be seamlessly combined and compared.

The biological issue is just as challenging. The main challenge is the difficulty to quantify the correctness of an alignment. For instance, if one assumes an evolutionary framework, a correct alignment can be defined as an alignment in which all residues corresponding to the same residue in the ancestral sequence are aligned to one another. Yet, estimating the evolutionary correctness would require knowing in advance the relationship among residues, something usually impossible. Likewise, if the alignment is computed in a structural framework, a correct alignment will be defined as an alignment where the aligned residues are all structurally homologous; it would therefore require the knowledge of the structure of each of the included sequences or a perfect understanding of the relationship between structure and sequence. As a consequence MSAs are usually estimated on the basis of sequence similarity, taking advantage of the evolutionary inertia. This approach works reasonably well for closely related sequences and it has been shown that structurally correct alignment can easily be inferred for sequences having more than 30 % identity; below this figure (the so-called twilight zone), direct pairwise comparison becomes much less informative. Nonetheless, MSA-based analysis can be used to align more distantly related sequences, provided that highly conserved featured positions can be used to estimate and validate the model. T-Coffee was designed to address both issues: first to deliver the most accurate possible MSAs, and second to estimate the accuracy of MSAs delivered by T-Coffee or alternative aligners.

2 Materials

2.1 T-Coffee Web Server

T-Coffee is available through a Web server http://tcoffee.crg.cat/apps/tcoffee/index.html [7] which contains the same modes as T-Coffee package; however it has limited capabilities regarding the size of your dataset (max 150 sequences) and the length of your sequences (2,500 characters). It is well suited for routine alignment but for more sophisticated or complex alignment, users should use the T-Coffee package. This chapter focuses only on a detailed explanation on how T-Coffee package can be used in its command line mode [8].

2.2 Computational Requirement

T-Coffee is distributed as precompiled binaries for Linux and Mac OS X platforms with a guided install procedure. This is the smoothest and quickest way to install T-Coffee on a local machine, since it comes with all required components and does not need any special user privilege. The Windows operating system is no longer supported; however users working on that environment can use a Linux station working in a virtual environment such as VMware. Most Linux distributions should work, especially Fedora, Debian, and Ubuntu being the most thoroughly tested (*see* **Note 1**). It is also possible to compile the source code under the Cygwin environment within a Windows operating system.

2.3 Installation

On Linux

- Download the installer package from the following URL: http://tcoffee.org/Packages/Stable/Latest/linux/
- Grant execution permission to the downloaded file: chmod +x T-COFFEE-installer-<version-xxx>.bin
- Launch the installation wizard with ./T-COFFEE-installer-<version-xxx>.bin
- Follow the wizard instructions and complete the installation.

On Mac OS X

- Download the installer package from the following URL: http://tcoffee.org/Packages/Stable/Latest/macosx/
- Double click on the DMG file to open it.
- Double click on the installer icon (within the mounted image).
- Follow the wizard instructions and complete the installation.

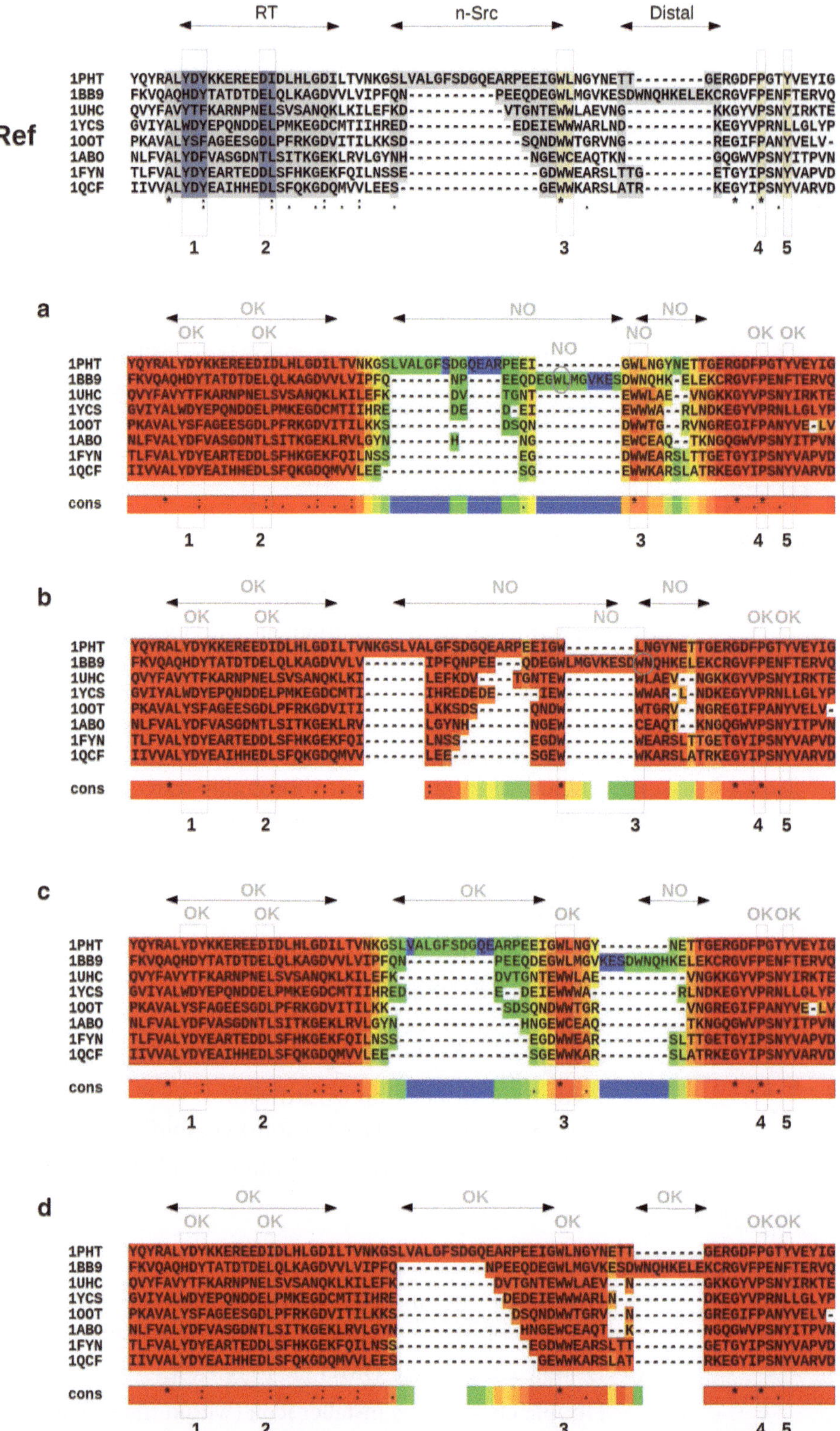

Fig. 1 Multiple sequence alignment of eight sh3 protein domains. "Ref" indicates a manually expert-curated MSA used as a benchmark. Functional regions are colored in *grayscale*. *Gray residues*: RT, n-Src, and distal regions;

2.4 Input Information

T-Coffee accepts as input virtually any type of sequences, nucleic acid or amino acid, given in a standard format (fasta or ClustalW only). You must have your sequence ready in order to use T-Coffee; however T-Coffee package comes along with reformatting options allowing you to change the format or manipulate your input sequences (cf. T-Coffee Tutorial at http://www.tcoffee.org/Projects/tcoffee/).

A few external tools or databases are required for specific T-Coffee modes (BLAST, PDB, nr) and are, by default, accessed via their respective servers. T-Coffee can use instead versions installed locally; however their installation and management are the user responsibility.

3 Methods

T-Coffee in its default mode runs on all types of sequences you provide (nucleic acids or amino acids). For all methods described here, T-Coffee always generate the following output files:

- The resulting alignment (you can choose your format for the output MSA using the flag -output).
- The resulting guide tree (tree generated to build the alignment; it is not a phylogenetic tree).
- An html file that can be visualized using any browser, showing the resulting alignment colored according to the T-Coffee color scheme. The color scheme is related to the consistency from blue (low-consistency regions) to red (high-consistency regions).

Depending on your biological problem and/or the nature of your sequences, you may choose a more adapted mode of T-Coffee rather than the default mode (*see* **Note 2**). For each case, additional output file will be detailed in each section.

3.1 Aligning Protein Sequences

The different T-Coffee modes were used to aligned proteins of the sh3 protein domain family as an example. All alignments generated here are shown in Fig. 1 and compared to a reference alignment manually curated to evaluate their respective accuracy, shown in Table 1. The CPU time requirement for each mode is indicated also in Table 1. While using T-Coffee on your own dataset, keep in mind

Fig. 1 (continued) blue-gray: aromatic and polar motifs of the RT loop, respectively; *beige*: conserved triad in the binding pocket. *Asterisks* indicate conserved amino acid in all sequences. *Colons* indicate position of the MSA composed of residues with similar physicochemical properties. *Dots* indicate columns of the MSA for which semi-conserved substitutions are observed. (**a–d**) MSA of the sh3 sequences with the T-Coffee default mode, fast M-Coffee, PSI-Coffee, and Expresso, respectively. The alignments are colored according to their consistency (CORE index). Regions in *red* have the highest consistency whereas regions in *blue* have the lowest consistency (*see* **step 1** in Subheading 3.3). Functional motifs correctly aligned or misaligned are tagged with the word "OK" or "NO," respectively

that T-Coffee performances will depend on the nature and size of your data (*see* **Note 3**).

1. M-Coffee
   M-Coffee [9] is a meta-method meant to combine the output of several alternative aligners into one final output, assuming that errors produced by independent prediction systems should not be consistent, therefore suggesting agreement as an indication of correctness. M-Coffee alignments are on overall less accurate than T-Coffee alignments; however they reveal inconsistent regions of your alignments. M-Coffee is run using the following command line:

   t_coffee -seq sh3.fasta -mode mcoffee

   By default, M-Coffee runs eight different aligners; however the users have the option to generate their own combination. The list of all alternative aligners available in the T-Coffee package can be displayed by simply running the command **t_coffee**. One of the most used variants is defined as Fast M-Coffee (fmcoffee), combining the current three fastest aligners (kalign, mafft, and muscle). Fast M-Coffee (Fig. 1b) is run using the following command line:

   t_coffee -seq sh3.fasta -mode fmcoffee

2. PSI-Coffee
   PSI-Coffee [8, 10] is designed for aligning distantly related proteins using evolutionary information by combining homology extension and T-Coffee consistency-based progressive alignment. Homology extension consists in generating a homology profile by running a BLAST against nr for each sequence within a given dataset (*see* **Note 4**). These profiles are then aligned rather than their corresponding sequences within the given dataset. PSI-Coffee is slower than T-Coffee or M-Coffee (Table 1), but it is the most accurate mode of T-Coffee using only sequence information. PSI-Coffee (Fig. 1c) is run using the following command:

   t_coffee -seq sh3.fasta -mode psicoffee

   A particular application of PSI-Coffee is TM-Coffee [10], a mode especially designed for transmembrane protein (TMP) in which the homology extension step uses a nonredundant database containing only TMP sequences. TM-Coffee achieves the same level of accuracy as PSI-Coffee while only requiring a tenth of the CPU time. TM-Coffee is run using the following command:

   t_coffee -seq sh3.fasta -mode psicoffee -template_file PSITM

   The PSITM template file is automatically generated and is used to display a colored MSA version (.tm_html output file)

Table 1
Comparisons between a manually curated alignment of eight sh3 domains and the different alignment generated with the different T-Coffee modes. CPU times are given for alignments computed with an 8 Intel(R) Xeon(R) CPU E5504 at 2.00 GHz

Method	Column score (CS)	CPU time (s)
T-Coffee	69.8	0.5
Fast M-Coffee	74.7	0.8
PSI-Coffee	80.0	21.7
Expresso	90.0	32.6

reflecting structural predictions (yellow: inside loop; red: TM helix; blue: outside loop).

3. Expresso/3D-Coffee
Expresso/3D-Coffee [11, 12] is a multiple structural aligner, using the 3D structures available in the Protein Data Bank (PDB) as templates to align the sequences within a given dataset. It is suited when all the sequences in the dataset have a close homolog with a known 3D structure (*see* **Note 5**). Expresso is similar to 3D-Coffee except that template file and the PDB files are automatically generated and fetched (*see* **Note 4**). If no template can be identified for a given sequence, alignment for this sequence will be achieved using the default T-Coffee mode. Expresso has been shown to be the most accurate aligner existing; however it is also the slowest (Table 1). Expresso (Fig. 1d) is run using the following command:

t_coffee -seq sh3.fasta -mode expresso

Expresso generates a template file in which the 3D structures associated with your query sequences are explicitly declared (*see* **Note 6**). The Expresso mode also comes along several filtering parameters related to the type of 3D structure (XRAY/NMR), the coverage and the identity between your query sequences and the identified templates. By default the structural aligner used is SAP; however the user can freely choose or combine other structural aligners using the tag -method. For instance the following command line will fetch 3D structures ("d" standing for XRAY and/or "n" standing for NMR) 95 % identical with your query sequences and will align them using the MUSTANG structural aligner:

t_coffee -seq sh3.fasta -mode expresso -method mustang_pair -pdb_type dn -pdb_min_sim 95

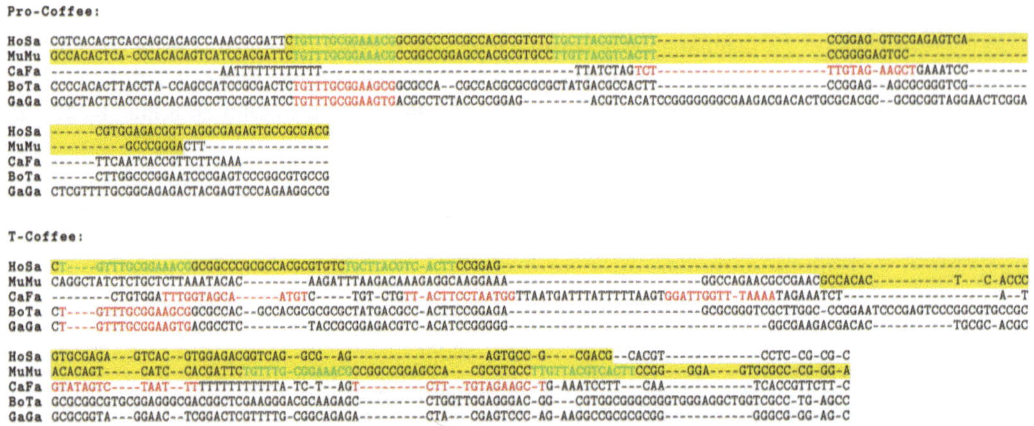

```
Pro-Coffee:

HoSa CGTCACACTCACCAGCACAGCCAAACGCGATTCTGTTTGCGGAAACGGCGGCCCGCGCCACGCGTGTCTGCTTACGTCACTT-----------------------CCGGAG-GTGCGAGAGTCA--------
MuMu GCCACACTCA-CCCACACAGTCATCCACGATTCTGTTTGCGGAAACGCCGGCCGGAGCCACGCGTGCCTTGTTACGTCACTT-----------------------CCGGGGAGTGC----------------
CaFa ------------------------AATTTTTTTTTTTT---------------------------------TTATCTAGTCT-----------------------TTGTAG-AAGCTGAAATCC--------
BoTa CCCCACACTTACCTA-CCAGCCATCCGCGACTCTGTTTGCGGAAGCGGCGCCA--CGCCACGCGCGCGCTATGACGCCACTT-----------------------CCGGAG--AGCGCGGTCG---------
GaGa GCGCTACTCACCCAGCACAGCCCTCCGCCATCCTGTTTGCGGAAGTGACGCCTCTACCGCGGAG---------ACGTCACATCCGGGGGGGCGAAGACGACACTGCGCACGC--GCGCGGTAGGAACTCGGA

HoSa ------CGTGGAGACGGTCAGGCGAGAGTGCCGCGACG
MuMu ----------GCCCGGGACTT-----------------
CaFa ------TTCAATCACCGTTCTTCAAA------------
BoTa ------CTTGGCCCGGAATCCCGAGTCCCGGCGTGCCG
GaGa CTCGTTTTGCGGCAGAGACTACGAGTCCCAGAAGGCCG

T-Coffee:

HoSa CT----GTTTGCGGAAACGGCGGCCCGCGCCACGCGTGTCTGCTTACGTC-ACTTCCGGAG-----------------------------------------------------------------------
MuMu CAGGCTATCTCTGCTCTTAAATACAC---------AAGATTTAAGACAAAGAGGCAAGGAAA--------------------------GGCCAGAACGCCGAACGCCACAC-----------T---C-ACCC
CaFa --------CTGTGGATTTGGTAGCA------ATGTC-----TGT-CTGTT-ACTTCCTAATGGTTAATGATTTATTTTTAAGTGGATTGGTT-TAAAATAGAAATCT---------------------A--T
BoTa CT----GTTTGCGGAAGCGGCGCCAC--GCCACGCGCGCGCTATGACGCC-ACTTCCGGAGA--------------------------GCGCGGGTCGCTTGGC-CCGGAATCCCGAGTCCCGGCGTGCCGC
GaGa CT----GTTTGCGGAAGTGACGCCTC---------TACCGCGGAGACGTC-ACATCCGGGGG-----------------------------------GGCGAAGACGACAC-----------TGCGC-ACGC

HoSa GTGCGAGA---GTCAC--GTGGAGACGGTCAG--GCG--AG------------------AGTGCC-G----CGACG--CACGT--------------CCTC-CG-CG-C
MuMu ACACAGT-----CATC--CACGATTCTGTTTG-CGGAAACGCCGGCCGGAGCCA---CGCGTGCCTTGTTACGTCACTTCCGG---GGA----GTGCGCC-CG-GG-A
CaFa GTATAGTC----TAAT--TTTTTTTTTTTTTA-TC-T--AGT---------CTT--TGTAGAAGC-TG-AAATCCTT---CAA-----------TCACCGTTCTT-C
BoTa GCGCGGCGTGCGGAGGGCGACGGCTCGAAGGGACGCAAGAGC---------CTGGTTGGAGGGAC-GG---CGTGGCGGGCGGGTGGAGGCTGGTCGCC-TG-AGCC
GaGa GCGCGGTA---GGAAC--TCGGACTCGTTTTG-CGGCAGAGA---------CTA---CGAGTCCC-AG-AAGGCCGCGCGCGG------------GGGCG-GG-AG-C
```

Fig. 2 Comparison of Pro-Coffee and T-Coffee alignments of the proximal promoter region of the human gene C18orf19. *Yellow boxes* indicate ChIP-seq regions for the transcription factor CEBPA. Predicted CEBPA-binding sites are shown in *green* when falling in ChIP-seq regions and in *red* when falling outside. Pro-Coffee aligns correctly the factor-binding regions and their binding sites while the default T-Coffee fails to do so

3.2 Aligning DNA/RNA Sequences

1. Functional DNA elements: Pro-Coffee

Aligning non-transcribed DNA is probably one of the most challenging tasks in the field of sequence alignment as a consequence of the reduced alphabet in nucleic acid sequences and the heterogeneity of functional features contained in genomic sequences. However, making use of footprints in homologous promoter or enhancer regions can increase your chances in motif finding or when scanning for known motifs using programs that accept alignments as input [13, 14]. Pro-Coffee [15] was designed to address this need. It makes use of a substitution matrix between dinucleotides, where the substitution counts were estimated from the seed alignments of TRANSFAC weight matrices. Pro-Coffee is run using the following command:

t_coffee -seq c18orf19.fasta -mode procoffee

The accuracy of a promoter alignment (Fig. 2) may increase when considering longer sequences. If the region of interest lays 500 bp upstream of the transcription start site, it can be beneficial to align a longer stretch, say from −1,500 bp to +500 bp relative to TSS, and then ignore the part of the resulting alignment you are not interested in. You can extract a block from, e.g., position 1,000 to position 1,500 with respect to a reference sequence "ref" in your aligned sequences using the command:

t_coffee -other_pg seq_reformat -in c18orf19.aln -action +extract_block 'ref' 1000 1500

```
                 T-Coffee                       R-Coffee
1F6X_A           GGAAGUCCGGUCUUCGGACCGGCUUCC    GGAAGUCCGGUCUUCGGACCGGCUUCC
1F78_A           GGAAGUCCGGUCUUCGGACCGGCUUCC    GGAAGUCCGGUCUUCGGACCGGCUUCC
1F7F_A           GGAAGUCCGGUCUUCGGACCGGCUUCC    GGAAGUCCGGUCUUCGGACCGGCUUCC
1F7H_A           GGAAGUCCGGUCUUCGGACCGGCUUCC    GGAAGUCCGGUCUUCGGACCGGCUUCC
1JOX_A           GG-----CGGUGCUGAGAU-GCCCGUC    GGCGG-----UGCUGAGAU-GCCCGUC
1JP0_A           GG-----CGGUGCUGAGAU-GCCCGUC    GGCGG-----UGCUGAGAU-GCCCGUC
1XSG_A           GGAAGACCGGUCUUCGGACCGGCUUCC    GGAAGACCGGUCUUCGGACCGGCUUCC
1XST_A           GGAAGACCGGUCUUCGGACCGGCUUCC    GGAAGACCGGUCUUCGGACCGGCUUCC
#=GC SS_cons     .........(((....)))........    (((((.....((....))....)))))
#=GC SS_analyze  .........NWI....IWN........    WNCCW.....WI....IW....WCCNW
```

Fig. 3 Comparison of R-Coffee and T-Coffee RNA alignments. The color code indicates the type of complementarities between matching columns. *Orange columns* marked with a "W" correspond to perfect Watson–Crick (WC) pairing, without mutation. *Green columns* marked with an "N" are columns containing either WC or GU pairs (neutral). *Blue columns* marked with a "T" indicate position containing non-WC pairs. *Red columns* correspond to perfect WC positions, including compensated mutations, marked with a "C"

The current version of Pro-Coffee uses a default gap-opening penalty (GOP) of −60 and a gap-extension penalty (GEP) of −1. Yet, our benchmarks suggest that good promoter alignments may be obtained using GOP values between −50 and −70 while keeping the GEP at the default value. To show how alternative values can be explored, here we give the command line that explicitly spells out the default parameters:

t_coffee -seq <your seq>-method promo_pair@EP@GOP@-60@GEP@-1

2. RNA sequences: R-Coffee
R-Coffee [16] is a special mode of T-Coffee for aligning noncoding RNAs with conserved secondary structures (the predicted secondary structures are generated as output files rfold_files). This mode can also be used to analyze compensated mutations in the resulting MSA using a homemade reference benchmark from DARTS, a database of RNA sequences with known 3D structure. R-Coffee (Fig. 3) is run using the following command:

t_coffee -seq RNA_rcoffee.fa -mode rcoffee -outfile RNA_rcoffee.aln

One of the main advantages of noncoding RNA in sequence analysis is the relative clear pattern left by compensated mutations when doing multiple sequence analysis. Given an RNA alignment with either a predicted or an estimated secondary structure, the seq_reformat option of T-Coffee can be used to estimate the number of columns showing compensated mutations by the following command:

t_coffee -other_pg seq_reformat -in RNA_rcoffee.aln -action +alifold2analyze color_html > RNA_tcoffee.html

3.3 Evaluating Alignments

A major issue with MSAs is accuracy estimation. It is well documented that MSA packages deliver significantly different alignments of the same dataset and thus can affect the results obtained from any downstream analysis. The comparison and evaluation of MSA is therefore an essential step in order to assess its quality and reliability. Currently, aligners are compared through external reference alignment expert-curated or benchmarks. However, it has been shown that no aligner performs better than all others on every available benchmark. T-Coffee provides three different scoring systems, allowing the comparison of alternative alignments, independently from such reference alignments.

1. Sequence only consistency-based accuracy evaluation: Consistency of Overall Residue Evaluation (CORE) index.
   The CORE index is one of the most versatile tools developed to display the agreement between a set of alignments and a given model. The CORE index is directly based on T-Coffee consistency estimation scheme. Every aligned residue is colored according to its consistency score (red for high and blue for low). This normalized score reflects the agreement between the actual alignment of the residue (column) and the alternative alignments. It can be displayed by running the following command:

   t_coffee sh3.fasta -method mafft_pair clustalw2_pair proba_pair poa_pair -output score_html, score_ascii

   In this example, the scores are displayed in ascii format and can be visualized using the corresponding html file. In this particular case, it is a measure of the agreement between the four considered methods mafft, clustalw, tcoffee, and poa and the final alignment. The CORE index is only meaningful for (1) datasets containing at least four sequences when running single aligner and (2) alignments combining at least three methods when using the M-Coffee mode.

2. Single-structure accuracy evaluation: STRIKE.
   The STRIKE [17] score is the latest scoring system developed within the T-Coffee package to assess and identify the most accurate MSA amongst alternative MSAs of the same sequence dataset. To assess protein MSA, the use of structure is often considered as a gold standard; however, such information is often not available or in low abundance. STRIKE's only requirement is a single homologous 3D structure to evaluate and rank alternative alignments of a given dataset. MSA accuracy is computed using a contact matrix estimated through residue–residue contact in a dataset of nonredundant high-quality protein structures from the ASTRAL database. STRIKE can be run using the following command:

t_coffee -other_pg strike -aln sh3.aln -template_file sh3.template

3. Multiple structure accuracy evaluation: iRMSD.
   The most accurate scoring system provided by T-Coffee is the iRMSD, which delivers a quantifiable and comparable score using 3D structural information. The iRMSD [18, 19] is an RMSD-like measure independent from structure superposition and thus unbiased by the process itself. The iRMSD is calculated using intramolecular distance matrixes (one per each sequence with an available 3D structure), distances being calculated between residues considered as equivalent as defined by the MSA. The limitation comes from the need to have a closely related structure for each sequence within the dataset; however the user is free to choose the identity threshold to identify homologous templates. The iRMSD is run using the following command:

   t_coffee -other_pg irmsd -aln sh3.aln -template_file sh3.template

   The template file corresponds to the explicit association between a query sequence and a structure file. It is for instance generated automatically when running Expresso. The output of the iRMSD will give two scores, the iRMSD score as described above and the NiRMSD, corresponding to the iRMSD score normalized by the length of the sequences.

4 Notes

1. T-Coffee package installation on Mac OS X and different Linux distributions have been heavily tested; however, it does not preclude especially for beta versions of T-Coffee to encounter installation problems. In such case, do not hesitate to contact the T-Coffee developers (tcoffee@googlegroups.com).
2. T-Coffee package is designed to address biological problems and as a consequence of the versatility of biological data, problems or limitations can occur. For this reasons, the T-Coffee developers can be always contacted for any problem you might encounter while using T-Coffee (tcoffee@googlegroups.com).
3. T-Coffee alignments depend on the T-Coffee mode, the complexity, and the size of your datasets, and thus can be quite expensive in terms of memory and computation. All different T-Coffee modes should not be used for dataset containing more than 1,000 sequences, and no more than 200 sequences when running structural mode. This limitation is of course empirical and is no more than an indication.

4. When using external database via their respective Web servers, the results can vary with the upgrade of these databases. Moreover, when these servers are unavailable, the T-Coffee modes requiring these databases will not be able to deliver an alignment. A solution would be to install locally the external servers you require (*see* corresponding external servers).
5. When using structure files as templates, PDB files are recognized only when following the PDB standard format. If the PDB structure files are modified, they might no longer be recognized by T-Coffee. If you want to do so, you should use T-Coffee reformatting options to modify the PDB file (cf. T-Coffee Tutorial). It is particularly adapted to extract chains, parts, or blocks using the reformatting option **-other_pg extract_from_pdb** specifying the chain and the boundaries of the sequence of interest (*see* T-Coffee Tutorial at http://www.tcoffee.org/Projects/tcoffee/).
6. When using structural aligner, having available structure files does not mean that the structural aligner will successfully structurally align your sequences with their corresponding templates; in such case, an error message will be printed out and the alignment will be performed using the T-Coffee default sequence aligner (*see* **step 3** in Subheading 3.1).

References

1. Edgar RC, Batzoglou S (2006) Multiple sequence alignment. Curr Opin Struct Biol 16(3):368–373
2. Kemena C, Notredame C (2009) Upcoming challenges for multiple sequence alignment methods in the high-throughput era. Bioinformatics 25(19):2455–2465
3. Just W (2001) Computational complexity of multiple sequence alignment with SP-score. J Comput Biol 8(6):615–623
4. Notredame C (2007) Recent evolutions of multiple sequence alignment algorithms. PLoS Comput Biol 3(8):e123
5. Notredame C, Higgins DG, Heringa J (2000) T-Coffee: a novel method for fast and accurate multiple sequence alignment. J Mol Biol 302 (1):205–217
6. Larkin MA et al (2007) CLUSTALW and CLUSTALX version 2.0. Bioinformatics 23: 2947–2948
7. Di Tommaso P, Moretti S, Xenarios L, Orobitg M, Montanyola A, Chang JM, Taly JF, Notredame C (2011) T-Coffee: a Web server for the multiple sequence alignment of protein and RNA sequences using structural information and homology extension. Nucleic Acids Res 39(Web Server issue):W13–W17
8. Taly JF, Magis C, Bussotti G, Chang JM, Di Tommaso P, Erb I, Espinosa-Carrasco J, Kemena C, Notredame C (2011) Using the T-Coffee package to build multiple sequence alignments of protein, RNA, DNA sequences and 3D structures. Nat Protoc 6(11):1669–1682
9. Wallace IM, O'Sullivan O, Higgins DG, Notredame C (2006) M-Coffee: combining multiple sequence alignment methods with T-Coffee. Nucleic Acids Res 34(6):1692–1699
10. Chang JM, Di Tommaso P, Taly JF, Notredame C (2012) Accurate multiple sequence alignment of transmembrane proteins with PSI-Coffee. BMC Bioinformatics 13(Suppl 4):S1
11. O'Sullivan O, Suhre K, Abergel C, Higgins DG, Notredame C (2004) 3DCoffee: combining protein sequences and structures within multiple sequence alignments. J Mol Biol 340 (2):385–395
12. Armougom F, Moretti S, Poirot O, Audic S, Dumas P, Schaeli B, Keduas V, Notredame C (2006) Expresso: automatic incorporation of structural information in multiple sequence alignments using 3D-Coffee. Nucleic Acids Res 34(Web Server issue):W604–W608
13. Siddharthan R, van Nimwegen E (2007) Detecting regulatory sites using PhyloGibbs. Methods Mol Biol 395:381–402

14. Arnold P, Erb I, Pachkov M, Molina N, van Nimwegen E (2012) MotEvo: integrated Bayesian probabilistic methods for inferring regulatory sites and motifs on multiple alignments of DNA sequences. Bioinformatics 28(4):487–494
15. Erb I, González-Vallinas JR, Bussotti G, Blanco E, Eyras E, Notredame C (2012) Use of ChIP-Seq data for the design of a multiple promoter-alignment method. Nucleic Acids Res 40(7):e52
16. Wilm A, Higgins DG, Notredame C (2008) R-Coffee: a method for multiple alignment of non-coding RNA. Nucleic Acids Res 36(9):e52
17. Kemena C, Taly JF, Kleinjung J, Notredame C (2011) STRIKE: evaluation of protein MSAs using a single 3D structure. Bioinformatics 27(24):3385–3391
18. Armougom F, Moretti S, Keduas V, Notredame C (2006) The iRMSD: a local measure of sequence alignment accuracy using structural information. Bioinformatics 22(14): e35–e39
19. O'Sullivan O, Zehnder M, Higgins D, Bucher P, Grosdidier A, Notredame C (2003) APBD: a novel measure for benchmarking sequence alignment methods without reference alignment. Bioinformatics 19(Suppl 1):215–221

Chapter 8

MAFFT: Iterative Refinement and Additional Methods

Kazutaka Katoh and Daron M. Standley

Abstract

This chapter outlines several methods implemented in the MAFFT package. MAFFT is a popular multiple sequence alignment (MSA) program with various options for the progressive method, the iterative refinement method and other methods. We first outline basic usage of MAFFT and then describe recent practical extensions, such as dot plot and adjustment of direction in DNA alignment. We also refer to MUSCLE, another high-performance MSA program.

Key words Multiple sequence alignment, Iterative refinement, Fast Fourier transform, Metagenome, Protein structure

1 Introduction

MAFFT [1] is a general purpose multiple sequence alignment (MSA) program for nucleotide or amino acid sequences. Due to its high performance [2–6], MAFFT is becoming popular in recent years. It has several different options depending on the size and type of alignment problem. In this chapter, first we outline various MAFFT options. Then, we describe several practical new features for web service and command-line versions. As a possible direction for future development, we discuss our ongoing efforts to use structural information in MSA calculations. We also clarify the limitation of MAFFT in actual analyses, by showing typical inappropriate usage. In addition, we refer to MUSCLE [7, 8], another high-performance MSA program.

2 Basic Algorithms

MAFFT has various options as discussed below. Generally, there is a trade-off between speed and accuracy. However, there are exceptional situations (*see* Subheading 6) that should be carefully considered when selecting an option to apply. Like most MSA methods,

David J. Russell (ed.), *Multiple Sequence Alignment Methods*, Methods in Molecular Biology, vol. 1079,
DOI 10.1007/978-1-62703-646-7_8, © Springer Science+Business Media, LLC 2014

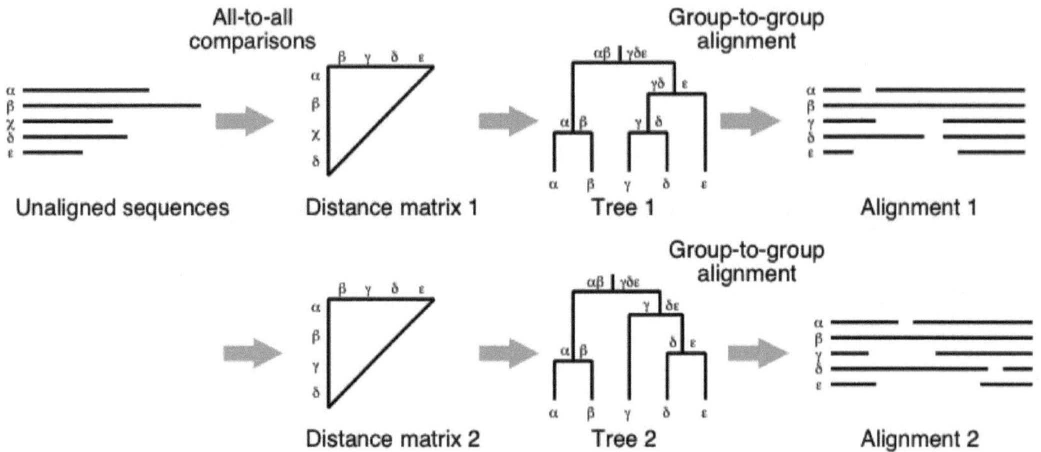

Fig. 1 Calculation procedure for the progressive options (reprinted from [57])

MAFFT assumes that the sequences involved in an MSA are homologous, that is, sharing a single common ancestral sequence. MAFFT always generates an MSA that has all the letters of the input sequences. The order of the letters in each sequence is identical to that of the input sequence, although the sequences can be reordered according to similarity.

2.1 Progressive Methods

The progressive method [9, 10] is the most basic MSA algorithm. A guide tree is created based on all-to-all pairwise comparisons, and an MSA is constructed using a group-to-group alignment algorithm at each node of the guide tree. To achieve a reasonable balance between speed and accuracy, MAFFT uses a two-cycle progressive method, FFT-NS-2, shown in Fig. 1. First, low-quality all-pairwise distances are rapidly calculated, based on the number of shared *k*mers, to build an initial tree [11]. *k* is 6 for both protein and nucleotide sequences as noted in [1]. A tentative MSA is computed using the initial guide tree, and then a refined tree is built using the tentative MSA. Finally, an MSA is constructed based on the refined tree. This is the default option

```
mafft input > output    (FFT-NS-2)
```

A faster option, FFT-NS-1, which performs the first cycle only, is also available.

```
mafft --retree 1 input > output   (FFT-NS-1)
```

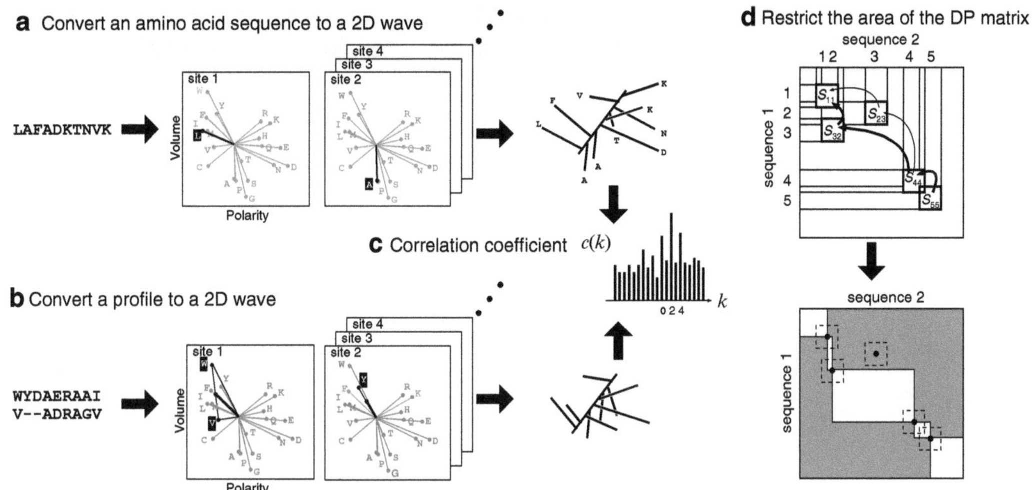

Fig. 2 Calculation procedure of the iterative refinement options (reprinted from [57])

In the group-to-group alignment calculation, each group can have gaps introduced in previous progressive steps. These gap positions are approximately considered [1].

In the FFT-NS-2 and FFT-NS-1 options, a group-to-group alignment algorithm based on the FFT approximation [1] is used (*see* Fig. 2). It enables fast calculation for long sequences. However, for handling many short sequences, sometimes the use of FFT is not preferable, because it requires additional memory space and conversion steps from sequence to wave. For such cases, standard dynamic programming (DP) can be selected. These options are called NW-NS-1 and NW-NS-2 for one-cycle and two-cycle versions, respectively.

```
mafft --nofft --retree 1 input > output   (NW-NS-1)
mafft --nofft            input > output   (NW-NS-2)
```

It is possible to use pairwise alignments by DP, instead of 6mer comparison, for calculating a guide tree.

```
mafft --localpair  input > output   (L-INS-1)
mafft --globalpair input > output   (G-INS-1)
```

L-INS-1 and G-INS-1 use local alignment and global alignment, respectively, for all the pairs to compute a distance matrix. These methods are slower but more accurate than FFT-NS-2 and NW-NS-2 in most benchmarks.

There are many MSA programs based on the progressive method, such as the Clustal series [6], PRANK [12], and Kalign [13].

2.2 Iterative Refinement Methods

The progressive method has a drawback in that once a gap is incorrectly introduced, especially at an early step (near a leaf of the guide tree), the gap is never removed in later steps. To overcome this drawback, the iterative refinement method was proposed [14–16]. MAFFT adopted this strategy (*see* Fig. 3). The iterative refinement method requires an objective function that represents the "goodness." An initial MSA, calculated by the progressive or other method, is subjected to an iterative process so that the objective function is maximized. Various objective functions and maximization strategies have been proposed to date [14–20]. Among them, Gotoh's method with the weighted sum-of-pairs (WSP) objective function is the most successful one. MAFFT adopted this method in the FFT-NS-i option. In this method, an initial MSA is partitioned into two groups, and the two groups are re-aligned using a group-to-group alignment algorithm. This process is repeated until no more improvements are made. The partitions of the MSA are restricted to those corresponding to the branches of the guide tree [21]. In one cycle of iterative refinement, all the branches in the guide tree are tried as partitioning points. To run this option, use

```
mafft --maxiterate 16 input > output    (FFT-NS-i)
```

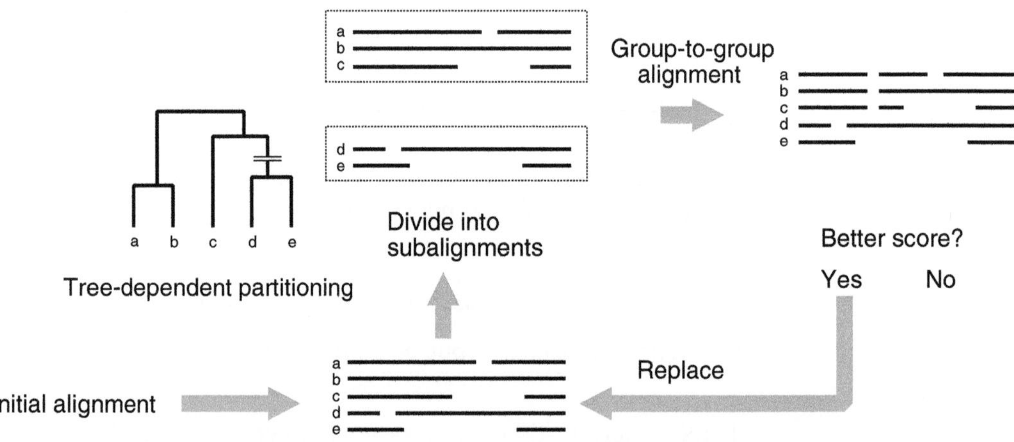

Fig. 3 Outline of a fast group-to-group alignment algorithm using FFT (reprinted from [30]). (**a**) A sequence is converted to a two-dimensional (2D) wave, arrangement of vectors. (**b**) A set of aligned sequences is also converted to a 2D wave. (**c**) The correlation between the two waves can be rapidly computed with FFT. (**d**) The highly conserved regions detected by FFT are used as anchors, and the area of the DP matrix is restricted

The number after `--maxiterate` specifies the number of cycles. In the current version, even if a number larger than 16 is specified, the calculation is finished at the 16th cycle. This is because major improvements in alignment quality are made in early cycles.

An option, NW-NS-i, without the FFT algorithm, is also available.

```
mafft --nofft --maxiterate 16 input > output    (NW-NS-i)
```

2.3 Consistency Criteria

In order to further improve accuracy, MAFFT partly uses the consistency criterion. Several different types of consistency criteria were described previously [22–26]. TCoffee [25] achieved a great improvement in accuracy. ProbCons [26] and other methods [27–29] are largely based on this idea. These methods generally require long computational time. Unlike them, three iterative refinement options of MAFFT, L-INS-i, G-INS-i, and E-INS-i use a consistency criterion similar to COFFEE [24], combining it with the WSP objective function, and maximize it using the iterative refinement process.

```
mafft --globalpair --maxiterate 16 input > output    (G-INS-i)
mafft --localpair  --maxiterate 16 input > output    (L-INS-i)
mafft --genafpair  --maxiterate 16 input > output    (E-INS-i)
```

These three options are designed for different types of input sequences. G-INS-i is suitable forsequences that have homology over the entire region, whereas L-INS-i and E-INS-iare suitable for sequences that have homology only in partial regions. *See* [30] formore detailed information.

2.4 RNA Alignments

The X-INS-i and Q-INS-i options are specially designed for RNA alignment, considering secondary structures [31]. In Q-INS-i, the base pairing probability is calculated by McCaskill's algorithm [32], incorporated into the objective function, and an iterative refinement method is applied to maximize the objective function. In X-INS-i, secondary structure is also considered in the pairwise comparison stage, by using SCARNA [33], in addition to the objective function.

```
mafft-qinsi input > output    (Q-INS-i)
mafft-xinsi input > output    (X-INS-i)
```

Both Q-INS-i and X-INS-i use source codes from the Vienna RNA package [34], MXSCARNA [35] and ProbconsRNA [26]. Some MAFFT packages do not contain these source codes and thus do not support these options.

These types of methods were intensively studied recently, and many alternative methods, such as PicXAA-RNA [5], CentroidAlign [36], and RCoffee [37], are available.

2.5 Profile Alignments

MAFFT has a subprogram to align two alignments.

```
mafft-profile alignment1 alignment2 > output
```

This program is useful only when two alignments are phylogenetically separated. Careless application of this method results in serious misalignments, as shown in [38] and Subheading 6.

We are preparing a safer option, `--addprofile`, to avoid such mistakes.

```
mafft --addprofile alignment1 alignment2 > output
```

This option does not return any result if the sequences in alignment1 do not form a monophyletic cluster. Thus this method is not always useful for every user and is still in the testing phase.

2.6 MSA of a Large Number of Sequences

To align a large number of sequences, MAFFT has an approximate option, PartTree [39], which skips the calculation of the full distance matrix consisting of $O(N^2)$ elements, where N is the number of sequences. Instead, n sequences are randomly selected and the distances between the n sequences and the remaining sequences are computed to classify the sequences into n groups. The n groups are recursively subjected to the same process, to create a tree-like classification. The time complexity of this processes is $O(N \log N)$. There are several subtypes of the PartTree option. The fastest one is

```
mafft --parttree --retree 1 input > output
```

in which distances are computed based on the number of shared 6mers. A more accurate subtype is also available.

```
mafft --dpparttree --retree 1 input > output
```

in which distances are computed based on DP. The application of DP to a large dataset might seem to be impractical, but as a result of the PartTree algorithm, we can drastically restrict the number of DP runs. Accordingly, this option is feasible and gives slightly better accuracy than the 6mer-based option in our tests. *See* [39] for details. The latest version of the Clustal series, Clustal Omega [6], provides an alternative method for large MSA, using the mBed algorithm [40].

3 Difference Between MAFFT and MUSCLE

MUSCLE [7, 8] is another high-performance MSA program. It adopted the overall design of the NW-NS-i option of MAFFT (*see* Subheading 2.2). Other options corresponding to NW-NS-1 and NW-NS-2 (*see* Subheading 2.1) can be selected by specifying the number of iterations. The accuracies of these options are close to the corresponding options of MAFFT. However, MUSCLE and MAFFT have several differences in the scoring system, the weighting system, and so on. Among these, MUSCLE made a great contribution to this area by introducing an approximate tree-building algorithm with a time complexity of $O(N^2)$, where N is the number of sequences. At that time, this algorithm was remarkably faster than those used by other programs. Then this algorithm was subsequently adopted by MAFFT [39] and the Clustal series [40]. MAFFT made a slight modification such that the resulting tree is exactly identical to that by the standard method. Due to this modification, the tree-building step is slightly faster in MUSCLE than in MAFFT without the PartTree option.

4 Dot Plot

All the options in MAFFT assume that there are no genomic rearrangements (translocations or inversions). By default, MAFFT uses an algorithm to accelerate a group-to-group alignment calculation with the FFT algorithm [1]. It first finds highly conserved regions and then aligns remaining regions using DP as shown in Fig. 2. Thus MAFFT can align long DNA sequences more efficiently than normal DP, if a number of highly conserved regions are found. Genomic rearrangements can result in conserved regions that appear in an inconsistent order. In such a case, DP has to be applied almost directly. It sometimes takes impractically long time, and the result does not make sense.

To avoid such cases, the web version of MAFFT displays dot plots between the first sequence and the remaining sequences, using the LAST local alignment program [41], for every nucleotide alignment run. By viewing the dot plots, a user can easily check for genomic rearrangements and the directions of input sequences.

4.1 Example

Some examples are shown in Fig. 4. If a plot like d is returned by the server, the calculation should be re-run with the "Adjust direction" option (for the web version) or with the `--adjust-direction` option (for the command-line version), as noted in the next section. If a more complicated plot, like e, is returned, other tools that assume genomic rearrangements should be applied,

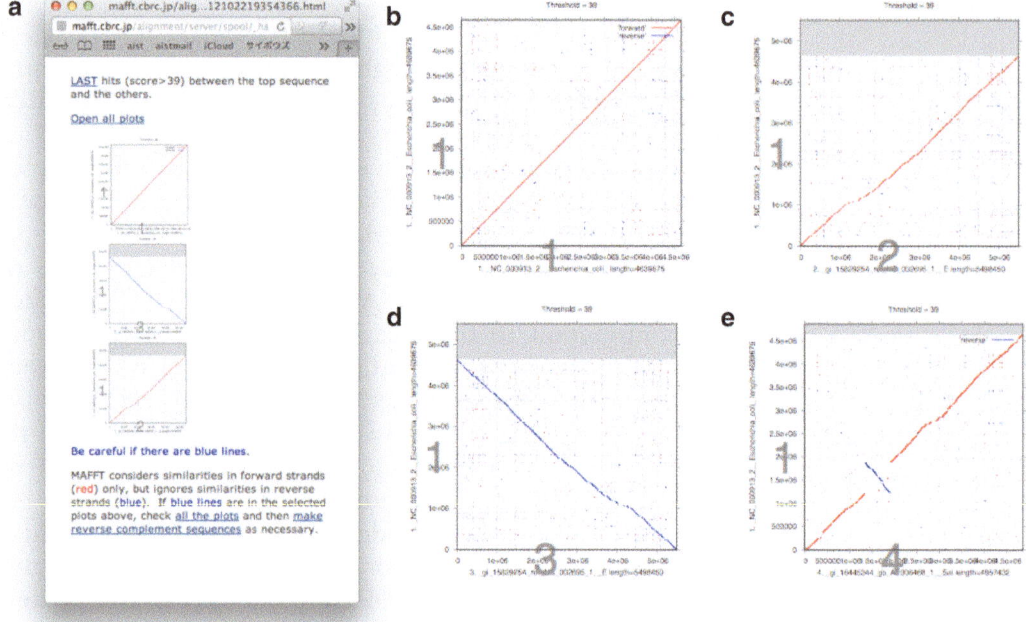

Fig. 4 Examples of dot plot. (**a**) Screenshot of dot plots in MAFFT service; (**b–d**) individual plots; (**b**) comparison of identical sequence; (**c**) no obvious large-scale genomic rearrangements; (**d**) reverse complementary sequences; (**e**) one large inversion; MAFFT cannot be applied to **e**

instead of MAFFT. Even if MAFFT returns an MSA for such a problem, the MSA is inappropriate for the region shown in blue in this plot.

5 Estimating the Direction of DNA Sequences

In the case of nucleotide alignments, if some of the input sequences have an entirely opposite direction to the other sequences, the directions can be automatically adjusted by the `--adjustdirection` option. This option is also available on the web version, with the "Adjust direction" button. There are several possible natural methods to determine the direction of nucleotide sequences.

Most naively, the direction of sequences can be estimated by comparing the first sequence and the other sequence with both forward and reverse directions. If the similarity score of forward–forward comparison is worse than that of forward–reverse comparison, the sequence is judged to have the opposite direction to the first sequence and its reverse complement replaces the sequence. This strategy works well in most cases, but, when the first sequence is phylogenetically isolated in the input data, the difference in similarity score between forward–forward comparison and forward–reverse comparison can be too small to judge the direction.

To give a more stable result, the current version of MAFFT uses the following procedure to determine the direction of each sequence. Suppose that the n input sequences are numbered from 0 to $n - 1$. For sequence i ($i = 1$ to $n - 1$) other than the first sequence,

1. Calculate the similarity scores, $S_f(j)$, between sequence i and sequences j ($j = 0$ to $i - 1$).
2. Calculate the similarity scores, $S_r(j)$, between the reverse complement of sequence i and sequences j ($j = 0$ to $i - 1$).
3. If $\max_j(S_f(j)) < \max_j(S_r(j))$, then sequence i is replaced with its reverse complement.

This procedure requires $O(n^2)$ comparisons and is slow when the scores are calculated with DP. However, when the scores are rapidly calculated based on the number of shared 6mers, the speed is practical.

To run this calculation on the command line, use

```
mafft --adjustdirection input > output
```

which computes the distances based on the number of shared 6mers. The slower but more exact calculation based on DP can be selected with

```
mafft --adjustdirectionaccurately input > output
```

Our preliminary assessment based on computer simulation showed that the difference between these two options is small unless the input sequences are highly divergent and short. Thus the `--adjustdirection` option is recommended in most cases.

6 Adding Unaligned Sequences into an MSA

The need for MSAs with a large number of sequences is increasing, as a result of advances in sequencing technologies. There are several different approaches to enable larger MSAs, e.g., rapid algorithms, and parallelization. MAFFT [1, 39, 42] and many other programs were recently developed or extended by incorporating these advances. In our opinion, another promising approach for large MSAs is the use of an existing alignment. A relatively small number of sequences have been carefully aligned and annotated in databases, e.g., [43–45]. Sometimes we align newly sequenced data into an existing MSA taken from such a database. This is more efficient than rebuilding the entire MSA from a set of ungapped sequences.

Moreover, biological knowledge is sometimes incorporated into MSAs in databases. Such information can be retained in a large alignment if the original alignment is kept. Based on such considerations, around 2010, we implemented an option, `--add`, to add unaligned sequences to an existing MSA. The implementation of the `--add` option was almost trivial; no change was necessary from the conventional progressive method, except that the alignment calculation is skipped at the nodes whose children are all in the existing alignment.

Several tools [46–48] for aligning short reads to existing alignment were developed between 2011 and 2012. Indeed such analysis is recently becoming important, along with the popularization of second-generation sequencers. For this purpose, a limitation of the `--add` option of MAFFT was pointed out in [48]. Thus we implemented a new option, `--addfragments`, which does not consider the relationship among the sequences to be added, for this purpose. Details of the `--add` and `--addfragments` options are described in [38].

6.1 Example: SSU rRNA

Here we use an example from Mirarab et al. [49]. They provide four datasets, M2, M3, M4, and 16S.B.ALL, for assessing the performance of phylogenetic placement. The first three are simulated datasets, which we used to assess the accuracy of alignments in [38]. Here we use the last one, which is based on actual data. It consists of a curated MSA of 13,822 bacterial SSU rRNA sequences, taken from the Gutell Comparative Ribosomonal Website (CRW) [50], and 13,821 fragmentary sequences, which are originally included in the CRW alignment but ungapped and artificially truncated.

Suppose a situation where we already have an MSA (`existingmsa`) consisting of 13,822 sequences, which are manually curated, and we have newly sequenced 13,821 fragments (`frags`) in a metagenomics project. Both files, `existingmsa` and `frags`, are in the multi-fasta format. To build a full alignment consisting of 27,643 sequences, use

```
mafft --addfragments frags existingmsa > output
```

in which full DP is used for computing the distances between the sequences in the existing MSA and new fragments. A faster option based on the number of shared 6mers is also available.

```
mafft --6merpair --addfragments frags existingmsa > output
```

The latter option is recommended unless the data is divergent.

If the new sequences were all from a single known species, this is a standard problem of mapping short reads to the (genomic)

Table 1
Comparison of different options using the 16S.B.ALL dataset [49]

Command	Accuracy	CPU time	Actual time‡
`mafft --addfragments frags existingmsa`	0.9969	6.67 days	18.3 h
`mafft --6merpair --addfragments frags existingmsa`	0.9949	3.77 h	36.2 min
`mafft --localpair --add frags existingmsa`	0.9707	39.7 days†	4.21 days†
`mafft --6merpair --add frags existingmsa`	0.9604	1.32 h	1.44 h
profile alignment	0.2779	14.8 h	1.53 h

The estimated alignments were compared with the CRW alignment to measure theaccuracy (the number of correctly aligned letters/the number of aligned letters inthe CRW alignment). Calculations were performed by MAFFT version 6.954, on aLinux PC with 2.67 GHz Intel Xeon E7-8837/256 GB RAM (for the case marked with †), oron a Linux PC with 3.47 GHz Intel Xeon X5690/48 GB RAM (for the othercases) ‡ Wall-clocktime with ten cores. Command-line argument for parallel processing is `--thread 10` [42]

sequence of the known species. However, in metagenomic analysis when new sequences are from multiple (and some novel) species, the phylogenetic position of the new sequences should be considered, like PaPaRa [46], PAGAN [48] and this option of MAFFT.

The accuracy of resulting MSAs was estimated by comparing them with the original CRW alignment (Table 1). CPU time and wall-clock time for each method are also listed in the table. Since the sequences in this dataset are highly conserved, the difference in accuracy between the default (`--addfragments`) and the faster option (`--6merpair --addfragments`) is small.

We also compared the performances of some subtypes of the `--add` option using the same dataset.

```
mafft --6merpair  --add frags existingmsa  > output (based on FFT-NS-1, faster)
mafft --localpair --add frags existingmsa  > output (based on L-INS-1, slower)
```

These options have no advantage for this problem, according to the third and fourth lines in Table 1. This is probably because the relationship among new fragments does not make sense, since most of them do not overlap with each other. In such cases, `--addfragments`, which does not consider this relationship, is more suitable than `--add`, which considers this relationship.
This observation suggests that the trade-off between accuracy and speed does not always hold. Rather, a method designed for the appropriate purpose should be applied. The application of a computationally expensive method based on L-INS-1 (`--localpair --add`) has no advantage, because the extra computational time is spent on the comparison of non-overlapping fragmentary sequences, which have no reasonable solutions.

Moreover, the last line in Table 1 shows results of profile alignment, in which the existing alignment is converted to a profile and each new sequence is separately aligned to the profile, equivalently to `mafft-profile`. This result clearly indicates that the application of profile alignment must be avoided in this case.

7 Portability

MAFFT is developed in a UNIX-like environment. Thus it runs natively on Linux and Mac OS X. However, previously, it did not smoothly run on Windows. We are now providing an all-in-one package, which includes SH and other necessary GNU utilities, for Windows. It runs almost like a native Windows program and can also be bundled with other packages or programs.

8 Use of Structural Information

We have been discussing alignments in terms of nucleotide or amino acid sequences. However, many amino acid sequences fold into unique tertiary structures. The use of such information in MSA construction was the basis of the 3DCoffee program [51], and subsequently PROMALS3D [52]. In this section we address several issues arising when incorporating protein structural information in MSA calculations. At the time of this writing, the number of sequenced proteins far exceeds the number of known structures. It would appear, then, that the scope of problems that can be addressed by sequence alignment far exceeds that of structure alignment. On the other hand, the number of sequence superfamilies is limited, and a large number of superfamilies contain members whose structures have been solved. Structural alignment represents a logical next step towards quantifying the similarity between remotely homologous families within a superfamily. However, to make practical use of sequence and structural information, a number of obstacles have to be overcome. Some of the obstacles are technical and result from the complexity and noisiness of structural information. While sequence information is discrete (i.e., 20 common amino acids) and compact (can be represented by a single letter), structural information is continuous (e.g., the position of a particular atom in space) and relatively large (there are between 4 and 13 heavy atoms in the 20 common amino acids). Moreover, due to the dynamic nature of proteins and limitations in experimental techniques, it is not uncommon for some atomic positions to be undefined or to have ambiguous positional assignments in typical protein structure database entries. The complexity of protein structure is reflected in the algorithms used to align them. Structural alignment methods are generally slow compared with sequence alignment methods, so any effort to combine the two must weigh the costs of such integration against the benefits.

There are also conceptual questions that need to be addressed. The biggest one is how to incorporate structural information into MSA calculations. A structural alignment is generally at least as accurate as a sequence alignment. However, not all parts of the alignment are equally reliable. For example, as a general rule "core" residues will align better than residues close to the molecular surface. When importing structural alignments into MSA calculations, we need a way of describing such variations in alignment quality. Below we will describe a particular structural alignment package, ASH, and discuss MAFFT-ASH integration at a conceptual level. ASH [53] is a pairwise protein structural alignment program that is based on the double dynamic programming (DDP) algorithm originally proposed by Orengo and Taylor [54, 55] and extended by Toh [56]. The source code of ASH is available from the Protein Data Bank Japan.

An essential feature of ASH is that the alignment is generated from a score matrix defined purely in terms of the structure of the two proteins. A particular element in the score matrix takes the form of a Gaussian-shaped function of the inter-residue distance

$$e_{ij} = \exp(-(d_{ij}/d_0)^2),$$

where d_{ij} is the distance between two alpha carbons i and j in the two input structures and d_0 is a parameter that defines tolerance in the score. The alignment results are fairly robust with respect to the particular choice of d_0, and the default behavior is to set the parameter to 4 Å. The distance between any two residues in the two input structures is obviously a function of their relative displacement and orientation. Thus the goal of ASH is to find the relative orientation that maximizes the equivalences when summed over the alignment. For domains that are topologically quite similar, minimization of the root-mean square deviation (RMSD) for a continuous subsequence of residues can provide a good initial guess. However, cases of repeating structural motifs can cause problems with convergence to a unique global maximum.

The residue-level equivalences, which form the basis of all ASH alignments, provide a convenient route for combining MAFFT and ASH. Given a set of input structures, we can compute structural alignments for all unique pairs. We can then set a threshold for the residue equivalence (e.g., .5), which we will define as "high confidence." MAFFT allows such "seed" alignments to be input as restraints [57].

```
mafft-linsi --seed pair1 --seed pair2 ··· sequences > output
```

where pair1, pair2, etc. are the high confidence structural alignments. If the sequences identities between the aligned structures are low, then we can expect an improvement in the resulting MSA relative to conventional MAFFT. Based on the approach outlined above, we are developing an integrative service for protein structure-informed MSA construction.

References

1. Katoh K, Misawa K, Kuma K, Miyata T (2002) MAFFT: a novel method for rapid multiple sequence alignment based on fast Fourier transform. Nucleic Acids Res 30:3059–3066
2. Nuin PA, Wang Z, Tillier ER (2006) The accuracy of several multiple sequence alignment programs for proteins. BMC Bioinformatics 7:471
3. Dessimoz C, Gil M (2010) Phylogenetic assessment of alignments reveals neglected tree signal in gaps. Genome Biol 11:R37
4. Letsch HO, Kuck P, Stocsits RR, Misof B (2010) The impact of rRNA secondary structure consideration in alignment and tree reconstruction: simulated data and a case study on the phylogeny of hexapods. Mol Biol Evol 27:2507–2521
5. Sahraeian SM, Yoon BJ (2011) PicXAA-R: efficient structural alignment of multiple RNA sequences using a greedy approach. BMC Bioinformatics 12(Suppl 1):S38
6. Sievers F, Wilm A, Dineen D, Gibson TJ, Karplus K, Li W, Lopez R, McWilliam H, Remmert M, Soding J, Thompson JD, Higgins DG (2011) Fast, scalable generation of high-quality protein multiple sequence alignments using Clustal Omega. Mol Syst Biol 7:539
7. Edgar RC (2004) MUSCLE: multiple sequence alignment with high accuracy and high throughput. Nucleic Acids Res 32:1792–1797
8. Edgar RC (2004) MUSCLE: a multiple sequence alignment method with reduced time and space complexity. BMC Bioinformatics 5:113
9. Feng DF, Doolittle RF (1987) Progressive sequence alignment as a prerequisite to correct phylogenetic trees. J Mol Evol 25:351–360
10. Higgins DG, Sharp PM (1988) CLUSTAL: a package for performing multiple sequence alignment on a microcomputer. Gene 73:237–244
11. Wilbur WJ, Lipman DJ (1983) Rapid similarity searches of nucleic acid and protein data banks. Proc Natl Acad Sci USA 80:726–730
12. Loytynoja A, Goldman N (2008) Phylogeny-aware gap placement prevents errors in sequence alignment and evolutionary analysis. Science 320:1632–1635
13. Lassmann T, Sonnhammer EL (2005) Kalign—an accurate and fast multiple sequence alignment algorithm. BMC Bioinformatics 6:298
14. Barton GJ, Sternberg MJ (1987) A strategy for the rapid multiple alignment of protein sequences. Confidence levels from tertiary structure comparisons. J Mol Biol 198:327–337
15. Berger MP, Munson PJ (1991) A novel randomized iterative strategy for aligning multiple protein sequences. Comput Appl Biosci 7:479–484
16. Gotoh O (1993) Optimal alignment between groups of sequences and its application to multiple sequence alignment. Comput Appl Biosci 9:361–370
17. Gotoh O (1995) A weighting system and algorithm for aligning many phylogenetically related sequences. Comput Appl Biosci 11:543–551
18. Ishikawa M, Toya T, Hoshida M, Nitta K, Ogiwara A, Kanehisa M (1993) Multiple sequence alignment by parallel simulated annealing. Comput Appl Biosci 9:267–273
19. Notredame C, Higgins DG (1996) Saga: sequence alignment by genetic algorithm. Nucleic Acids Res 24:1515–1524
20. Gotoh O (1996) Significant improvement in accuracy of multiple protein sequence alignments by iterative refinement as assessed by reference to structural alignments. J Mol Biol 264:823–838
21. Hirosawa M, Totoki Y, Hoshida M, Ishikawa M (1995) Comprehensive study on iterative

algorithms of multiple sequence alignment. Comput Appl Biosci 11:13–18

22. Vingron M, Argos P (1989) A fast and sensitive multiple sequence alignment algorithm. Comput Appl Biosci 5:115–121
23. Gotoh O (1990) Consistency of optimal sequence alignments. Bull Math Biol 52:509–525
24. Notredame C, Holm L, Higgins DG (1998) COFFEE: an objective function for multiple sequence alignments. Bioinformatics 14:407–422
25. Notredame C, Higgins DG, Heringa J (2000) T-Coffee: a novel method for fast and accurate multiple sequence alignment. J Mol Biol 302:205–217
26. Do CB, Mahabhashyam MS, Brudno M, Batzoglou S (2005) ProbCons: probabilistic consistency-based multiple sequence alignment. Genome Res 15:330–340
27. Roshan U, Livesay DR (2006) Probalign: multiple sequence alignment using partition function posterior probabilities. Bioinformatics 22:2715–2721
28. Pei J, Grishin NV (2007) PROMALS: towards accurate multiple sequence alignments of distantly related proteins. Bioinformatics 23:802–808
29. Liu Y, Schmidt B, Maskell DL (2010) MSAProbs: multiple sequence alignment based on pair hidden markov models and partition function posterior probabilities. Bioinformatics 26:1958–1964
30. Katoh K, Toh H (2008) Recent developments in the MAFFT multiple sequence alignment program. Brief Bioinform 9:286–298
31. Katoh K, Toh H (2008) Improved accuracy of multiple ncRNA alignment by incorporating structural information into a MAFFT-based framework. BMC Bioinformatics 9:212
32. McCaskill JS (1990) The equilibrium partition function and base pair binding probabilities for RNA secondary structure. Biopolymers 29:1105–1119
33. Tabei Y, Tsuda K, Kin T, Asai K (2006) SCARNA: fast and accurate structural alignment of rna sequences by matching fixed-length stem fragments. Bioinformatics 22:1723–1729
34. Hofacker IL, Fekete M, Stadler PF (2002) Secondary structure prediction for aligned RNA sequences. J Mol Biol 319:1059–1066
35. Tabei Y, Kiryu H, Kin T, Asai K (2008) A fast structural multiple alignment method for long RNA sequences. BMC Bioinformatics 9:33
36. Hamada M, Sato K, Kiryu H, Mituyama T, Asai K (2009) CentroidAlign: fast and accurate aligner for structured RNAs by maximizing expected sum-of-pairs score. Bioinformatics 25:3236–3243
37. Wilm A, Higgins DG, Notredame C (2008) R-Coffee: a method for multiple alignment of non-coding RNA. Nucleic Acids Res 36:e52
38. Katoh K, Frith MC (2012) Adding unaligned sequences into an existing alignment using MAFFT and LAST. Bioinformatics 28:3144–3146
39. Katoh K, Toh H (2007) PartTree: an algorithm to build an approximate tree from a large number of unaligned sequences. Bioinformatics 23:372–374
40. Blackshields G, Sievers F, Shi W, Wilm A, Higgins DG (2010) Sequence embedding for fast construction of guide trees for multiple sequence alignment. Algorithms Mol Biol 5:21
41. Kiełbasa SM, Wan R, Sato K, Horton P, Frith MC (2011) Adaptive seeds tame genomic sequence comparison. Genome Res 21:487–493
42. Katoh K, Toh H (2010) Parallelization of the MAFFT multiple sequence alignment program. Bioinformatics 26:1899–1900
43. Punta M, Coggill PC, Eberhardt RY, Mistry J, Tate J, Boursnell C, Pang N, Forslund K, Ceric G, Clements J, Heger A, Holm L, Sonnhammer EL, Eddy SR, Bateman A, Finn RD (2012) The Pfam protein families database. Nucleic Acids Res 40:D290–D301
44. Sigrist CJ, Cerutti L, deCastro E, Langendijk-Genevaux PS, Bulliard V, Bairoch A, Hulo N (2010) PROSITE, a protein domain database for functional characterization and annotation. Nucleic Acids Res 38:D161–D166
45. Cole JR, Wang Q, Cardenas E, Fish J, Chai B, Farris RJ, Kulam-Syed-Mohideen AS, McGarrell DM, Marsh T, Garrity GM, Tiedje JM (2009) The Ribosomal Database Project: improved alignments and new tools for rRNA analysis. Nucleic Acids Res 37:D141–D145
46. Berger SA, Stamatakis A (2011) Aligning short reads to reference alignments and trees. Bioinformatics 27:2068–2075
47. Sun H, Buhler JD (2012) PhyLAT: a phylogenetic local alignment tool. Bioinformatics 28:1336–1344
48. Löytynoja A, Vilella AJ, Goldman N (2012) Accurate extension of multiple sequence alignments using a phylogeny-aware graph algorithm. Bioinformatics 28:1684–1691
49. Mirarab S, Nguyen N, Warnow T (2012) SEPP: SATé-Enabled phylogenetic placement. Pac Symp Biocomput 17:247–258

50. Cannone JJ, Subramanian S, Schnare MN, Collett JR, D'Souza LM, Du Y, Feng B, Lin N, Madabusi LV, Muller KM, Pande N, Shang Z, Yu N, Gutell RR (2002) The comparative RNA web (CRW) site: an online database of comparative sequence and structure information for ribosomal, intron, and other RNAs. BMC Bioinformatics 3:2
51. O'Sullivan O, Suhre K, Abergel C, Higgins DG, Notredame C (2004) 3DCoffee: combining protein sequences and structures within multiple sequence alignments. J Mol Biol 340:385–395
52. Pei J, Kim BH, Grishin NV (2008) PROMALS3D: a tool for multiple protein sequence and structure alignments. Nucleic Acids Res 36:2295–2300
53. Standley DM, Toh H, Nakamura H (2004) Detecting local structural similarity in proteins by maximizing number of equivalent residues. Proteins 57:381–391
54. Taylor WR, Orengo CA (1989) Protein structure alignment. J Mol Biol 208:1–22
55. Orengo CA, Taylor WR (1993) A local alignment method for protein structure motifs. J Mol Biol 233:488–497
56. Toh H (1997) Introduction of a distance cut-off into structural alignment by the double dynamic programming algorithm. Comput Appl Biosci 13:387–396
57. Katoh K, Asimenos G, Toh H (2009) Multiple alignment of DNA sequences with MAFFT. Methods Mol Biol 537:39–64

Chapter 9

Multiple Sequence Alignment Using Probcons and Probalign

Usman Roshan

Abstract

Sequence alignment remains a fundamental task in bioinformatics. The literature contains programs that employ a host of exact and heuristic strategies available in computer science. Probcons was the first program to construct maximum expected accuracy sequence alignments with hidden Markov models and at the time of its publication achieved the highest accuracies on standard protein multiple alignment benchmarks. Probalign followed this strategy except that it used a partition function approach instead of hidden Markov models. Several programs employing both strategies have been published since then. In this chapter we describe Probcons and Probalign.

Key words Sequence alignment, Expected accuracy, Hidden Markov models, Partition function

1 Introduction

Multiple protein sequence alignment is one of the most commonly used tasks in bioinformatics [1]. It has widespread applications that include detecting functional regions in proteins [2] and reconstructing complex evolutionary histories [1, 3]. Techniques for constructing accurate alignments are therefore of great interest to the bioinformatics community.

ClustalW [4] is one of the earliest multiple sequence aligners and remains popular to date. Other programs include Dialign [5], T-Coffee [6], MUSCLE [7], and MAFFT [8]. Given the importance of multiple sequence alignment, several protein alignment benchmarks have been created for unbiased accuracy assessment of alignment quality. Of these, BAliBASE [9–11] is by far the most commonly used. The BAliBASE benchmark alignments are computed using superimposition of protein structures.

Prior to Probcons [12] most programs optimized the sum-of-pairs score of a multiple alignment or computed the Viterbi alignment [3]. Probcons computes the maximal expected accuracy alignment instead. The expected accuracy of an alignment is based upon posterior probabilities of residues [3, 12–14].

David J. Russell (ed.), *Multiple Sequence Alignment Methods*, Methods in Molecular Biology, vol. 1079,
DOI 10.1007/978-1-62703-646-7_9,

Probcons computes these probabilities using a Hidden Markov Model (HMM) for pairwise sequence alignment. The HMM parameters are learned using unsupervised learning on the BAliBASE 2.0 benchmark.

Probalign [13] on the other hand estimates amino acid posterior probabilities from the partition function of alignments as described by Miyazawa [14]. It then proceeds to compute the maximal expected accuracy multiple sequence alignment by following the strategy of Probcons. We first describe both methods of computing posterior probabilities in detail below. We then describe the Probcons alignment algorithm that makes use of the probabilities to output a final alignment. Probalign follows the same approach.

2 Methods

2.1 Posterior Probabilities for Expected Accuracy Sequence Alignment

The expected accuracy of an alignment is based upon the posterior probabilities of aligning residues in two sequences. Consider sequences x and y and let a^* be their true alignment. Following the description in Do [12] the posterior probability of residue x_i aligned to y_j in a^* is defined as

$$P(x_i \sim y_j \in a^*|x,y) = \sum_{a \in A} P(a|x,y)\ 1\{x_i \sim\ y_j \in a\}, \qquad (1)$$

where A is the set of all alignments of x and y and $1(expr)$ is the indicator function which returns 1 if the expression $expr$ evaluates to true and 0 otherwise. $P(a|x,y)$ represents the probability that alignment a is the true alignment a^*. This can easily be calculated using a pairwise HMM if all the parameters are known (described below). From here on we represent the posterior probability as $P(x_i \sim y_j)$ with the understanding that it represents the probability of x_i aligned to y_j in the true alignment a^*.

According to Eq. 1 as long as we have an ensemble of alignments A with their probabilities $P(a|x,y)$ we can compute the posterior probability $P(x_i \sim y_j)$ by summing up the probabilities of alignments where x_i is paired with y_j. Probcons uses hidden Markov models while Probalign uses the partition function of sequence alignments to generate the ensemble.

2.2 Posterior Probabilities by Hidden Markov Models

Probcons uses a basic sequence alignment hidden Markov model (HMM) shown in Fig. 1.

The emission probabilities for the hidden states M, I_x, and I_y are given by $p(x_i, y_j)$, $q(x_i)$, and $q(y_j)$ where x_i is the ith residue of sequence x and y_j defined correspondingly. The terms δ and ε represent transition probabilities for gap open and gap extensions. The probability of a sequence alignment under this model is well-defined and the one with the highest probability can be found with

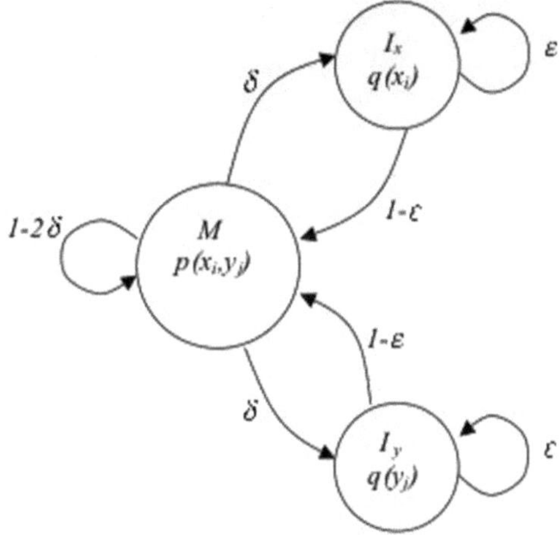

Fig. 1 Hidden Markov model for pairwise sequence alignment

the Viterbi algorithm [3]. The posterior probabilities can then be obtained by

$$P\left(x_i \sim \; y_j \in a^* | x, y\right) \frac{f(i,j)b(i,j)}{P(x,y)}. \tag{2}$$

In the above equation $f(i,j)$ is the sum of all probabilities of all alignments of $x_{1...i}$ and $y_{1...j}$ where $x_{1...i}$ are the first i characters of sequence x and $y_{1...j}$ is defined the same way. The term $b(i,j)$ is the sum of all probabilities of all alignments of $x_{i+1...m}$ and $y_{j+1...n}$ where m and n are the lengths of sequences x and y respectively. And finally $P(x,y)$ is the sum of the probabilities of all alignments of x and y under the model. These can be obtained by modifying the Viterbi algorithm to add instead of taking the max as shown in Durbin [3].

2.3 Posterior Probabilities by Partition Function

Amino acid scoring matrices that are normally used for sequence alignment are represented as log-odds scoring matrices as defined by Dayhoff [15]. The commonly used sum-of-pairs score of an alignment [3] is defined as the sum of residue-residue pairs and residue-gap pairs under an affine penalty scheme.

$$S(a) = T \sum_{(i,j) \in a} \ln\left(M_{ij}/f_i f_j\right) + (\text{gap\_penalties}). \tag{3}$$

Here T is a constant and set according to the scoring matrix, M_{ij} is the mutation probability of residue i changing to j and f_i and f_j are background frequencies of residues i and j. In fact, it can be shown that any scoring matrix corresponds to a log odds matrix [16, 17]. Miyazawa [14] proposed that the probability of alignment $P(a)$ of sequences x and y can be defined as

$$P(a) \propto e^{S(a)/T}, \tag{4}$$

where $S(a)$ is the score of the alignment under the given scoring matrix. In this setting one can then treat the alignment score as negative energy and T as the thermodynamic temperature, similar to what is done in statistical mechanics. Analogous to the statistical mechanical framework Miyazawa [14] defined the partition function of alignments as

$$Z(T) = \sum_{a \in A} e^{S(a)/T}, \tag{5}$$

where A is the set of all alignments of x and y. With the partition function in hand the probability of an alignment a can now be defined as

$$P(a, T) = e^{S(a)/T}/Z(T). \tag{6}$$

As T approaches infinity all alignments are equally probable, whereas at small values of T only the nearly optimal alignments have the highest probabilities. Thus, the temperature parameter T can be interpreted as a measure of deviation from the optimal alignment.

The alignment partition function can be computed using recursions similar to the Needleman–Wunsch dynamic algorithm. Let Z_{ij}^M represent the partition function of all alignments of $x_{1...i}$ and $y_{1...j}$ ending in x_i paired with y_j, and $S_{ij}(a)$ represent the score of alignment a of $x_{1...i}$ and $y_{1...j}$. According to Eq. 2.

$$Z_{i,j}^M = \sum_{a \in A_{ij}} e^{S_{ij}(a)/T} = \left(\sum_{a \in A_{ij} = ij-1} e^{S_{i-1,j-1}(a)/T} \right) e^{s(x_i, y_j)/T}, \tag{7}$$

where A_{ij} is the set of all alignments of $x_{1...i}$ and $y_{1...j}$, and $s(x_i,y_j)$ is the score of aligning residue x_i with y_j. The summation in the bracket on the right hand side of the above equation is precisely the partition function of all alignments of $x_{1...i-1}$ and $y_{1...j-1}$. We can thus compute the partition function matrices using standard dynamic programming.

$$\begin{aligned}
Z_{i,j}^M &= \left(Z_{i-1,j-1}^M + Z_{i-1,j-1}^E + Z_{i-1,j-1}^F \right) e^{s(x_i, y_j)/T} \\
Z_{i,j}^E &= Z_{i,j-1}^M e^{g/T} + Z_{i,j-1}^E e^{ext/T} \\
Z_{i,j}^F &= Z_{i-1,j}^M e^{g/T} + Z_{i-1,j}^F e^{ext/T} \\
Z_{i,j} &= Z_{i,j}^M + Z_{i,j}^E + Z_{i,j}^F.
\end{aligned} \tag{8}$$

Here $s(x_i,y_j)$ represents the score of aligning residue x_i with y_j, g is the gap open penalty, and *ext* is the gap extension penalty. The matrix Z_{ij}^M represents the partition function of all alignments ending in x_i paired with y_j. Similarly Z_{ij}^E represents the partition function of all alignments in which y_j is aligned to a gap and Z_{ij}^F all alignments in which x_i is aligned to a gap. Boundary conditions and further details can be obtained from Miyazawa [14].

Once the partition function is constructed, the posterior probability of x_i aligned to y_j can be computed as

$$P(x_i \sim y_j) = \frac{Z^M_{i-1,j-1} Z'^M_{i+1,j+1}}{Z} e^{s(x_i, y_j)/T}, \tag{9}$$

where $Z'^M{}_{i,j}$ is the partition function of alignments of subsequences $x_{i...m}$ and $y_{j...n}$ beginning with x_i paired with y_j and m and n are lengths of x and y respectively. This can be computed using standard backward recursion formulas [3]. In the above equation $Z^M_{i-1,j-1}/Z$ and $Z'^M_{i+1,j+1}/Z$ represent the probabilities of feasible suboptimal alignments (as determined by the T parameter) of $x_{1...i-1}$ and $y_{1...j-1}$, and $x_{i+1...m}$ and $y_{j+1...n}$ respectively, where m and n are lengths of x and y respectively. Thus, the equation weighs alignments according to their partition function probabilities and estimates $P(x_i \sim y_j)$ as the sum of probabilities of all alignments where x_i is paired with y_j.

2.4 Maximal Expected Accuracy Alignment

Given the posterior probability matrix $P(x_i \sim y_j)$, we define the expected accuracy of the alignment of x and y as

$$\frac{1}{\min\{|x||y|\}} \sum_{x_i - y_j \in a} P(x_i \sim y_j \in a^* | x, y). \tag{10}$$

The maximum expected accuracy alignment score is computed by dynamic programming using the following recurrence described in Durbin [3].

$$\text{for } i = 1 \text{ to } |x|$$
$$\text{for } j = 1 \text{ to } |y|$$
$$A(i,j) = \max \left\{ \begin{array}{l} A(i-1, j-1) + P(x_i \sim y_j) \\ A(i-1, j) \\ A(i, j-1) \end{array} \right\}. \tag{11}$$

The first row and column of A are set to 0. The alignment score is given by $A(|x|, |y|)$ where $|x|$ and $|y|$ denote the lengths of sequences x and y. The actual alignment of x and y can be recovered by keeping track of which cell the maximum value is obtained from for each entry of A [3].

Both Probcons and Probalign first estimate posterior probabilities for amino acid residues for all pairs of protein sequences in the input. Probcons introduced a number of new approaches for constructing a multiple alignment with posterior probabilities for all pairs of sequences. It first performs a probabilistic consistency transformation to improve posterior probabilities with the aid of a third sequence [12]. It then adapts three standard approaches in multiple sequence alignment, namely construction of a guide-tree, progressive alignment, and iterative refinement to the expected accuracy alignment approach. The guide-tree construction is similar to UPGMA [18] except that expected accuracies are used to measure distance between clusters [12]. Profile-profile alignment [3], another standard technique in multiple sequence alignment, is extended to incorporate expected accuracies which facilitates the progressive and iterative alignment strategies. Probalign follows all of these procedures for constructing its multiple alignment.

3 Practical Issues

Probalign is freely available at http://probalign.njit.edu [19] with gap penalties optimized for standard protein and RNA alignment benchmarks and Probcons is available from its authors. In terms of running time both Probcons and Probalign are slower than several previous approaches and so the alignment of thousands of sequences remains a challenge. Some runtime improvements have been made to Probalign and the most recent version 1.4 (at the time of writing this chapter) is considerably faster than earlier ones.

The Probalign webserver, also called eProbalign, provides a useful tool for eliminating poorly aligned columns. The problem of determining reliably aligned columns frequently comes up in practice. eProbalign provides one solution by averaging pairwise posterior probabilities in each column and displaying them in different shades of red. The server also allows the alignment to be saved in text and pdf formats.

In practice Probalign outperforms existing programs by large margins when the data contains sequences of varying lengths [13]. Thus it is particularly suitable for protein and RNA datasets where the sequence length variation is high.

The alignment of genomic length DNA sequences pose a runtime challenge to Probalign and Probcons. Both work best for protein and RNA sequences. However, the program Pecan [20] and webserver plastrna.njit.edu [21] adapt the expected accuracy approach for genome analysis. The former is for genome alignment while the latter searches for evolutionary related RNAs in genomes.

References

1. Notredame C (2002) Recent progresses in multiple sequence alignment: a survey. Pharmacogenomics 3(1):131–144
2. La D, Sutch B, Livesay DR (2005) Predicting protein functional sites with phylogenetic motifs. Proteins 58:309–320
3. Durbin R, Eddy S, Krogh A, Mitchison G (1998) Biological sequence analysis: probabilistic models of proteins and nucleic acids. Cambridge University Press, Cambridge
4. Thompson JD, Higgins DG, Gibson TJ (1994) ClustalW: improving the sensitivity of progressive multiple sequence alignment through sequence weighting, position specific gap penalties, and weight matrix choice. Nucleic Acids Res 27(13):2682–2690

5. Subramanian AR, Weyer-Menkhoff J, Kaufmann M, Morgenstern B (2005) Dialign-T: an improved algorithm for segment-based multiple sequence alignment. BMC Bioinformatics 6:66
6. Notredame C, Higgins D, Heringa J (2000) T-Coffee: a novel method for multiple sequence alignments. J Mol Biol 302:205–217
7. Edgar RC (2004) MUSCLE: multiple sequence alignment with high accuracy and high throughput. Nucleic Acids Res 32(5):1792–1797
8. Katoh K, Misawa K, Kuma K, Miyata T (2005) MAFFT version 5: improvement in accuracy of multiple sequence alignment. Nucleic Acids Res 33:511–518
9. Thompson JD, Plewniak F, Poch O (1999) A comprehensive comparison of multiple sequence alignment programs. Nucleic Acids Res 27(13):2682–2690
10. Bahr A, Thompson JD, Thierry JC, Poch O (2001) BAliBASE (Benchmark Alignment dataBASE) enhancements for repeats, transmembrane sequences, and circular permutations. Nucleic Acids Res 29(1):323–326
11. Thompson JD, Koehl P, Ripp R, Poch O (2005) BAliBASE 3.0: latest developments of the multiple sequence alignment benchmark. Proteins 61:127–136
12. Do CB, Mahabhashyam MSP, Brudno M, Batzoglou S (2005) PROBCONS: probabilistic consistency based multiple sequence alignment. Genome Res 15:330–340
13. Roshan U, Livesay DR (2006) Probalign: multiple sequence alignment using partition function posterior probabilities. Bioinformatics 22(22):2715–2721
14. Miyazawa S (1995) A reliable sequence alignment method based upon probabilities of residue correspondences. Protein Eng 8(10): 999–1009
15. Dayhoff MO, Schwartz RM, Orcutt BC (1978) A model for evolutionary change in proteins. In: Dayhoff MO (ed) Atlas of protein sequence and structure, vol 5. National Biochemical Research Foundation, Washington, DC, pp 345–352
16. Karlin S, Altschul SF (1990) Methods for assessing the statistical significance of molecular sequence features by using general scoring schmes. Proc Nat Acad Sci USA 87(6): 2264–2268
17. Altschul SF (1993) A protein alignment scoring system sensitive at all evolutionary distances. J Mol Evol 36(3):290–300
18. Sneath PHA, Sokal RR (1973) Numerical taxonomy. Freeman, San Francisco, CA
19. Chikkagoudar S, Roshan U, Livesay DR (2010) PLAST-ncRNA: partition function Local Alignment Search Tool for non-coding RNA sequences. Nucleic Acids Res 38: W59–W63
20. Paten B, Herrero J, Beal K, Birney E (2009) Sequence progressive alignment, a framework for practical large-scale probabilistic consistency alignment. Bioinformatics 25(3): 295–301
21. Roshan U, Chikkagoudar S, Livesay DR (2008) Searching for evolutionary distant RNA homologs within genomic sequences using partition function posterior probabilities. BMC Bioinformatics 9:61

Chapter 10

Phylogeny-aware alignment with PRANK

Ari Löytynoja

Abstract

Evolutionary analyses require sequence alignments that correctly represent evolutionary homology. Evolutionary and structural homology are not the same and sequence alignments generated with methods designed for structural matching can be seriously misleading in comparative and phylogenetic analyses. The phylogeny-aware alignment algorithm implemented in the program PRANK has been shown to produce good alignments for evolutionary inferences. Unlike other alignment programs, PRANK makes use of phylogenetic information to distinguish alignment gaps caused by insertions or deletions and, thereafter, handles the two types of events differently. As a by-product of the correct handling of insertions and deletions, PRANK can provide the inferred ancestral sequences as a part of the output and mark the alignment gaps differently depending on their origin in insertion or deletion events. As the algorithm infers the evolutionary history of the sequences, PRANK can be sensitive to errors in the guide phylogeny and violations on the underlying assumptions about the origin and patterns of gaps. These issues are discussed in detail and practical advice for the use of PRANK in evolutionary analysis is provided. The PRANK software and other methods discussed here can be found from the program home page at http://code.google.com/p/prank-msa/.

Key words Phylogeny-aware alignment, Evolutionary sequence analysis, Insertions and deletions, Character homology

1 Introduction

Multiple sequence alignment has a central role in evolutionary sequence analysis, in some studies so central that the alignment and evolutionary inferences should be performed simultaneously or at least in a tightly coupled manner. The connection between alignment and phylogeny inference was noticed early [1], but the evolutionary consequences of it are still largely ignored by mainstream alignment methods. A probable explanation for this oversight is the historical focus on protein alignments and extensive use of structural benchmarks in the development and comparison of the analysis methods. The use of these benchmarks has produced great alignments for structural studies of proteins but, as noticed by many users of the resulting methods, the very same alignments may be unsuitable for evolutionary analyses.

David J. Russell (ed.), *Multiple Sequence Alignment Methods*, Methods in Molecular Biology, vol. 1079,
DOI 10.1007/978-1-62703-646-7_10,

This chapter focuses on evolutionary sequence alignment and the use of a phylogeny-aware alignment algorithm to infer alignments for evolutionary studies. The definition of evolutionary homology is central and we will start by discussing that and its correct representation in multiple sequence alignments. We will then introduce—with lots of figures and no equations—the phylogeny-aware alignment algorithm implemented in PRANK. After detailing the strengths and weaknesses of the method, we will see what this means in practice and give advice for the use of PRANK. We will finish with a brief discussion on the future plans for PRANK and related methods.

In the following sections, some methods based on the classical progressive algorithm are criticized and shown to perform poorly. This criticism is based on their performance in evolutionary analyses only and, as demonstrated in other chapters of this book, the alignments they produce can be suitable for other types of analyses. Similarly, the phylogeny-aware algorithm may perform poorly in non-evolutionary alignment tasks and alternative methods should be used.

2 Evolutionary Homology in Sequence Alignment

A multiple sequence alignment represents sitewise homology among the characters in different sequences. The type of homology denoted by the alignment depends on the application and the intended use of the data: when the alignment is used for evolutionary analyses, the characters placed in the same column are believed to be evolutionarily homologous and share a common ancestor. Evolutionary homology is not the same as structural homology and the difference between the two homology types is most clearly evidenced by the role of insertions. Two independent insertions at the same position can lead to identical changes in the structure and the characters inserted independently may thus be considered structurally homologous; in contrast, independent insertions—even at exactly the same position—do not share a common ancestor and can never be evolutionarily homologous. To correctly indicate the evolutionary homology of insertions, the characters descending from different insertions events should be placed in separate alignment columns (Fig. 1).

If one restricts the analysis to relatively short sequence fragments, one can assume that the sequences evolve by substitutions, insertions, and deletions only. The three processes can be assumed to occur at relatively constant (although for different processes distinct) rates, substitutions typically being at least an order of magnitude more common than insertions and deletions [2]. The three processes differ greatly in their effect on the sequences and on one's ability to infer the events from the data: (1) a character at a

True alignment: **Re-alignment with ClustalW:**

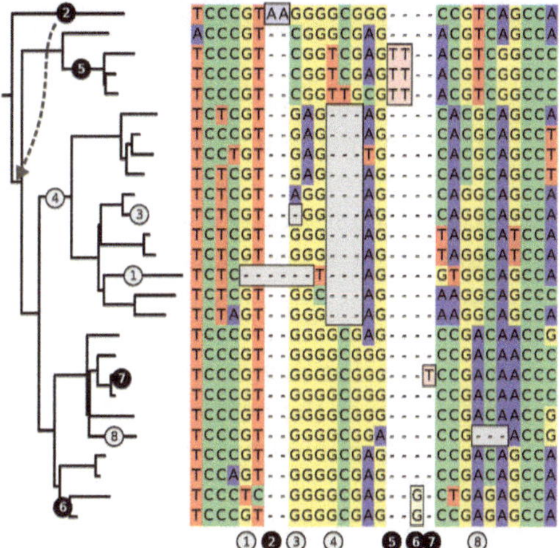

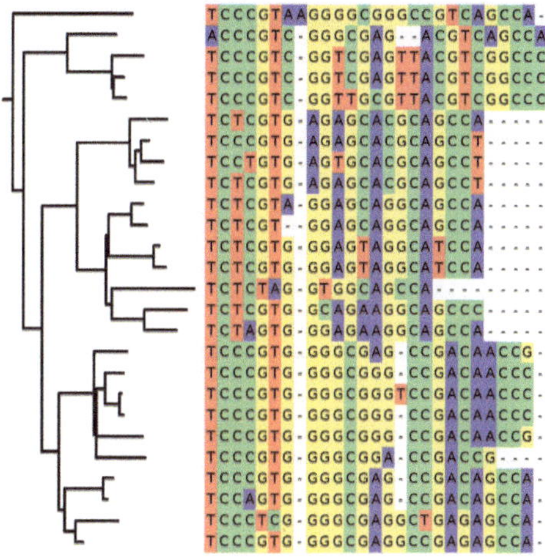

Fig. 1 The gap patterns in a true alignment reflect the underlying phylogeny of the sequences. Insertions and deletions create gap patterns (*boxes* in the alignment; numbered at the bottom) that reflect the phylogenetic locations of the events (*black* and *gray bubbles* in the tree). If multiple parallel insertions occur at homologous positions (events 5, 6 and 7), inserted columns can be placed in any order without effect on the homology statement. In the case of more than two parallel insertions, some inserted fragments are disconnected from the rest of the alignment (here, event 6). The phylogenetic location and type of event 2 is uncertain and it could also be a deletion in the sister branch. Whereas the phylogeny-aware algorithm would re-align the sequences correctly, the classical progressive algorithm (here ClustalW [7]) fails to resolve the true insertion and deletion events

certain site can be substituted several times, subsequent substitutions may turn the character state back to an earlier one and characters in different evolutionary lineages may independently obtain the same new character state, all still remaining homologous to each other; (2) an insertion adds new characters to the sequence and subsequent insertions may be nested inside a fragment inserted by a preceding insertion event but, as mentioned above, insertions in different lineages are never homologous and evolve independently; and (3) a deletion removes characters permanently and the characters once deleted cannot be reverted, a potential insertion at the same position introducing new characters that are not homologous with the deleted ones.

By a comparison of two homologous sequences we can detect sites that have undergone substitutions but, without more information, we cannot tell which sequence has changed at which position. In contrast to this, length differences between two sequences can be explained by deletions of existing characters in one sequence or insertions of new characters in the other. The evolutionary lineage on which the substitution-differences between two sequences have occurred can be resolved using outgroup sequences and by inferring the character states for the ancestor of the two.

Similarly to this, the evolutionary lineages—and thus the types—of insertion/deletion events creating length differences between two sequences can be inferred using phylogenetic information from related sequences.

Over time, sequences accumulate changes. Despite some differences in genome sizes, we can assume that sequences tend to retain their approximate length and insertion of new characters is counterbalanced by deletion of others. After a split from a common ancestor, the number of substitution-differences at homologous positions in descendant sequences increases until the sequence identity drops to the level expected by random sequences. The effect of insertions and deletions is very different. Assuming that each new insertion is not immediately followed by the deletion of the newly inserted characters, the total number of independent homologous sites within a set of sequence keeps increasing. With more than few sequences in the set, the increase in the number of independent homologous sites—and thus the number of columns in the alignment representing them—is not significantly affected by deletions as the chances of the same sites being independently deleted in all evolutionary lineages are small. Thus, the total length of the sequence alignment correctly representing the evolutionary homology among the characters is expected to grow roughly linearly with the evolutionary time covered by the different sequence lineages. Over long periods of time, the ancestral characters of a neutrally evolving sequence (or sequence region) are expected to be completely replaced by new characters through combinations of insertions and deletions: as a result, the correct evolutionary alignment of highly-diverged descendant sequences should not match a single character. Typically, the more freely-evolving sequence regions are flanked by conserved regions (e.g., loops and coils vs. core region in protein sequences) and the alignment is both possible and meaningful.

In practice, the alignment length rarely grows linearly with the evolutionary divergence. If the alignment is performed with methods based on the classical progressive algorithm [3, 4], the alignment length may grow linearly with the number of substitution changes for a while, but the growth curves of the two then separate and the alignment length increases only slowly, if at all (Fig. 2). The reason for this is that the classical algorithm does not distinguish insertions from deletions and, inherently, considers all length differences as deletion events. The use of such biased alignments in evolutionary analysis is likely to lead to erroneous conclusions.

3 Phylogeny-Aware Alignment

Independent insertions at the same position are not homologous and have to be identified to allow for their correct placement in different alignment columns. This alone demonstrates that an evolutionarily accurate alignment cannot be generated without

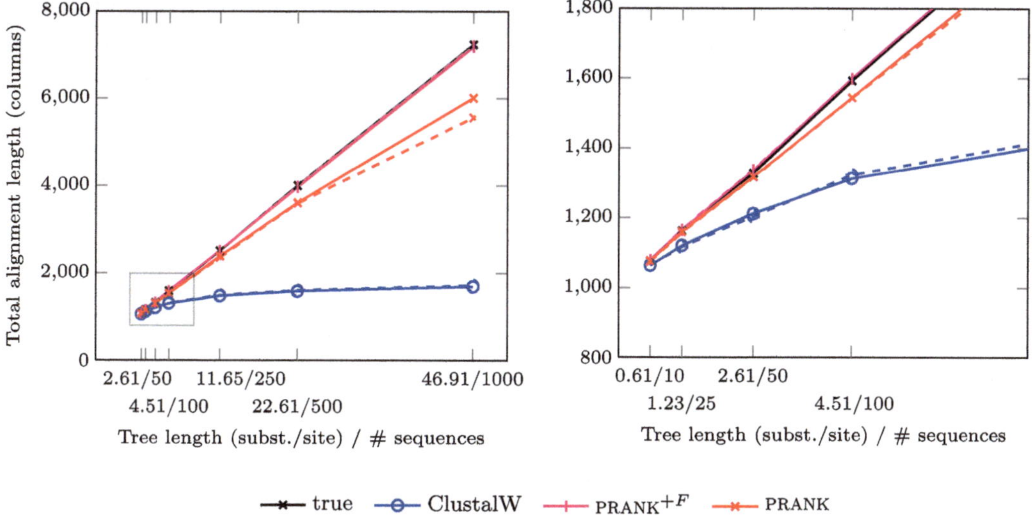

Fig. 2 The length of the alignment is expected to grow linearly with the evolutionary divergence contained within the sequences. One thousand sequences were simulated under a random tree with the maximum root-to-tip distance of 0.1 substitutions per site [17]. Subsets of 10, 25, 50, 100, 250, and 500 sequences as well as the full datasets were re-aligned with ClustalW [7] and PRANK, and the length of the resulting alignments is plotted as a function of the length of the tree relating the included sequences. As the insertion–deletion process used for simulation is time-dependent and defined relative to the substitution rate, the correlation between the two values for the true alignment (*black line*) is perfect. The two variants of the phylogeny-aware function (PRANK and PRANK^{+F}) produce alignments with lengths close to the true length whereas a method based on the classical progressive alignment algorithm (ClustalW) over-aligns the sequences and the length of the alignment is seriously underestimated. *Solid* and *dashed lines* indicate alignments based on the true and estimated guide trees, respectively. The *rectangle* in the left plot indicates the area shown on the right

considering the phylogeny of the sequences included. In practice, not only are insertions at the same position problematic but the correct alignment of sequences with insertions and deletions at near-by positions requires the identification of distinct evolutionary events and their subsequent correct handling.

Progressive alignment algorithms exploit the sequence phylogeny and align the sequences pairwise in the reverse order, starting from the most closely-related ones and, at each step clustering the aligned subsets, progressing towards the root of the tree. A major reason for the use of progressive algorithms is the prohibitive computational complexity of the exact multiple sequence alignment algorithm: with progressive algorithms, the complexity of aligning n sequences of length l is reduced from $O(l^n)$ to $O((n-1)l^2)$. The additional beauty of the approach is that the algorithm starts with alignments that are expected to be easiest and thus minimizes the chances of early alignment errors in its greedy processing of sequences. The classical algorithm does not use the phylogeny for anything else, however, and the placement of gaps—that is, the inference of which characters have been inserted or deleted in the evolutionary past—in the resulting alignments is often phylogenetically implausible [5].

Assuming that the alignment guide tree is correct and that the sequences are relatively closely-related, the progressive alignment approach provides the information necessary to identify insertion and deletion events. The phylogeny-aware progressive algorithm implemented in PRANK [5, 6] uses outgroup information from the next alignment step to decide if the length difference observed between the aligned sequences (representing either true extant sequences or internal nodes representing an aligned subset) was caused by an insertion or a deletion (Fig. 3). By identifying the true evolutionary event, the phylogeny-aware algorithm can handle insertions correctly and avoid penalizing the single event multiple times in later stages of the alignment.

The phylogeny-aware algorithm flags sites that contain an alignment gap in the immediately preceding stage of the progressive alignment, allowing for free placement of new gaps at flagged positions in the very next round. For an insertion, a new gap is created at exactly the same position and the flags indicating the gap are retained; for a gap caused by a deletion, a better alignment is obtained by matching the sites and the flags are removed (Fig. 3). The algorithm keeps the inserted sites at the later stages of the progressive alignment and the sequences it reconstructs for the internal nodes of the alignment tree may not reflect the true length of the ancestral sequences. Despite that, the identification and marking of the insertion events avoids penalizing for the same events multiple times and provides a significant improvement over the classical algorithm that, in practice, considers all length differences as deletions.

Penalization of a single event multiple times seems an insignificant error if the procedure nevertheless reconstructs the correct alignment. In trivial alignment tasks that may be the case but in more complex ones the classical algorithm will allow for the matching of insertions with non-homologous characters, the resulting alignments indicating false homologies (Fig. 4). The heuristics proposed to correct for insertion events by lowering the gap cost at sites already containing gaps (e.g., [7, 8]) cannot prevent this; in contrast, they typically cause further errors by moving gaps caused by deletion events at near-by sites to the same columns and produce block-like alignments with alternating gappy and conserved regions (*see* Fig. 1). The basic version of the phylogeny-aware algorithm greatly reduces the problem but even that cannot completely avoid the matching of independent insertions, especially in the alignment of large datasets in which the chances of mutation events at near-by positions is significant (Fig. 2).

As discussed above, the phylogeny-aware algorithm identifies the type of insertion–deletion event and then handles the event accordingly, either creating a new gap or removing the flags indicating the gap. A variant of the phylogeny-aware algorithm, known as PRANK^{+F}, uses this information to mark sites at which the flagged

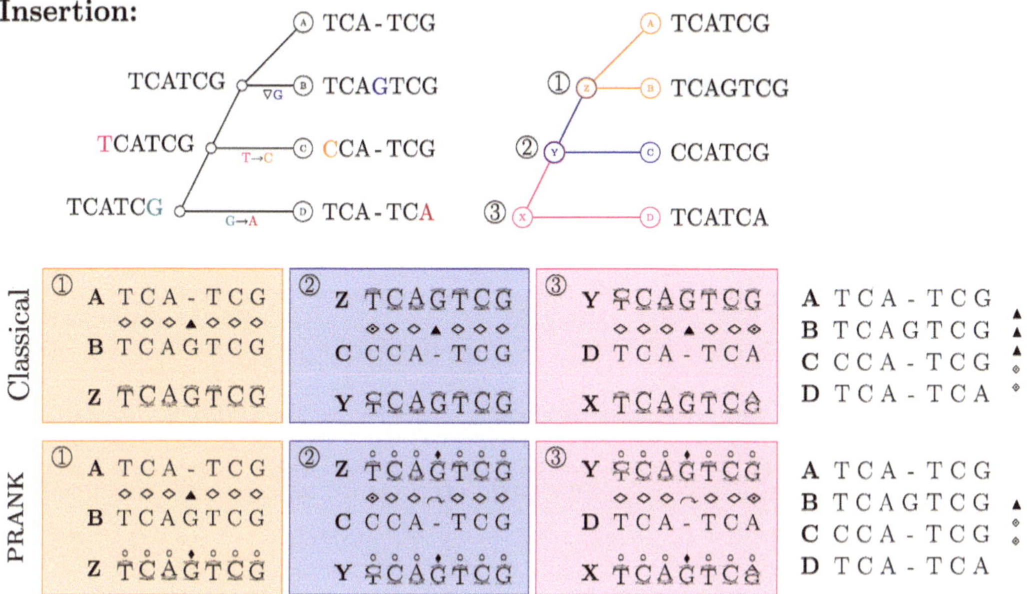

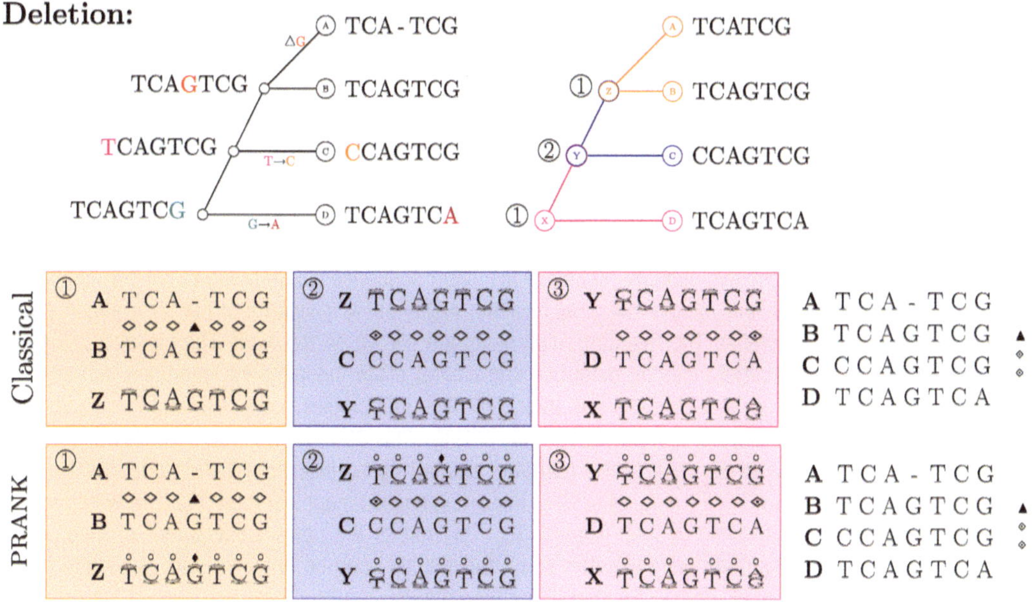

Fig. 3 The phylogeny-aware algorithm distinguishes insertions from deletions and treats them differently. The trees on the *left* represent the evolutionary histories of four short sequences undergoing two substitutions and either an insertion (*top*) or a deletion (*bottom*). The trees on the *right* indicate how the alignment of sequences is divided into three pairwise alignments, each creating an ancestral sequence (**Z**, **Y**, **X**) that is placed at the corresponding internal node and then aligned pairwise with the next sequence. The classical alignment algorithm penalizes the single insertion three times (indicated with *filled triangle*; *open diamond* and *white diamond with black dot* denote match and mismatch, respectively); in contrast, the phylogeny-aware algorithm implemented in PRANK flags the gapped site after the first alignment (indicated by *filled diamond* above the sequence) and can then open a new gap at the flagged position without a further penalty (indicated by *curved right arrow*). For a deletion, the gap needs to be created only once and the phylogeny-aware algorithm removes the flag indicating the gap after the second alignment

Two insertions, two deletions:

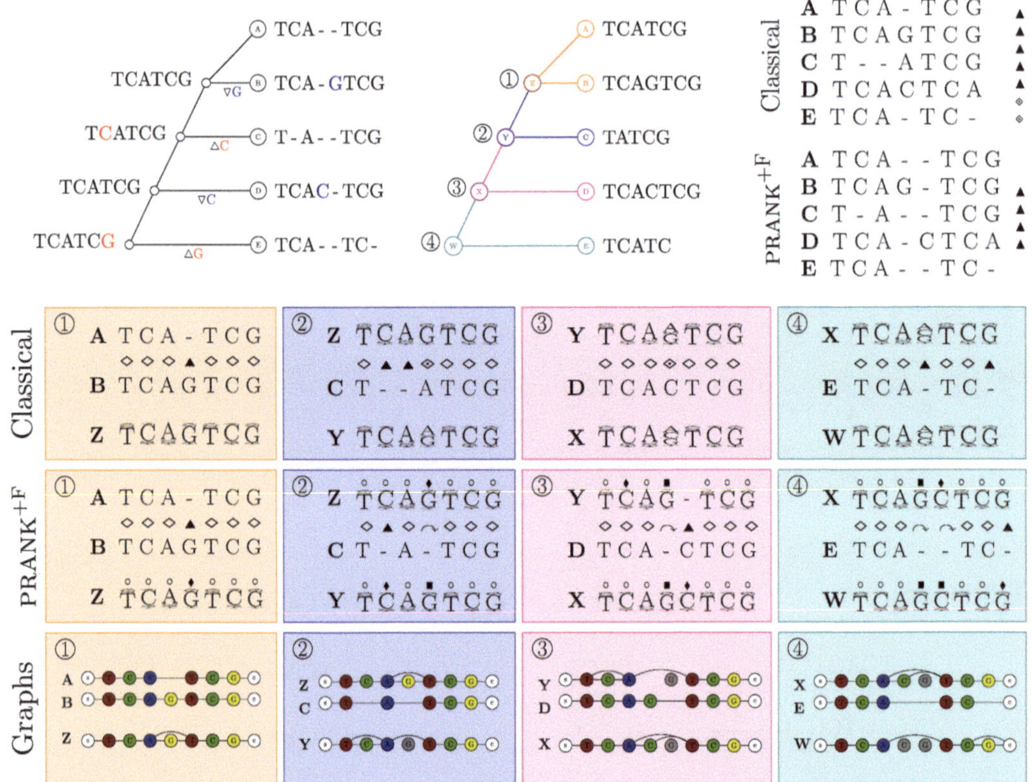

Fig. 4 The phylogeny-aware algorithm can distinguish and correctly align near-by insertion and deletions. The tree on the *left* represents the evolutionary history of five short sequences undergoing two insertion and two deletion events. The tree in the *middle* indicates how the alignment is divided into pairwise alignments of sequences (or sequence graphs). The resulting alignments are shown on the *right*. The classical alignment algorithm considers length differences as deletions and cannot place independent insertions into separate columns; often it also moves near-by gaps and indicates false homologies, here resulting in substitutions. A variant of the phylogeny-aware algorithm with greedy calling of insertions, known as PRANK^{+F}, considers the re-use of a flagged gap as evidence that the gap was created by an insertion. It then changes the flags indicating a pre-existing gap (*filled diamond*) to ones indicating a permanent insertion (*filled square*) and does not allow matching of these sites at later alignments. This forces the correct placement of independent insertions into separate alignment columns. The same functionality can be obtained with sequence graphs and greedy pruning of the graph edges. *See* Fig. 3 for the notation

gap is re-used as permanent insertions that cannot be matched at the later stages of the progressive alignment; to prevent overlapping deletions from confirming embedded insertions, the re-use of a gap has to be done for its full length with matching characters at the flanking sites. This approach can separate multiple insertions at the same position to independent events without effect on the placement of gaps caused by deletions (Fig. 4). When the order of aligning the sequences is correct and the sequence sampling is dense enough to call near-by gaps as separate events, PRANK^{+F}

works very well and can reconstruct alignments with lengths very close to the true length (Fig. 2). When the underlying assumptions hold, the method in principle scales up to any number of sequences.

The phylogeny-aware algorithm reconstructs ancestral sequences with information about sites that are believed to be insertions. The ancestral sequences are required for the alignment but they can be useful otherwise, too: PRANK allows for outputting inferred ancestral sequences, using gaps to indicate sites that are believed to have been later inserted and not present in the ancestors, along with the alignment of the extant sequences. Such alignments are unique and enable studying the process of change and timing certain events to specific evolutionary branches. In addition to ancestral sequences, the algorithm also infers the type of mutation events that have caused the length differences between the sequences and can provide this information in the output. Although an experienced user may distinguish insertions and deletions from the gap patterns they create, the marking of gaps caused by insertions and deletions with different symbols, as can be done with PRANK, is helpful.

The explanation and illustration of the flagging approach used by PRANK is slightly simplified and only considers one level of flagging. In practice, the algorithm marks the gaps in the immediately preceding alignments and, for the sites not cleared of flags, for the one before that. This procedure prevents long deletions in one branch from masking overlapping insertions in the descendants of its sister branch. For details, *see* [5, 6].

4 Limitations of the Phylogeny-Aware Algorithm

Unlike typical progressive alignment algorithms, the phylogeny-aware algorithm does not align sub-alignments to each other but reconstructs ancestral sequences to represent the parents of sets of aligned descendant sequences and then aligns pairwise these ancestral sequences. Accurate representation of the ancestral sequences, including the detection of inserted and deleted sites, is required for the correct distinction between insertion and deletion events in the subsequent stages of alignment. Correct reconstruction of sequences naturally requires that such ancestral sequences really existed and were true ancestors for the given sets of descendant sequences.

As the ancestral sequences are reconstructed for the internal nodes of the alignment phylogeny, it is crucial that the phylogeny accurately reflects the evolutionary history of the sequences. The role of alignment phylogeny is especially central in the calling of permanent insertions (PRANK^{+F}) that considers the re-use of a flagged gap as a confirmation that the gap has been created by an insertion. With the wrong order of aligning the sequences,

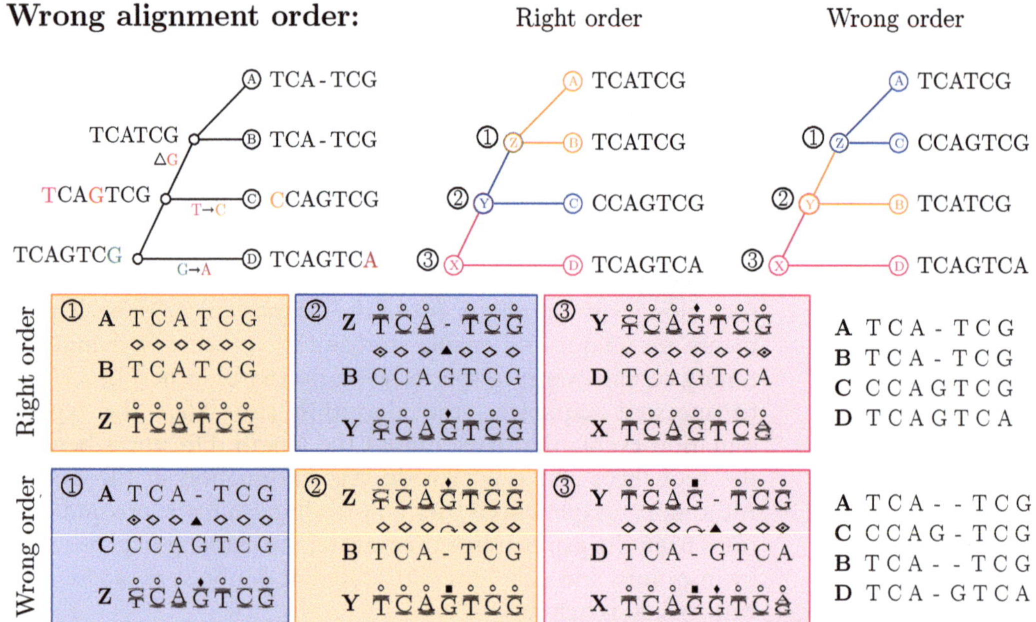

Fig. 5 The phylogeny-aware algorithm can be sensitive to errors in the guide phylogeny. The tree on the *left* represents the true evolutionary history and the trees in the *middle* and on the *right* indicate the right and a wrong order of aligning the sequences. The greedy calling of insertions (PRANK^{+F}) marks flagged gaps that are re-used (*curved right arrow*) as permanent insertions. When the alignment order is correct (*top*), the algorithm works perfectly. If **A** and **C** are incorrectly aligned first (*bottom*), the subsequent alignment of **B** appears to confirm an insertion in **C** although the true event is a deletion shared by **A** and **B**. As the insertion in column 4 is marked permanent (*filled square*), the site belonging to that columns has to placed in a columns of its own. The resulting alignment is too long and gappy. *See* Fig. 3 for the notation

a deletion may appear as an insertion and, by marking sites incorrectly as a permanent insertion, the algorithm has to place characters truly homologous to that to separate columns (Fig. 5). Although the resulting alignment is too long and gappy, small errors in the alignment order may not be too serious in typical evolutionary analyses: an incorrect alignment such as that in Fig. 5 does not indicate all true homologies but neither does it contain false homology statements.

In addition to the wrong alignment order, missing data can cause errors with the PRANK^{+F} variant. The algorithm assumes that alignment gaps are caused by insertions and deletions and then chooses the most plausible explanation of the two. One isolated gap caused by missing data may not be serious but if several sequences lack data at the same region, the gap pattern created may look like an insertion in the complete sequences; when this region is falsely marked as a permanent insertion, the subsequent alignment must place the affected region in separate columns. As sequences are often truncated at their ends, the marking of terminal gaps as permanent insertions is by default disabled by PRANK.

Dense vs. sparse sampling:

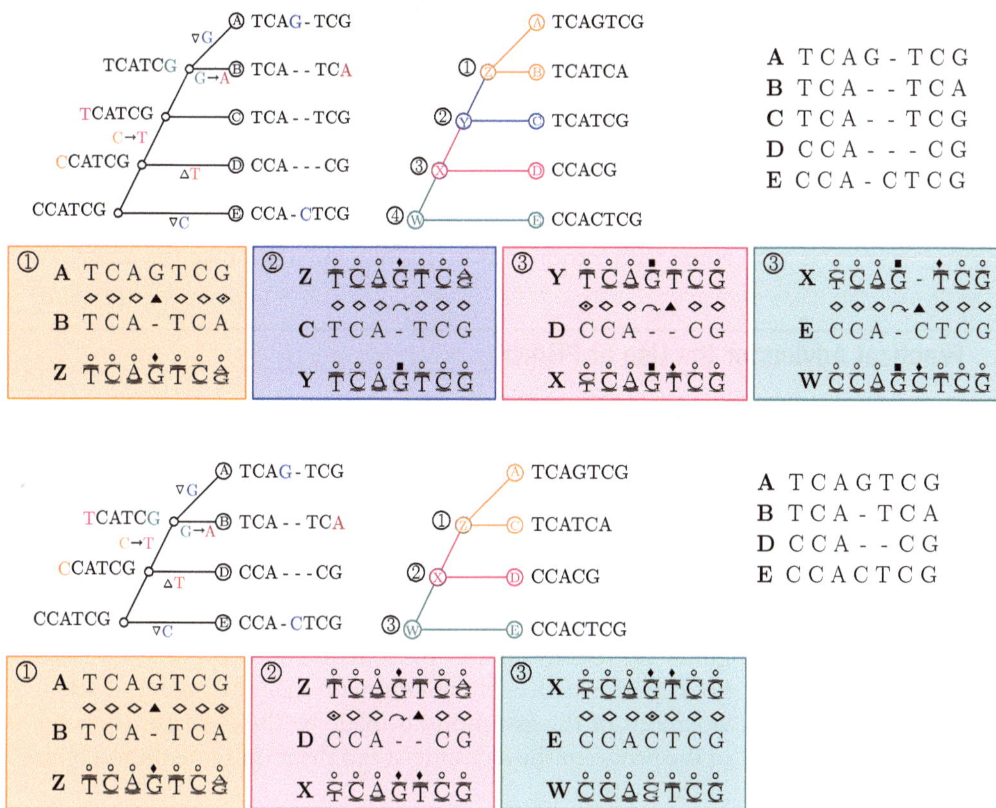

Fig. 6 Correct identification of independent insertion and deletion events requires closely-related sequences. With a dense sampling of sequences (*top*) each insertion and deletion event can be identified using the outgroup information from the next alignment and the correct homology is recovered. With a sparser sampling (*bottom*), the insertion in **A** cannot be identified because of a deletion at an adjacent position in **D**. As a result, the independent insertions in **A** and **E** are incorrectly matched

Similar heuristics unfortunately cannot be provided for missing data in other parts of the sequences.

The phylogeny-aware alignment algorithm assumes that each alignment gap is caused by one insertion or deletion event and that the very next alignment provides information to distinguish between the two types of events. When the sequences are relatively closely related (and, as stated previously, the alignment order is correct), these assumptions are typically valid. If the sequences are more diverged, the chances of independent insertion and deletions events at near-by positions in the adjacent evolutionary branches become significant. As a result of this, either the gap created in the first alignment may be a combination of two or more separate events, or the subsequent alignment of an outgroup sequence fails to confirm the event as an insertion or a deletion due to an overlapping independent event in the neighboring branch (Fig. 6).

Some of the limitations of the approach and the measures to overcome them can be contradicting. Accurate calling of insertion and deletion events requires densely-sampled sequence sets but the inference of alignment phylogeny for a large dataset is prone to errors [9] and the alignment may therefore suffer. Furthermore, incomplete lineage sorting is more likely among closely-related sequences and possibly no single phylogeny correctly reflects the evolutionary history of all sites of a very densely-sampled sequence set. As discussed below, some of these contradictions can be overcome with a more advanced implementation of the algorithm.

5 Practical Advice for the Use of PRANK

Evolutionary sequence analysis is based entirely on multiple sequence alignment and the accuracy of the downstream analysis depends on the correctness of the underlying alignment. Alignments produced with PRANK have been shown to provide exceptionally accurate inferences of selection on protein-coding sequences [10, 11] and perform well in phylogenetic analyses [12], although the latter finding is somewhat controversial due to the role of the guide phylogeny in the phylogeny-aware alignment. Despite its good performance in evolutionary analyses, PRANK is sensitive to violations of the assumptions made by the algorithm and the users of the program should understand the requirements and limitations of the method.

Alignment phylogeny: The phylogeny-aware alignment algorithm uses the alignment guide phylogeny to distinguish insertions from deletions. The algorithm is therefore sensitive to errors in the guide phylogeny, the variant with permanent insertions (PRANK^{+F}) being especially sensitive. Any PRANK alignment should be performed using an accurate guide phylogeny: if a high-quality phylogeny is available for the sequence set, it should be used instead of the heuristic phylogeny inferred by the program. (See below for future plans on integrated phylogeny search with PRANK.)

Evolutionary distances: PRANK uses the branch lengths provided by the guide phylogeny to re-compute the substitution and gap scoring for each alignment step. Depending on the expected evolutionary divergence, a region with several dissimilarities may be considered homologous and matched (distant sequences), or non-homologous and placed in separate columns (close sequences). Although the algorithm is not sensitive to small deviations in the branch lengths provided, a guide phylogeny with accurate distance estimates should be used when available.

Option +F: Given that the alignment guide phylogeny is correct and the sequence sampling is dense, the variant with permanent insertions (PRANK^{+F}) has been shown to outperform the basic algorithm [5]. If the alignment guide phylogeny is likely to contain errors or the input sequences are incomplete (i.e., contain missing data), the option +F can be problematic and the resulting alignment should at least be compared to one produced without it.

Reproducibility: Most pairwise alignments have several equally good solutions. In progressive alignment, the choice between these alternative solutions may trigger larger changes in the later stages of the process and lead to very different multiple alignments. Most alignment methods are deterministic and always pick the same solution and thus guarantee to produce the same final alignment. This practice hides the uncertainty in the data and has led to post-processing methods to recover the hidden variation [13]. By default, PRANK picks randomly one of the alternative solutions and may produce different results on independent runs of the very same data. This behavior may be disabled if reproducibility is required.

Sequence alphabet: PRANK represents sites at ancestral sequences with vectors of conditional likelihoods for the descendant sub-tree given different character states at the parent. This requires $O(A^2)$ computations for each cell in the dynamic programming matrix, where A is the size of the character alphabet, and makes the alignment of sequences with a large alphabet relatively slow. For protein-coding sequences, the alignments performed on codon level has been shown to outperform those done on protein sequences [10, 11]. Despite its slower computation, the use of codon alignment is recommended whenever possible. In general, protein-coding DNA sequences should not be aligned as DNA without good reason. If codon alignment is found to be too slow, PRANK provides an option to translate protein-coding DNA sequences to protein, perform the alignment on protein sequences, and back-translate the resulting alignment to DNA.

Sequence sampling: Given that the alignment guide phylogeny is correct and the sequence sampling is dense, PRANK is unbiased and scales up to any number of sequences. Even if the question in hand would not require an alignment of a large number of sequences, the quality of the resulting alignment is expected to be better when it is performed for many closely-related sequences than for a small number of distantly-related ones. Unneeded sequences can be removed after the alignment without affecting the statement of homology among the remaining sequences. PRANK is not suitable for the alignment of highly diverged sequences.

6 Future Directions

PRANK has been shown to perform well in benchmarks assessing the suitability of sequence alignments generated with various methods to different types of evolutionary analyses [10–12]. Despite its good performance in phylogenetic analyses, the method should be used with caution in analyses where the guide phylogeny is unknown prior to alignment. It is possible that the alignment generated by PRANK is influenced by the guide phylogeny and, in the case that the guide phylogeny is seriously wrong, the subsequent phylogenetic analysis based on the resulting alignment may be biased. On the other hand, if the problem with the guide phylogeny can be sorted out, PRANK is expected to provide superior alignments for evolutionary analyses, often closely approximating alignments produced with computationally much heavier statistical methods [14, 15].

Co-estimation of alignment and phylogeny with an iterative approach is a promising new idea [9] that can greatly reduce the problems caused by the inter-dependency of the two inferences. PRANK can merge two alignments using the very same phylogeny-aware approach that is used for de novo alignment and the method is in principle ready to be embedded in a similar framework, splitting the alignment task to speed up the search. We are in the process of studying the best practices to divide the problem to subtasks and then to measure the goodness of the resulting alignments. User-friendly tools to use PRANK for phylogenetic analyses will be provided shortly.

Although an iterative search strategy should help PRANK to greatly reduce the problems caused by an incorrect start guide phylogeny, iteration does not decrease the greediness of the algorithm nor can it solve the phylogeny for datasets that have no unique phylogeny, e.g., due to incomplete lineage sorting. Both challenges can be tackled by re-implementing the phylogeny-aware algorithm for the alignment sequence graphs [16]. By using additional edges to indicate unresolved gaps and then pruning the unused edges after the alignment of the next sequences, one can implement an algorithm very similar to that of PRANK^{+F} (Fig. 4). The advantages of the graph approach are greater, though, if the edges are not pruned but given weights or probabilities based on the evidence for the different mutation types. Such edge-weighting approach makes the method far less sensitive to errors in the guide phylogeny or different sites evolving under slightly different phylogenies. We are actively implementing the features of PRANK still missing from the graph approach and believe that the new method, called PAGAN, will soon replace PRANK.

As discussed above and evidenced by methods for co- and joint-estimation of alignment and phylogeny, the multiple sequence alignment should always be seen with the associated phylogeny. Understanding the alignment and drawing right conclusions from it is much easier when the relationships between the sequences are indicated next to the alignment. For a method such as PRANK, the phylogeny is also needed to indicate the relative positions of the ancestral sequences and to visualize the changes happening in different evolutionary branches. We are developing Wasabi, a browser-based graphical user interface, that integrates these ideas

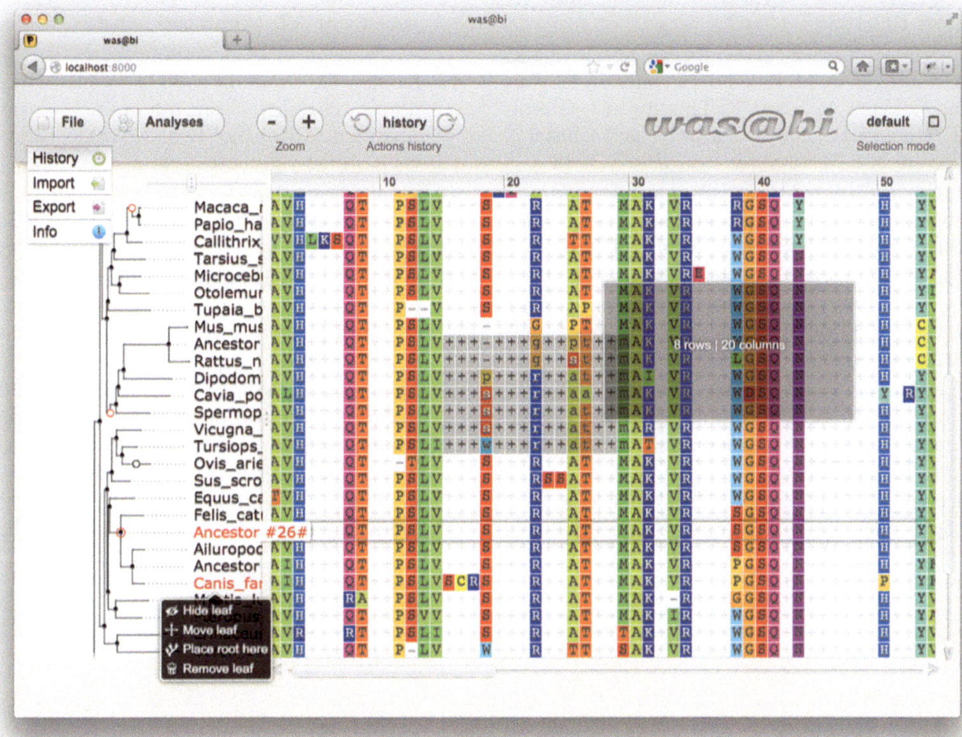

Fig. 7 Wasabi provides a web browser-based graphical interface to PRANK and related evolutionary sequence analysis programs. The visualization and analysis of multiple sequence alignments should be performed in the context of the phylogeny relating the sequences. In addition to integrated alignment and analysis with PRANK and related tools, the Wasabi browser interface allows for manipulating the sequence phylogeny, showing and hiding the inferred ancestral sequences and both automated and manual masking of parts of the alignment. Wasabi can be found from the PRANK home page at http://code.google.com/p/prank-msa/

and provides an easy-to-use access to PRANK and related methods (Fig. 7). This tool and all other methods discussed here can be found from the PRANK home page at http://code.google.com/p/prank-msa/.

References

1. Sankoff D (1975) Minimal mutation trees of sequences. SIAM J Appl Math 28:35–42
2. Ogurtsov A, Sunyaev S, Kondrashov A (2004) Indel-based evolutionary distance and mouse-human divergence. Genome Res 14:1610–1616
3. Hogeweg P, Hesper B (1984) The alignment of sets of sequences and the construction of phyletic trees: an integrated method. J Mol Evol 20:175–186
4. Feng D, Doolittle R (1987) Progressive sequence alignment as a prerequisite to correct phylogenetic trees. J Mol Evol 25:351–360
5. Löytynoja A, Goldman N (2008) Phylogeny-aware gap placement prevents errors in sequence alignment and evolutionary analysis. Science 320:1632–1635
6. Löytynoja A, Goldman N (2005) An algorithm for progressive multiple alignment of

sequences with insertions. Proc Natl Acad Sci USA 102:10557–10562

7. Larkin M, Blackshields G, Brown N, Chenna R, McGettigan P, McWilliam H, Valentin F, Wallace I, Wilm A, Lopez R, Thompson J, Gibson T, Higgins D (2007) Clustal W and Clustal X version 2.0. Bioinformatics 23:2947–2948
8. Edgar R (2004) MUSCLE: multiple sequence alignment with high accuracy and high throughput. Nucleic Acids Res 32:1792–1797
9. Liu K, Raghavan S, Nelesen S, Linder C, Warnow T (2009) Rapid and accurate large-scale coestimation of sequence alignments and phylogenetic trees. Science 324:1561–1564
10. Fletcher W, Yang Z (2010) The effect of insertions, deletions, and alignment errors on the branch-site test of positive selection. Mol Biol Evol 27:2257–2267
11. Jordan G, Goldman N (2012) The effects of alignment error and alignment filtering on the sitewise detection of positive selection. Mol Biol Evol 29:1125–1139
12. Dessimoz C, Gil M (2010) Phylogenetic assessment of alignments reveals neglected tree signal in gaps. Genome Biol 11:R37
13. Landan G, Graur D (2007) Heads or tails: a simple reliability check for multiple sequence alignments. Mol Biol Evol 24:1380–1383
14. Suchard M, Redelings B (2006) BAli-Phy: simultaneous Bayesian inference of alignment and phylogeny. Bioinformatics 22:2047–2048
15. Novák A, Miklós I, Lyngsø R, Hein J (2008) StatAlign: an extendable software package for joint Bayesian estimation of alignments and evolutionary trees. Bioinformatics 24:2403–2404
16. Löytynoja A, Vilella A, Goldman N (2012) Accurate extension of multiple sequence alignments using a phylogeny-aware graph algorithm. Bioinformatics 28:1684–1691
17. Löytynoja A, Goldman N (2010) webPRANK: a phylogeny-aware multiple sequence aligner with interactive alignment browser. BMC Bioinformatics 11:579

Chapter 11

GramAlign: Fast alignment driven by grammar-based phylogeny

David J. Russell

Abstract

Multiple sequence alignment involves identifying related subsequences among biological sequences. When matches are found, the associated pieces are shifted so that when sequences are presented as successive rows—one sequence per row—homologous residues line-up in columns. Exact alignment of more than a few sequences is known to be computationally prohibitive. Thus many heuristic algorithms have been developed to produce good alignments in an efficient amount of time by determining an order by which pairs of sequences are progressively aligned and merged. GRAMALIGN is such a progressive alignment algorithm that uses a grammar-based relative complexity distance metric to determine the alignment order. This technique allows for a computationally efficient and scalable program useful for aligning both large numbers of sequences and sets of long sequences quickly. The GRAMALIGN software is available at http://bioinfo.unl.edu/gramalign.php for both source code download and a web-based alignment server.

Key words Multiple sequence alignment, Progressive alignment, Relative complexity measure, Abstract grammar, GramAlign

1 Introduction

Generation of meaningful multiple sequence alignments (MSAs) of biological sequences is a well-studied NP-complete problem, which has significant implications for a wide spectrum of applications [1, 2]. In general, the challenge is aligning N sequences of varying lengths by inserting gaps in the sequences so that in the end all sequences have the same length. Of particular interest to computational biology are DNA/RNA sequences and amino acid sequences, which are comprised of nucleotide and amino acid residues, respectively.

Advances in sequencing technology continue to provide vast amounts of data in need of multiple alignment. In the case of large sequencing projects, high numbers of fragments that lead to longer contigs to be combined are generated with much less time and money [3]. In addition, as more organisms' genomes are

David J. Russell (ed.), *Multiple Sequence Alignment Methods*, Methods in Molecular Biology, vol. 1079,
DOI 10.1007/978-1-62703-646-7_11, © Springer Science+Business Media, LLC 2014

sequenced, approaches that require MSA of the same gene in different organisms now find a more populated data set. In both cases computational time in MSA is becoming an important issue that needs to be addressed.

Given a scoring scheme to evaluate the fitness of an MSA, calculating the best MSA is an NP-complete problem [1]. Variances in scoring schemes, need for expert-hand analysis in most applications, and many-to-one mappings governing elements-to-functionality (codon mapping and function) make MSA a more challenging problem when considered from a biological context as well [4].

Generally, three approaches are used to automate the generation of MSAs. The first uses a brute-force method of multidimensional dynamic programming [5], which may result in a good alignment but is generally computationally expensive and, therefore, usable only for a small number of sequences. Another method uses a probabilistic approach where Hidden Markov Models (HMMs) are approximated from unaligned sequences. The final method, progressive alignment, is possibly the most commonly used approach for obtaining MSAs [6].

A progressive alignment algorithm begins with an optimal alignment of two of the N sequences. Then, each of the remaining N sequences are aligned to the current MSA, either via a consensus sequence or one of the sequences already in the MSA. Variations on the progressive alignment method are presented throughout the chapters in this text. In most cases, the algorithms attempt to generate accurate alignments while minimizing computational time or space. This chapter presents GRAMALIGN, a computationally efficient progressive alignment method. In particular, the natural grammar inherent in biological sequences is estimated to determine the order in which sequences are progressively merged into the ongoing MSA. The following sections focus on the somewhat obscure idea of using a grammar-based relative complexity measure to estimate the distance matrix which is used to guide the progressive alignment process. This is followed by a section containing several details regarding the usage of GRAMALIGN.

2 Distance Matrix Calculation

As shown in Fig. 1, the first significant step in GRAMALIGN, and progressive alignment in general, is to determine the order in which sequences are added to the ongoing alignment. To do so it is typical for an algorithm to generate a distance matrix whose elements are inter-sequence distances which often reflect phylogeny. Unfortunately, many of these methods are known to be computationally expensive and sometimes impose a paradoxical situation in which an alignment is required prior to estimating the

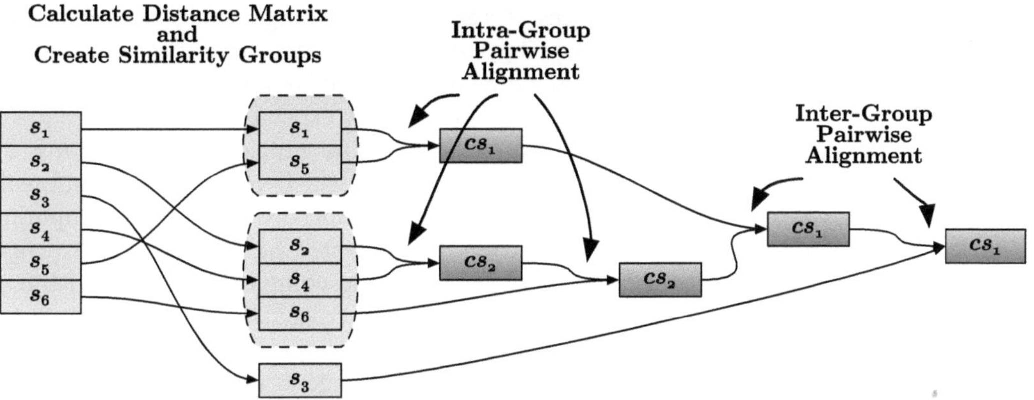

Fig. 1 Algorithm overview. Given the sequences to be aligned, $\{s_1, \ldots, s_6\}$, a distance matrix is calculated based on relative grammar complexity. The sequences are subsequently placed into similarity groups based on their relative distance compared to a user-specified similarity threshold (option -T). Sequences within each group are progressively aligned to form a group consensus sequence, cs_i. Finally, the consensus sequences are progressively aligned to result in the final alignment

phylogeny which in turn is required for determining the order in which the sequences are aligned. Some alignment algorithms overcome this by introducing a number of iterations to continuously refine both the distance estimations and the alignment. As will be described, GRAMALIGN generates a distance matrix based on a measure of relative grammar complexity naturally inherent in biological sequences which does not require knowledge of the underlying phylogeny. This section is dedicated to describing this computationally efficient way of creating a valid alignment order prior to performing the progressive alignment.

2.1 Phylogeny Estimation

Measures of distances between sequences are central to the development of molecular phylogeny. The measures of distance can be divided into two main classes, distances whose computation requires that the sequences be aligned, and methods that can compute distances without the need for alignment. Some of the measures that require sequence alignment [7–11] use a nucleotide or amino acid substitution model to develop a distance matrix which is then used to build a phylogenetic tree. Others use parsimony and maximum likelihood to evaluate the fitness of different topologies for the phylogenetic trees [12–19]. All these approaches assume the existence of homologous versions of a particular gene in each of the organisms which need to be aligned in order for the distances to be computed. The requirement of MSAs imposes a substantial computational cost which can become prohibitive when the number of sequences is large. Alignment free methods were first introduced by Sankoff et al. [20] to deal with situations where the comparisons were not with homologous genes but rather with whole genomes. Since then a number of alignment free methods that use edit distances, breakpoints, rearrangements, recombination, comparative mapping, and gene order have been introduced [21–30]. While these methods do not require multiple alignments they still tend to be computationally expensive.

The relative complexity measure utilized in GRAMALIGN is an alignment free distance measure which is also computationally efficient. The use of the relative complexity measure for computing phylogenetic distance was demonstrated via successful construction of phylogenies for eutharians using complete mitochondrial genomes [31]. This measure was further validated [32] by using it for phylogenetic analysis of medically relevant fungal species using the cytochrome b gene, the 18S rDNA gene and the ITS region. The topology constructed using this measure was robust to random removal of significant portions (up to 40 %) of the genes. We have made use of the computational efficiency of this distance metric in both GRAMALIGN and to develop a clustering algorithm that was validated by clustering large 16S data sets [33, 34]. GRAMALIGN itself has been validated and incorporated into the primer design pipeline UniPrime2 [35]. Finally, this measure was used to cluster protein families into functional subtypes [36]. In all these applications the distance measure is used to obtain the phylogenetic topology and thus what has been important has been the ordering of the distances between sequences.

2.2 Relative Complexity Measure

The relative complexity measure of distance is based on an approximation to Kolmogorov complexity [37] by Lempel–Ziv complexity [38]. The Lempel–Ziv (LZ) complexity of a sequence is the number of distinct phrases in the left-to-right parsing of a sequence. A distinct phrase is one which has not occurred in the history of a sequence. Consider the sequence $Q = gagacagt$. Initially the history of the sequence is the empty string. As g is not in the empty string g becomes the first distinct phrase. The history of the sequence now consists of the single letter g. As a is not in the history, a becomes the second unique phrase. The history of the sequence is now ga. The next two letters are g and a. As the sequence ga exists in the history we keep building the phrase until we get to gac which is not in the history of the sequence. Thus, gac is the third unique phrase. The history is now the sequence $gagac$. The next unique phrase in the sequence is agt, ag being available in the history. Thus the sequence $gagacagt$ can be represented by four unique phrases and therefore has an LZ complexity of four. We can represent the parsing by

$$Q = g \cdot a \cdot gac \cdot agt$$

and the LZ complexity by $c(Q)$. We can see that sequences that have more diversity will result in higher LZ complexity and sequences that are more uniform will result in smaller LZ complexity. If we concatenated this string with another similar to it, the increase in complexity will be quite small. For example, consider the sequence $R = gagacat$ and examine the parsing of the concatenated sequence QR:

$$QR = g \cdot a \cdot gac \cdot agt \cdot gagacat$$

The concatenated sequence QR while being almost twice as long as the sequence Q has an LZ complexity of five, which is only one more than the LZ complexity of Q. If, on the other hand, we concatenate a sequence with a very different composition to Q, we will get a greater increase in LZ complexity. Consider the sequence *tcgctta* and the parsing of the concatenation

$$QT = g \cdot a \cdot gac \cdot agt \cdot tc \cdot gc \cdot tta$$

with an LZ complexity of seven. In this case the LZ complexity has grown in proportion to the length of the sequence.

We have discussed the analysis of the sequence as simply a parsing. We can also view this same analysis as a way of discovering grammar rules which can be used to generate a sequence. Thus the entry *gac* in our parsing can be viewed as a rule for generating the letter following *ga*. From this perspective, the elements in our parsing can be viewed as a dictionary of grammar rules which can be used to generate sequences belonging to the same grammatical family as the sequence being analyzed. Thus, any increase in complexity when analyzing a concatenation of two sequences is a result of the need to include additional grammar rules for generating the latter sequence. The more different two sequences are the more the need for additional grammar rules. As this is a more general framework we will use this in our subsequent discussion. We introduce some notation in the following to formalize this approach.

2.3 Grammar-Based Distance Estimation

The distance calculation begins by generating a dictionary of grammar rules for each sequence. Let the notation G_m define the grammar dictionary for the mth sequence. Further, while the dictionary is under construction, the sequence will be scanned one residue at a time from left-to-right. We add a superscript on the dictionary notation, G_m^k, to indicate the current dictionary up to the kth residue with sequence m. Let f be the current substring "fragment" of the sequence that has not yet been added to the dictionary. A fragment is composed via concatenated symbols taken one at a time from the sequence being scanned. After each iteration, the sequence is checked for a matching fragment. If it is not found, the fragment is unique and added to the dictionary before it is reset to the empty string. As in the case of the dictionary, we use the superscript f^k to indicate the fragment location within the sequence. Finally, we use the term "visible sequence" to mean the left-most windowed portion of a sequence that has already been processed. So, at the kth iteration, the visible sequence would include only the first k residues of a scanned sequence.

Initially, each dictionary ${G_m}^1 = \emptyset$ is empty, a fragment $f^1 = s_m(1)$ is set to the first residue of the corresponding sequence, and only the first element $s_m(1)$ is visible to the algorithm. At the kth iteration of the procedure, the kth residue is appended to the $k - 1$ fragment and the visible sequence is checked. If $f^k \notin s_m(1, \ldots, k - 1)$ then f^k is considered a new rule, and so added to the dictionary $G_m^k = G_m^{k-1} \cup \{f^k\}$, and the fragment is reset, $f^k = \emptyset$. However, if $f^k \in s_m(1, \ldots, k - 1)$, then the current dictionary contains enough rules to produce the current fragment, i.e., $G_m^k = G_m^{k-1}$. In either case, the iteration completes by appending the kth residue to the visible sequence. This procedure continues until the visible sequence is equal to the entire sequence, at which time the size of the dictionary $| G_m |$ is recorded for the next step. As was described in [31], calculating the distance between sequences requires only the number of entries in the dictionary.

For the next step in generating the distance matrix, each sequence is compared with all other sequences. In particular, consider the process of comparing sequences m and n. Initially, the dictionary ${G_{m,n}}^1 = G_m$ is set to that of sequence m, a fragment $f^1 = s_n(1)$ is set to the first residue of the nth sequence, and the visible sequence is all of s_m. The algorithm operates as described previously, resulting in a new dictionary size $|G_{m,n}|$. When complete, more grammatically-similar sequences will have a new dictionary size with fewer entries as compared to sequences that are less grammatically-similar. Therefore, the size of the new dictionary $|G_{m,n}|$ will be close to the size of the original dictionary $|G_m|$ for grammatically similar sequences.

The distance between the sequences is then estimated using combinations of the old and new dictionary sizes. Five different distance measures were suggested in [31]. GRAMALIGN uses the distance measure

$$d_{m,n} = \frac{|G_{m,n}| - |G_m| + |G_{n,m}| - |G_n|}{\frac{|G_{m,n}| + |G_{n,m}|}{2}}, \tag{1}$$

where $m, n \in \{1, \ldots, N\}$ are indices of two sequences being compared. This particular metric accounts for differences in sequence lengths and normalizes accordingly. Thus, the final distance matrix D is composed of grammar-based distance entries given by Eq. 1. Smaller entries in D indicate a stronger similarity, at least in terms of the LZ-based grammar estimate.

2.4 Sequence Alphabets

The distance between sequences m and n as determined by Eq. 1 is based on how many additional rules need to be added to each grammar in order to generate both s_m and s_n. Because the real grammars are unknown, G_m and G_n are approximated by scanning the only observations available (i.e., s_m and s_n). The grammar approximation improves as the length of the observed sequences increases. And so, the distance calculations are a function of sequence lengths, becoming more accurate as the sequences increase in length. In practice, this calculation works well for DNA/RNA sequences, even of shorter lengths, because the approximated grammar of a DNA/RNA sequence can only contain rules involving words composed of combinations of elements from the alphabet {"A,""C,""G,""T/U"}. This small alphabet allows for a rapid generation of a reasonable grammar since there are a relatively small number of permutations of letters.

From a grammar perspective, amino acid sequences are generally much more difficult to process correctly using Eq. 1. The reason being the alphabet contains 23 letters, where each element is not equally different from all other elements. Due to the relatively large alphabet size, much longer sequences are necessary to generate a reasonable grammar approximation. Thus, the accuracy of distances calculated for sets of short amino acid sequences is diminished. Consider the substitution scores of "L" and "M" as taken from the GONNET250 and BLOSUM62 substitution matrices in Fig. 2. Notice in (a) and (c), that "L" receives a relatively high positive value when aligned with any of {"I,""L,""M,""V"}. In (b) and (d), "M" receives a relatively high positive value when aligned with any of the same set. Additionally, both "L" and "M" generally receive high negative values when compared to letters other than {"I,""L,""M,""V"}. When taking this type of scoring into account, the elements "L" and "M" could be considered the same letter in a grammatical sense.

Thus, GRAMALIGN offers the option to use a "Merged Amino Acid Alphabet" when calculating the distance matrix. The merged alphabet contains 11 elements corresponding to the 23 amino acid letters grouped into the sets {"A,""S,""T,""X"}, {"B,""D,""N"}, {"C"}, {"E,""K,""Q,""R,""Z"}, {"F"}, {"G"}, {"H"}, {"I,""L," "M,""V"}, {"P"}, {"W"}, and {"Y"}. These groupings were determined by considering all 23 rows of the BLOSUM45, BLOSUM62, BLOSUM80, and GONNET250 substitution matrices, and only grouping elements that had a strong similarity across the entire row in all four matrices. The merged alphabet has the benefit of containing fewer elements allowing for more accurate distance estimates based upon shorter observed sequences. In practice, average alignment scores increase when aligning the same data sets using the merged alphabet within the distance calculation, as compared to using the actual alphabet.

2.5 Progressive Alignment

Once the distances have been calculated, a minimal spanning tree based on these distances is used to determine the order in which sequences should be pairwise aligned. At the core of most progressive MSA algorithms is some method for performing pairwise

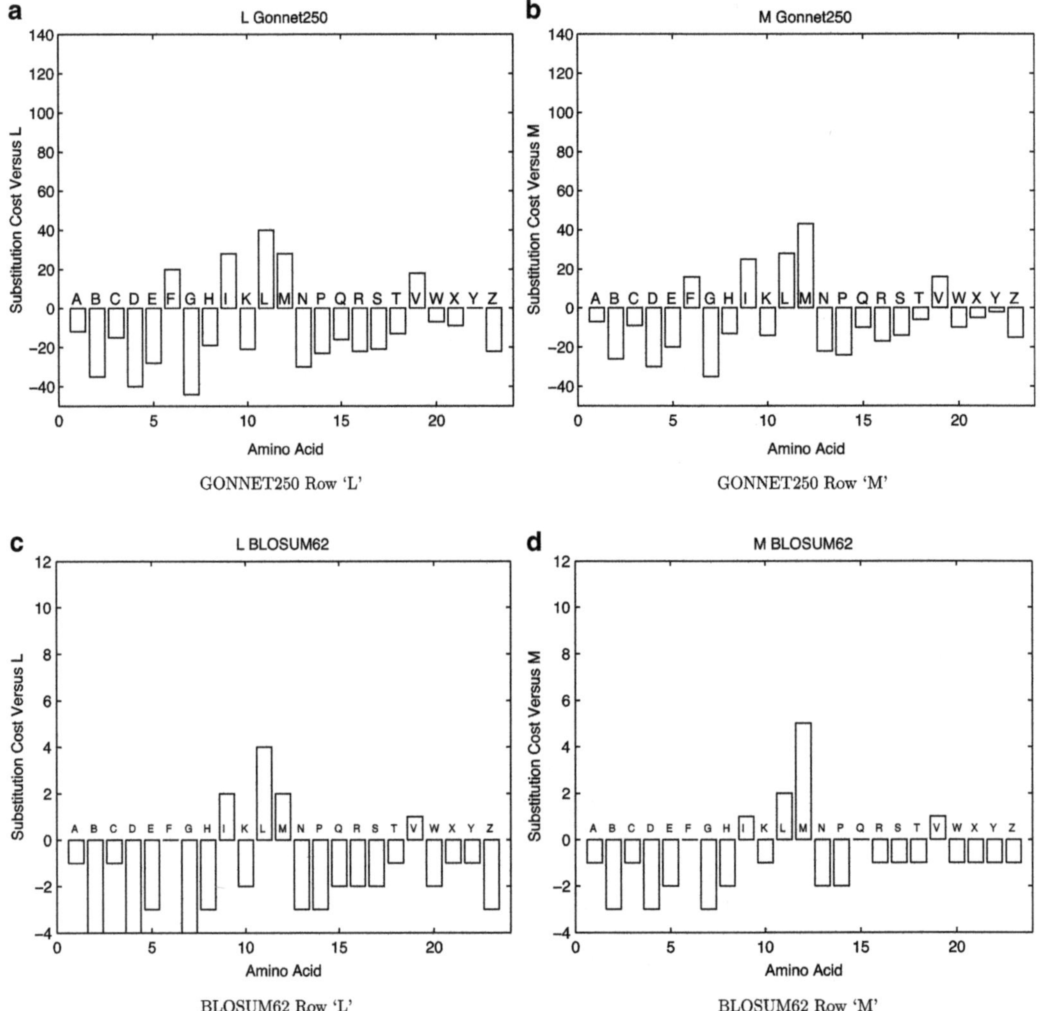

Fig. 2 Bar graphs of the substitution scores for amino acid "L" and "M" as taken from the GONNET250 and BLOSUM62 substitution matrices. The scores are shown based on an alphabetical ordering of amino acid letters from the leftmost "A" to rightmost "Z." (**a**) GONNET250 row "L;" (**b**) GONNET250 row "M;" (**c**) BLOSUM62 row "L;" (**d**) BLOSUM62 row "M"

alignments between two sequences. This work uses a version of the Needleman–Wunsch dynamic programming algorithm with affine gap scores as discussed in [2] to generate each pairwise alignment, the result of which is merged back into the ensemble alignment of all previously aligned sequences.

3 GramAlign Usage

The GRAMALIGN software is available at http://bioinfo.unl.edu/gramalign.php for both source code download and a web-based alignment server.

3.1 Install

GRAMALIGN is written in ANSI-C, and so should build without error on any platform with an ANSI-C compiler. Use the following commands to build the program.

```
cd src
make clean
make
```

At this point, the console-based executable `GramAlign` (on linux/MacOSX) or `GramAlign.exe` (on Windows) will reside in the src directory as well. Copy the executable to anywhere in your path.

3.2 Usage

To run GRAMALIGN from the command-line window, use the following command:

```
/path/to/executable/GramAlign [options]
```

Examples: The most basic example performs an alignment for the sequences in the file "input.fasta," which is located on the user's desktop. The output is written to a file named "output.txt," also on the desktop.

```
GramAlign -i ~/Desktop/input.fasta -o ~/Desktop/output.txt
```

An example of applying two additional options creates a distance matrix for the sequences in the file "input.fasta." Supplying the "-C" option tells the program to compute the full distance matrix, and the program's output is written to "output.txt." As in the previous example, the input and output files are accessed on the desktop based on the provided path.

```
GramAlign -f 0 -C -i ~/Desktop/input.fasta -o ~/Desktop/
output.txt
```

3.3 General Options

The most generic options include the following:

Option -h: Help! Display the command-line options.

Option -q: Turn on "quiet mode," which will prevent any intermediate text from being displayed by GRAMALIGN. The default is verbose which will output various progress during the alignment procedure.

Option -S <value>: Specify the maximum trace-back matrix size before temporary paging begins. The largest one-time memory requirement for GRAMALIGN occurs during each pairwise alignment process, at which time the trace-back matrix—an $|s_i| \times |s_j|$ matrix of bytes, where $|s_k|$ is the length of sequence k—is necessary for the backward portion of the dynamic programming procedure. If both sequences being aligned are large relative to the computer's physical available memory, then this block of memory can become so large that the computer spends a significant amount of processing time paging physical memory to and from the system virtual memory (i.e., hard-drive area). The value specified with this option determines a maximum size of the trace-back matrix in bytes before GRAMALIGN will (much more) efficiently copy pieces of the trace-back matrix to "page files" within the current directory. At the end of the trace-back procedure, these temporary page files will be removed. *Suggestion:* For systems with larger physical memory, this amount should be increased to improve performance. If this option is not specified, the default value is 100 (i.e., 100 million bytes). Note, the temporary page files are named `_ga_temp.page`*xxxxx*, where *xxxxx* is replaced with the proper page number. These files are safe to delete as long as GRAMALIGN is not running.

3.4 File Options

GRAMALIGN provides the following options for manipulating the input and output files and their formats.

Option -i <filename>: Specify the input file name, which needs to be in FASTA format. If this option is not used, the default name is "infile." The type of input sequence (nucleotide or amino acid) is determined by the input file type command-line option (`-F`).

Option -o <filename>: Specify the output file name. If this option is not used, the default file name is "outfile."

Option -f <Value>: Specify the output file format. A value of 0 will output the grammar-based distance matrix. A value of 1 will output the alignment in PHYLIP format. A value of 2 will output the alignment in Aligned FASTA format. A value of 3 will output the alignment in MSF/GCG format. A value of 4 will output the consensus sequence in html including gaps in the alignment. A value of 5 will output the consensus sequence in html ignoring any gap elements in the alignment. If this option is not specified, the default file format is PHYLIP.

Option -F <Value>: Specify the input file type. A value of 0 will cause GRAMALIGN to automatically detect if the input file contains amino acid sequences. The auto-detection is based on if a base other than A, C, G, T, U, or X is part of the sequence. Should any other character appear in any of the input sequences, the program will align the sequences as though the input file contains all amino acid sequences. A value of 1 will force the alignment to assume all sequences are either DNA or RNA. A value of 2 will force the alignment to assume all sequences are amino acid sequences. If this option is not specified, the default is to automatically detect the input file type.

3.5 Distance Matrix Options

The following options provide a means for adjusting the way in which GRAMALIGN creates the distance matrix which guides the order in which sequences are progressively aligned.

Option -C: Force GRAMALIGN to generate a complete distance matrix prior to determining the alignment order. The default allows GRAMALIGN to generate a partial distance matrix with a time complexity on the order of $N\log(N)$. Using this option will ensure the most accurate grammar-based alignment order, but requires a time complexity on the order of N^2. In creating the partial distance matrix, one initial column is completely filled in and divided into two clusters—one with the smallest distances and the other with the largest distances. Then each cluster is recursively processed, whereby one sequence is compared to all others in the cluster, the subset of which is further divided into two clusters, and so on. The underlying basis for this to work is the transitivity of grammars; if a sequence has a short grammar distance to two other sequences, then those two sequences should likely have a short grammar distance to each other. *Suggestion:* If you are using GRAMALIGN to output a distance matrix—say for studying phylogeny—then you should enable this option. Otherwise, you should not include this option in order to greatly decrease computation time, especially for many input sequences.

Option -M: Disable use of the merged amino acid alphabet. As discussed in Subheading 2.4, we developed a merged alphabet whereby certain amino acid characters were found to have similar row scores within the substitution matrices. We were able to reduce the original 23 characters into a set of 11 characters. This ability is particularly useful for the grammar-based distance calculation. This option will disable using the merged alphabet. This option is ignored for nucleotide sequences. *Suggestion:* Because this option only affects the distance matrix for amino acids, you should not use it unless you have a good reason to believe the grammar present in the original alphabet is significant to the alignment order. This option does not directly affect the pairwise alignment scoring.

Option -T <value>: Specify the relative grammar-based similarity threshold. Referring to the left half of Fig. 1, all sequence pairs that have a relative complexity measure below this threshold will be grouped together prior to alignment. Sequences within each group will be aligned to each other first. Then a consensus sequence for each group will be aligned to the overall alignment ensemble. Lower thresholds will force sequences to be more identical before they will be grouped together. If this option is not specified, the default value is 0.10. *Suggestion:* The default value is quite low, thereby ensuring that sequences need to be very similar before being grouped together. We have performed a series of classification comparisons on known 16S Ribosomal RNA sequences, the

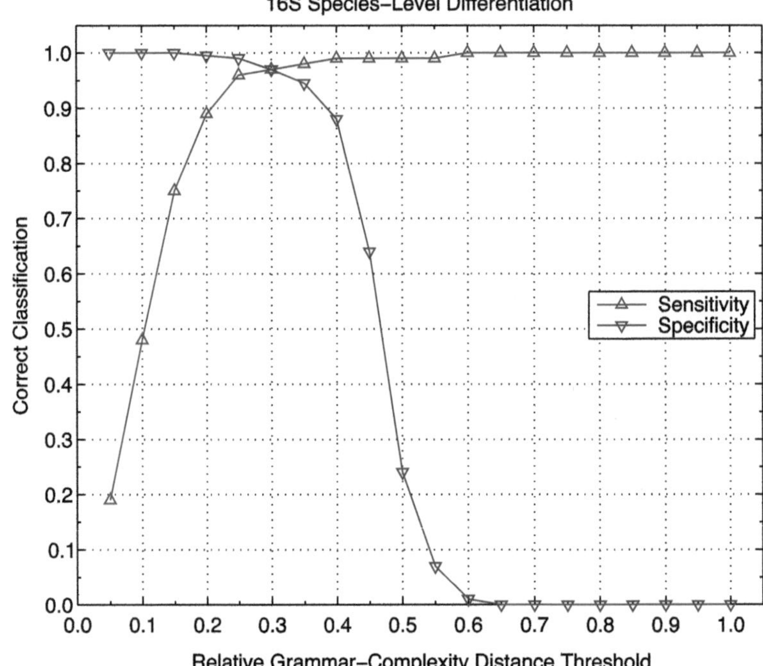

Fig. 3 Relative grammar-based distance classification thresholds at the species level. Given a known set of 16S Ribosomal RNA sequences in which multiple sequences belong to the same species, a binary classification is performed for varying grammar-based complexity thresholds. The classification procedure is to compare the pairwise distance between two sequences against a threshold. If the distance is less than the threshold, the two sequences are classified as belonging to the same species. If the distance is greater than the threshold, the two sequences are classified as belonging to different species

results of which are presented in Figs. 3 and 4. Given a known set of 16S Ribosomal RNA sequences in which multiple sequences belong to the same species Fig. 3 or genus Fig. 4, a binary classification was performed for varying thresholds. The classification procedure was to compare the pairwise distance between two sequences against a threshold. If the distance was less than the threshold, the two sequences were classified as belonging to the same species/genus. If the distance was greater than the threshold, the two sequences were classified as belonging to different species/genera. So, a reasonable threshold might be more on the order of 0.30. Dissimilar sequences start exhibiting relative distance scores above 0.45.

3.6 Alignment Heuristic Options

After the alignment order has been determined, progressive alignment is accomplished by performing repeated pairwise alignment. Referring to the right half of Fig. 1, the sequences being aligned may be: (1) two sequences that belong to the same similarity group, or (2) two consensus sequences that represent the agreed-upon average sequence of two dissimilar groups. The following options provide a means for adjusting the amount of the alignment matrix

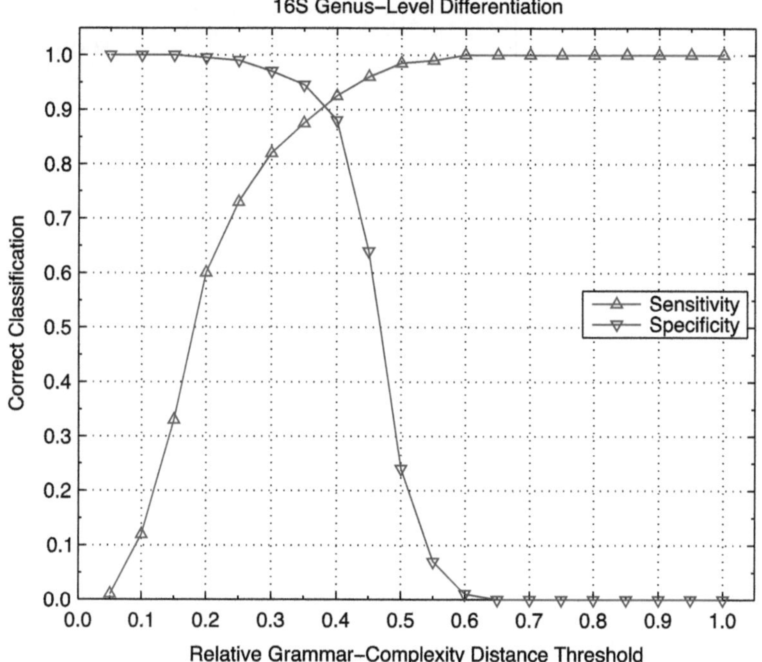

Fig. 4 Relative grammar-based distance classification thresholds at the genus level. Given a known set of 16S Ribosomal RNA sequences in which multiple sequences belong to the same genera, a binary classification is performed for varying grammar-based complexity thresholds. The classification procedure is to compare the pairwise distance between two sequences against a threshold. If the distance is less than the threshold, the two sequences are classified as belonging to the same genus. If the distance is greater than the threshold, the two sequences are classified as belonging to different genera

actually calculated, as more similar sequences are less likely to require the entire matrix be created.

Option -v <value>: Depicted in Fig. 5, specify the alignment overhang percentage for similar sequences. Normally in pairwise alignment every residue of one sequence is compared to all other residues of the second sequence. Especially for nearly identical sequences, this is unnecessary. This value represents the percent of residues shifted past both ends of the longer sequence. Smaller values will result in quicker alignments but at a risk of increased mismatches. If this option is not specified, the default value is 0.10. *Suggestion:* This value can be lowered to decrease computation time, but attention must be paid to the `-T` option. Increasing the -T option will result in more dissimilar sequences being grouped together, and then be pairwise aligned using this parameter. You should only lower this value if you also lower the -T value as well.

Option -V <value>: Similar to the previous option and depicted in Fig. 6, specify the alignment overhang percentage for dissimilar sequences. The alignment of more dissimilar sequences will cause the algorithm to deviate away from the diagonal of the matrix by a

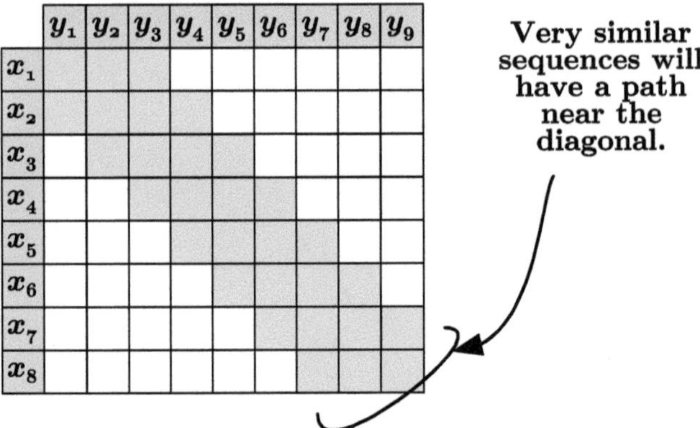

Fig. 5 Effect of option -v, the alignment overhang percentage for similar sequences. During pairwise alignment, normally every residue of one sequence is compared to all other residues of the second sequence. Especially for nearly identical sequences, this is unnecessary since the sequences should align along the diagonal of the matrix. The -v option represents the percent of residues shifted past both ends of the longer sequence. Smaller values will cause a narrower swath requiring fewer cells to be calculated. This will result in quicker alignments but at a risk of increased mismatches

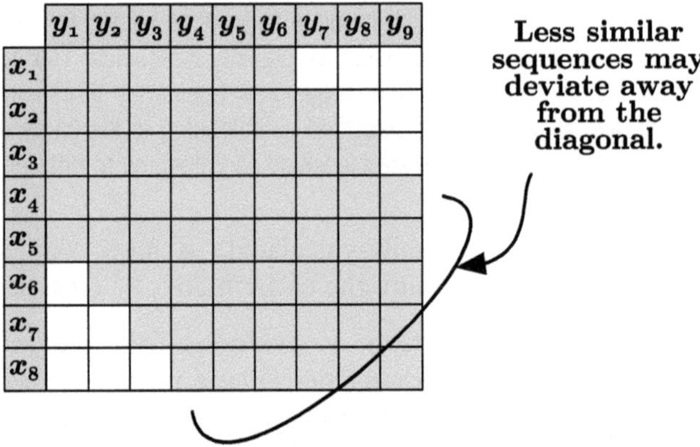

Fig. 6 Effect of option -V, the alignment overhang percentage for dissimilar sequences. During pairwise alignment, normally every residue of one sequence is compared to all other residues of the second sequence. The alignment of more dissimilar sequences will cause the algorithm to deviate away from the diagonal of the matrix by a larger number of residues. The -V option represents the percent of residues shifted past both ends of the longer sequence. Smaller values will cause a narrower swath requiring fewer cells to be calculated. This will result in quicker alignments but at a risk of increased mismatches

larger number of residues. This value represents the percent of residues shifted past both ends of the longer sequence. Smaller values will result in quicker alignments but at a risk of increased mismatches. If this option is not specified, the default value is 0.25.

Suggestion: This value determines the amount of the pairwise matrix that is calculated when aligning consensus sequences. Thus, they are less likely to be nearly identical, and so more of the matrix should be calculated.

3.7 Alignment Scoring Options

GRAMALIGN uses the following parameters for the pairwise alignment process, which is an affine-gap version of Needleman–Wunsch.

Option -g <value>: Specify the gap-open cost. At the core of the alignment algorithm is the pairwise alignment algorithm used to progressively align each sequence not already in the alignment. This pairwise process is the well-known Needleman–Wunsch dynamic programming algorithm modified for affine gap penalties. The value specified in this option represents the cost assigned anytime a gap in a sequence is started. If this option is not specified, the default value is 15.2 for protein sequences or 8.7 for DNA sequences. In the case the GONNET250 substitution matrix is used, this value is multiplied by 10.

Option -G <value>: Specify the tail gap-open cost. The value specified in this option represents the cost assigned if a gap is started at either the beginning or the ending of the alignment. If this option is not specified, the default value is 15.2 for protein sequences or 8.7 for DNA sequences. In the case the GONNET250 substitution matrix is used, this value is multiplied by 10.

Option -e <value>: Specify the gap-extension cost. The value specified in this option represents the cost assigned each time a gap in a sequence is extended by an additional character. If this option is not specified, the default value is 0.6 for protein sequences or 0.8 for DNA sequences. In the case the GONNET250 substitution matrix is used, this value is multiplied by 10.

Option -E <value>: Specify the tail gap-extension cost. The value specified in this option represents the cost assigned each time a gap is extended at either the beginning or the ending of the alignment. If this option is not specified, the default value is 0.3 for protein sequences or 0.4 for DNA sequences. In the case the GONNET250 substitution matrix is used, this value is multiplied by 10.

Option -m <value>: Specify the amino acid substitution matrix. Another important piece of the Needleman–Wunsch pairwise alignment procedure is the substitution scoring matrix. A value of 0 will use the GONNET250 matrix. A value of 1 will use the BLOSUM45 matrix. A value of 2 will use the BLOSUM62 matrix. A value of 3 will use the BLOSUM80 matrix. Note, when using the GONNET250 matrix, the values specified by -g, -G, -e, and -E are all multiplied by 10 to account for the relative differences between the GONNET and BLOSUM matrices. If this option is not specified, the default substitution matrix is the GONNET250. This option is ignored for nucleotide sequences, which use a simple matrix of positive diagonal entries and negative off-diagonal entries.

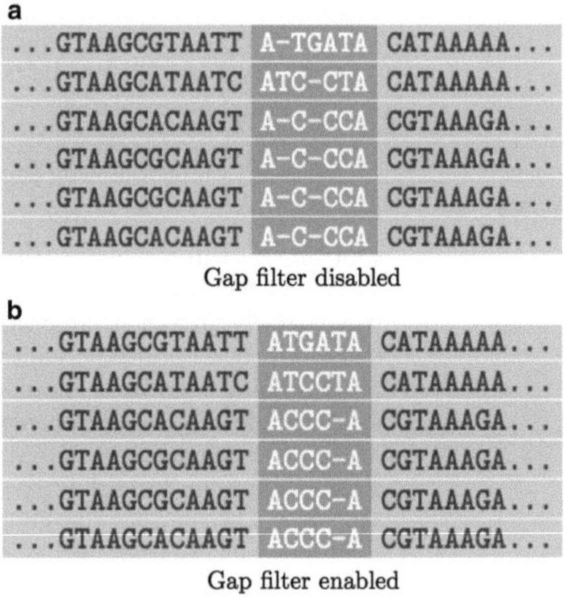

a

```
...GTAAGCGTAATT A-TGATA CATAAAAA...
...GTAAGCATAATC ATC-CTA CATAAAAA...
...GTAAGCACAAGT A-C-CCA CGTAAAGA...
...GTAAGCGCAAGT A-C-CCA CGTAAAGA...
...GTAAGCGCAAGT A-C-CCA CGTAAAGA...
...GTAAGCACAAGT A-C-CCA CGTAAAGA...
```

Gap filter disabled

b

```
...GTAAGCGTAATT ATGATA CATAAAAA...
...GTAAGCATAATC ATCCTA CATAAAAA...
...GTAAGCACAAGT ACCC-A CGTAAAGA...
...GTAAGCGCAAGT ACCC-A CGTAAAGA...
...GTAAGCGCAAGT ACCC-A CGTAAAGA...
...GTAAGCACAAGT ACCC-A CGTAAAGA...
```

Gap filter enabled

Fig. 7 Effect of post-alignment processing with the gap filter. Sometimes non-homologous regions appear with an interleaving of gap-abundant columns. This is due to a combination of alignment order and pairwise scoring parameter selection. Many times the non-homologous regions are not of much interest, in which case this doesn't matter. However, we have provided a post-processing step that scans the MSA in search for columns composed of gaps and with neighboring columns also with gaps, but in different rows. When found, the gaps are blindly shift in order to compress these non-homologous regions together

3.8 Alignment Gap Filter Options

After the MSA has been created, GRAMALIGN provides a post-processing step designed to perform a blind shift of gaps. The goal is to reduce non-homologous regions containing interleaved columns composed largely of gaps. For example *see* Fig. 7, which depicts an original alignment in (a) and the result of enabling the filter in (b). The locally-affected region is highlighted. By default this post-processing gap adjustment step is disabled based on the default values of the following options.

Option -t <value>: Specify the percentage of gaps in a column before a blind adjustment can occur. At the end of the MSA algorithm, the alignment is scanned for columns containing at least as many gaps as specified via this percentage (e.g., 0 = 0 % = zero gaps in the column, 1.0 = 100 % = column with all gaps, 0.5 = 50 % = at least half of the column entries are gaps). If any column contains at least this many gaps, a surrounding window (specified in the -w option) of columns is checked for possible gaps that may be shifted into the current column. To disable this action, simply set this value to 1.0, thus setting the threshold to be columns that contain nothing but gaps, which are non-existent based on the pairwise alignment process. If this option is not specified, the default value is 1.0 (i.e., 100 %).

Option -w <value>: Specify the number of columns in the gap-adjustment window. Regarding the process discussed in the (-t) option, when a column in the initial MSA is found to have at least the necessary number of gaps to be adjusted, this value determines the number of neighboring columns on either side to be scanned for gaps that may be shifted into the current column. Another way to disable this blind shifting is setting this value to 0. If this option is not specified, the default value is 0.

3.9 Secondary Structure Options (Experimental)

The next three command-line options are available, but only useful with the secondary-structure output from IVS, a secondary structure grammar inference program not currently released. So, it is recommended to ignore these parameters.

Option -p <filename>: Specify the grammar piece file, which is output by IVS. By default, there is no grammar piece file used. A grammar piece file is meant to help guide an alignment based on secondary structure present in DNA/RNA sequences. This option is ignored for amino acid sequences.

Option -c <value>: Specify the grammar piece mismatch cost. This is the subtractive penalty applied when either a residue position being aligned is not within a structural piece but the ensemble is, or a residue position is within a structural piece but the ensemble is not. This amount is subtracted from the regular pairwise substitution cost. If this option is not specified, the default cost is 0.0.

Option -s <value>: Specify the grammar piece match score. This is the additive benefit applied when both a residue position being aligned is contained within a structural piece and the ensemble location is also within a structural piece. This amount is added to the regular pairwise substitution score. If this option is not specified, the default score is 0.0.

4 Conclusion

This chapter presented GRAMALIGN, a computationally efficient progressive alignment method. The most interesting and novel aspects of the algorithm involve the generation of the distance matrix which guides the order of alignment. However, for many readers, perhaps the most practical and useful information was presented in Subheading 3. That is, an in-depth detailing of the various options that may be adjusted in order to modify the behavior and computation time necessary to complete an alignment procedure. The GRAMALIGN software is available at http://bioinfo.unl.edu/gramalign.php for both source code download and a web-based alignment server, as shown in Fig. 8.

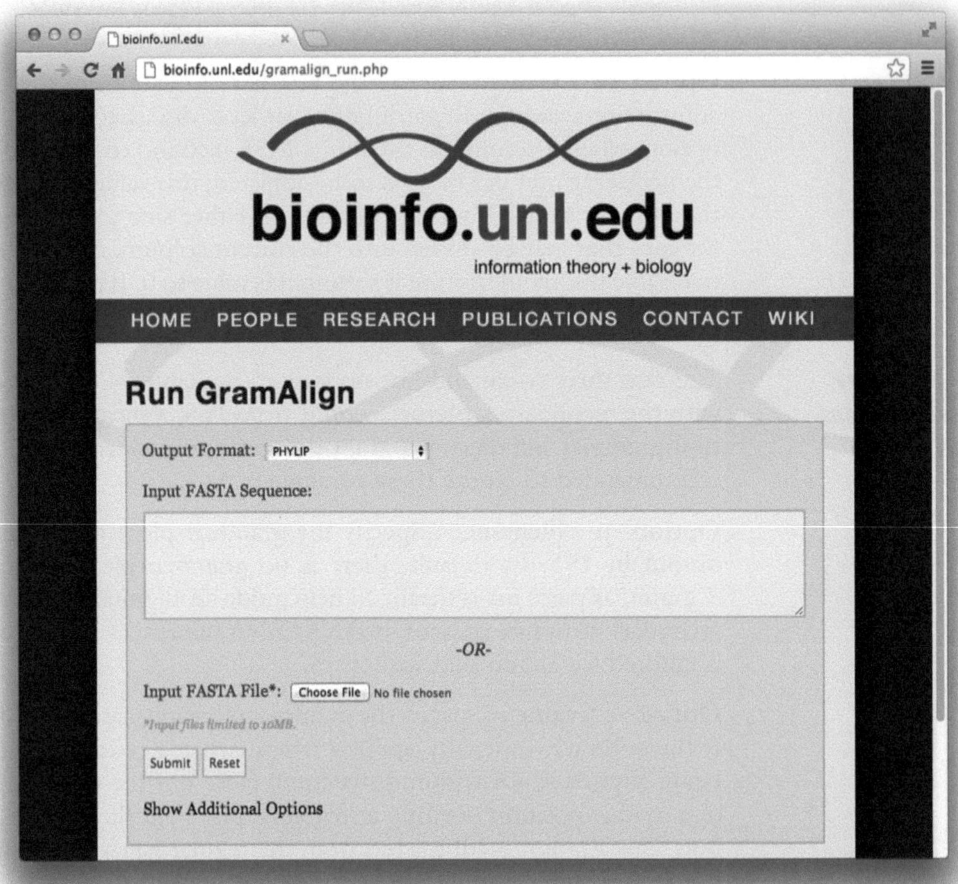

Fig. 8 GramAlign web-server screen shot. The GRAMALIGN software is available at http://bioinfo.unl.edu/gramalign.php for both source code download and a web-based alignment server

References

1. Clote P, Backofen R (1998) Computational molecular biology, an introduction. Cambridge University Press, New York
2. Durbin R, Eddy S, Krogh A, Mitchison G (1998) Biological sequence analysis, probabilistic models of proteins and nucleic acids. Cambridge University Press, New York
3. Sundquist A, Ronaghi M, Tang H, Pevzner P, Batzoglou S (2007) Whole-genome sequencing and assembly with high-throughput, short-read technologies. PLoS ONE 2, pp 1–14
4. Mitrophanov AY, Borodovsky M (2006) Statistical significance in biological sequence analysis. Brief Bioinform 7:2–24
5. Lipman DJ, Altschul SF, Kececioglu JD (1989) A tool for multiple sequence alignment. Proc Natl Acad Sci USA 86:4412–4415
6. Notredame C (2007) Recent evolutions of multiple sequence alignment algorithms. PLoS Comput Biol 3:1405–1408
7. Jukes TH, Cantor CR (1969) Evolution of protein molecules. In: Munro HN (ed) Mammalian protein metabolism. Academic, New York, pp 21–132
8. Kimura M (1980) A simple method for estimating evolutionary rates of base substitutions through comparative studies of nucleotide sequences. J Mol Evol 16:111–120

9. Barry D, Hartigan JA (1987) Asynchronous distance between homologous DNA sequences. Biometrics 43:261–276
10. Kishino H, Hasegawa M (1989) Evaluation of the maximum likelihood estimate of the evolutionary tree topologies from DNA sequence data, and the branching order in Hominoidea. J Mol Evol 29:170–179
11. Lake JA (1994) Reconstructing evolutionary trees from DNA and protein sequences: paralinear distances. Proc Natl Acad Sci USA 91:1455–1459
12. Camin JH, Sokal RR (1965) A method for deducing branching sequences in phylogeny. Evolution 19:311–326
13. Cavalli-Sforza LL, Edwards AWF (1967) Phylogenetic analysis: models and estimation procedures. Evolution 21:550–570
14. Fitch WM (1971) Toward defining the course of evolution: minimum change for a specific tree topology. Syst Zool 20:406–416
15. Adachi J, Hasegawa M (1996) MOLPHY version 2.3: programs for molecular phylogenetics based on maximum likelihood. Number 28 in computer science monographs. Institute of Statistical Mathematics, Tokyo
16. Guindon S, Gascuel O (2003) A simple, fast, and accurate algorithm to estimate large phylogenies by maximum likelihood. Syst Biol 52:696–704
17. Felsenstein J (1981) Evolutionary trees from DNA sequences: a maximum likelihood approach. J Mol Evol 17:368–376
18. Felsenstein J (1989) PHYLIP—phylogeny inference package (version 3.2). Cladistics 5:164–166
19. Swofford DL (1998) PAUP: phylogenetic analysis using parsimony (*and other methods). Sinauer Associates, Sunderland
20. Sankoff D, Leduc G, Antoine N, Paquin B, Lang BF, Cedergren R (1992) Gene order comparisons for phylogenetic inference: evolution of the mitochondrial genome. Proc Natl Acad Sci USA 89:6575–6579
21. Gramm J, Niedermeier R (2002) Breakpoint medians and breakpoint phylogenies: a fixed-parameter approach. Bioinformatics 18(Suppl 2):S128–S139
22. Lin Y, Rajan V, Swenson KM, Moret BME (2010) Estimating true evolutionary distances under rearrangements, duplications, and losses. BMC Bioinformatics 11(Suppl 1):1–11
23. Moret BME, Tang J, Wang L, Warnow T (2002) Steps toward accurate reconstructions of phylogenies from gene-order data. J Comput Syst Sci 65:508–525
24. Hannenhalli S, Pevzner PA (1995) Towards a computational theory of genome rearrangements. Lect Notes Comput Sci 1000:184–202
25. Kececioglu J, Sankoff D (1995) Exact and approximation algorithms for sorting by reversals, with application to genome rearrangement. Algorithmica 13:180–210
26. Kececioglu J, Gusfield D (1998) Reconstructing a history of recombinations from a set of sequences. Discrete Appl Math 88:239–260
27. Kececioglu J, Ravi R (1995) Of mice and men: algorithms for evolutionary distances between genomes with translocation. In: Proceedings of the 6th ACM-SIAM symposium on discrete algorithms, Society for Industrial and Applied Mathematics, Philadelphia, PA, USA, pp 604–613
28. Boore JL, Brown WM (1998) Big trees from little genomes: mitochondrial gene order as a phylogenetic tool. Curr Opin Genet Dev 8:668–674
29. Sankoff D, Blanchette M (1998) Multiple genome rearrangement and breakpoint phylogeny. J Comput Biol 5:555–570
30. Sankoff D (1999) Genome rearrangement with gene families. Bioinformatics 15:909–917
31. Otu HH, Sayood K (2003) A new sequence distance measure for phylogenetic tree construction. Bioinformatics 19:2122–2130
32. Bastola DR, Otu HH, Doukas SE, Sayood K, Hinrichs SH, Iwen PC (2004) Utilization of the relative complexity measure to construct a phylogenetic tree for fungi. Mycol Res 108:117–125
33. Russell DJ, Otu HH, Sayood K (2008) Grammar-based distance in progressive multiple sequence alignment. BMC Bioinformatics 9:1–13
34. Russell DJ, Way SF, Benson AK, Sayood K (2010) A grammar-based distance metric enables fast and accurate clustering of large sets of 16S sequences. BMC Bioinformatics 11:1–14
35. Boutros R, Stokes N, Bekaert M, Teeling EC (2009) UniPrime2: a web service providing easier universal primer design. Nucleic Acids Res 37(Web Server issue):W209–W213
36. Albayrak A, Otu HH, Sezerman UO (2010) Clustering of protein families into functional subtypes using relative complexity measure with reduced amino acid alphabets. BMC Bioinformatics 11:1–10
37. Li M, Vitanyi P (1997) An introduction to Kolmogorov complexity and its applications. Springer, Berlin
38. Lempel A, Ziv J (1976) On the complexity of finite sequences. IEEE Trans Inf Theory 22:75–81

Chapter 12

Multiple Sequence Alignment with DIALIGN

Burkhard Morgenstern

Abstract

DIALIGN is a software tool for multiple sequence alignment by combining global and local alignment features. It composes multiple alignments from local pairwise sequence similarities. This approach is particularly useful to discover conserved functional regions in sequences that share only local homologies but are otherwise unrelated. An anchoring option allows to use external information and expert knowledge in addition to primary-sequence similarity alone. The latest version of DIALIGN optionally uses matches to the *PFAM* database to detect weak homologies. Various versions of the program are available through *Göttingen Bioinformatics Compute Server (GOBICS)* at http://www.gobics.de/department/software.

Key words Motif discovery, Local alignment, Anchored alignment, Protein domain

1 Introduction

When the first software tools for multiple sequence alignment (MSA) were developed, alignment methods were generally classified as either *global* or *local*. Global alignment methods align sequences over their entire length, regardless of the real extension of sequence homology. Local methods return only one or several regions of sequence similarity, ignoring the remainder of the sequences. The classical *progressive* alignment methods proposed in the late 1980s were global methods [1–3]. Until today, many of the most advanced MSA approaches rely on this concept, e.g., *MAFFT* [4], *MUSCLE* [5], T-COFFEE [6], or *CLUSTAL Omega* [7]. In addition to these purely global approaches, algorithms for *local* multiple alignment have been proposed since the early days of MSA. Pioneering local MSA methods are *MEME* [8], *PIMA* [9], or *TEIRESIAS* [10]. These tools are limited to extract conserved motifs from a set of sequences, even if detectable homology is not restricted to one conserved motif.

The program *DIALIGN* was first released in 1996 [11], the main idea behind this program was to combine local and global alignment features. In *DIALIGN*, this is implemented by first

David J. Russell (ed.), *Multiple Sequence Alignment Methods*, Methods in Molecular Biology, vol. 1079,
DOI 10.1007/978-1-62703-646-7_12, © Springer Science+Business Media, LLC 2014

searching for local pairwise sequence similarities, and by then combining these "partial" similarities in a final multiple alignment. The idea was that pairwise or MSA should be limited to those parts of the sequences that share some statistically significant similarity, but one should not try to align non-related parts of the sequences.

In this sense, *DIALIGN* is a sort of hybrid approach, between local and global alignment methods. If sequence homology extends over the entire length of the input sequences, it produces global alignments similar to the output of strictly global methods such as *CLUSTAL*, but if sequences share only one region of homology, this region is aligned and the rest of the sequences is ignored. A main advantage of *DIALIGN* is that the program itself decides which parts of the sequences are to be aligned, so the user does not need to decide if a local or a global methods is suitable for a particular sequence set under study. One single program can be applied to both global *and* local homologies. More importantly, *DIALIGN* can deal with sequences that share regions of detectable homology that are separated by unrelated regions. Such sequences can neither be aligned by purely global MSA methods nor by purely local motif finders. Nevertheless, although *DIALIGN* integrates local and global MSA features in a single program, the main strength of the program is *local* homology detection.

DIALIGN can be applied to nucleic acid and protein sequences. Originally, the program was developed and tested with a focus on *protein* sequences. However, when more and more genomic DNA sequences became available around the turn of the millennium, the hybrid strategy of *DIALIGN* combining local and global alignment was found particularly useful to study genomic sequences where conserved functional regions such as exons or regulatory sites are separated by large non-related parts of the genome.

With the increasing number of partially or fully sequenced genomes, cross-species alignment of genomic sequences became a valuable tool to discover functional elements in genome sequences, *see*, e.g., [12] or [13] for a review. The idea is that functionally conserved elements of the genome are more conserved than non-functional parts, so local conservation in an alignment of syntenic genomic sequences usually indicates biological functionality.

Obviously, neither traditional methods for global alignment such as *CLUSTAL W* nor purely local methods as *BLAST* are suitable to align genomic sequences. Although *DIALIGN* was originally designed as a generic MSA program, not specialized for genomic alignment, it was the first MSA method that could produce meaningful alignments of genomic sequences. In a number of pioneering studies, *DIALIGN* was used to discover non-coding functional elements in genome sequences that could later be confirmed in experimental studies, e.g., [14, 15].

If large sequences are to be aligned, program run time becomes an issue. Similar as with more traditional alignment methods, the run time of *DIALIGN* for pairwise alignment is proportional to the product of the lengths of the input sequences [16]. This is too slow to align large genomic sequences. To speed up the program run time, a previously developed *anchoring* option proved to be useful.

2 Anchored Alignment

Most MSA methods are fully automated and do not require any human intervention. The input from the user is restricted to selecting a set of input sequences and to choose the necessary parameter values, e.g., for *gap penalties.* In most cases, default parameter values are used which have been found useful in the typical situations where a program is used.

Automated alignment is clearly required where no further information about the input sequences is available. Also, if large data sets are to be processed and manual intervention would be too time consuming, automated MSA is mandatory. It should be clear, however, that the accuracy of automatic methods for sequence analysis is fundamentally limited. At best, they can produce alignments with a (near-)optimal *quality score* in some mathematical sense. But there can be no guarantee that mathematically optimal or high-scoring alignments are biologically meaningful.

The standard version of *DIALIGN* is fully automated, i.e. like other MSA methods, it works without human intervention. The only input parameter is a *threshold* T for the quality of the local similarities considered for alignment. Often, however, an expert user has already some information about (putative) homologies among the input sequences. In this case, it is desirable to *force* an MSA program to align these homologies and to align only the remainder of the sequences in the usual automatic fashion.

For this reason, *DIALIGN* has an option for *anchored alignment* where MSAs are produced in a *semi-automatic* way [17, 18]. With this option, the user can select parts of the input sequences that are to be aligned to each other. The final alignment produced by *DIALIGN* can then be seen as an *extension* of this user-specified alignment *anchor.* In more detail, the user selects equal-length pairs of sequence segments that will end up aligned to each other without gaps. Such pairs of segments are called *anchor points.* In general, it may not be possible to align all of the specified anchor points in one single output alignment, so it may be necessary to discard some of the user-defined anchor points. Therefore, the user has to assign *score* to each anchor point determining their priority in case not all anchor points can be used.

In addition to including expert knowledge in otherwise automatically produced MSAs, the anchored-alignment option in

DIALIGN can be used to speed up the alignment procedure. Indeed, if an anchor point enforces alignment of two selected sequence segments, this reduces the search space of the remaining automatic alignment procedure (e.g., if the middle positions of two sequences are used as anchor point, the search spaced for the pairwise alignment is reduced by a factor of two).

Therefore, the ***anchoring*** option was also used to align long genomic sequences [19, 20]. Here, a fast method for local homology detection such as *BLAST* [21] can be used to find strong sequence homologies that can then be used as ***anchor points*** for *DIALIGN*. This approach has been implemented and made available on our web server [19]. Our anchored-alignment approach to genomic sequence comparison has also been used to improve the performance of ***gene-finding*** methods in eukaryotes [22]. Other applications of anchored multiple alignment are the possibility to study the behavior of alignment methods in detail, or the integration of new algorithmic approaches for multiple alignment instead of the greedy heuristic used in the standard version of *DIALIGN* [23].

3 *DIALIGN-T* and *DIALIGN-TX*

Studies have shown that *DIALIGN* is often superior to other MSA tools where sequences with local homologies are aligned. On globally related sequences with weak primary-sequence similarity, however, it tends to be outperformed by strictly global methods such as *CLUSTAL W* [24], *MUSCLE* [5, 25], *MAFFT* [4], or *PROBCONS* [26]. One might think that a possible reason for this relative weakness could be the greedy optimization method used for multiple alignment in *DIALIGN*. Indeed, it is easy to see that the heuristic in *DIALIGN* can produce MSAs with scores far below the possible optimal MSA. If that would be the reason for the relative weakness of the program on global, weak homologies, one would make efforts to find more efficient optimization algorithms, leading to higher-scoring MSAs in the sense of the fragment-based scoring function used in *DIALIGN*. This has been done in the past, e.g., in [27, 28]. More recent results based on anchored alignments indicate, however, that the relative weakness of *DIALIGN* on global homologies with low similarity at the primary-sequence level is caused by the underlying objective function, and not so much by the greedy optimization algorithm. Thus, MSAs with mathematically higher scores may not necessarily be more meaningful of biologically. We therefore adopted other approaches to improve the performance of *DIALIGN* on those sequence families where strictly global MSA methods were still superior. This resulted in the development of *DIALIGN-T* and *DIALIGN-TX*.

DIALIGN-T is a complete re-implementation of *DIALIGN* [29]. As the first implementation of *DIALIGN*, it starts with calculating all pairwise alignments of the input sequences [16, 30]. That is, an optimal chain of *fragments* is calculated for each pair of input sequences. The difference to previous versions of the program is in the way, these similarities are integrated into a final multiple alignment. Like in the first implementation, a greedy heuristic is used, but *DIALIGN-T* uses a various tricks to prevent the algorithm from aligning spurious, isolated random similarities which might prevent a greedy method from finding a biologically correct global alignment. *DIALIGN-T*, for example, does not only consider the local degree of similarity in a fragment, but also its *context* within the two aligned sequences. Fragments that belong to a high-scoring pairwise alignment are preferred to isolated fragments. Together with some other heuristics, this led to a considerable improvement of the performance compared with the original implementation of *DIALIGN*.

In *DIALIGN-TX* [31], more sophisticated methods were used to reduce the influence of isolated local similarities. This implementation relies on the traditional *progressive* approach to multiple alignment [1–3] and adapts this approach to the focus on local similarities that is used in *DIALIGN*. In a first step, a *guide tree* is calculated for the input sequences. This is done by transforming the fragment-based similarities in the pairwise alignments into distance values. As in more traditional progressive methods, sequences and groups of previously aligned sequences are aligned, going from the tips to the root of the guide tree.

In progressive methods such as *CLUSTAL*, a group of previously aligned sequences is represented as a *profile*, i.e. as a matrix containing the residue frequencies for each alignment column. This cannot be generalized to the segment-based approach where an alignment is seen as a set of local homologies, and parts of the sequences may remain unaligned. *DIALIGN-TX* therefore uses a different approach to align two groups G_1 and G_2 of previously aligned sequences. Fragments are selected, each of which aligns one sequence from G_1 with another sequence from G_2. To remove fragments that are inconsistent with the previously selected fragments aligning sequences from G_1 and G_2, respectively, to each other, a graph algorithm is used [32].

4 Using Matches to Pfam for Improved Protein Alignment

Traditional alignment approaches are based on primary-sequence information only. In one way or the other, they define an *alignment score* based on detectable primary-sequence similarity and then try to calculate optimal or near-optimal alignments in the sense of this scoring scheme. Such approaches are clearly reasonable where no further information about the input sequences is available. But, as mentioned above, there is no guarantee that a mathematically optimal alignment makes also sense from a biological point of view. Where possible, it is advisable to exploit additional information that may be available about the sequences to be aligned.

With more and more known genes and proteins, it is likely that sequences under study have known homologs in a database. Information about these homologs can also be used for improved alignment. For example, the program *DbCLUSTAL* [33] uses *BLAST* [21] to search for homologs of the input sequences in databases. These homologs are then aligned together with the original sequence set using *CLUSTAL W*, and finally the database hits are removed again to obtain a MSA of the original set of input sequences. If local similarities to database sequences are found, they are used as a sort of anchor points. It could be shown that this approach increases the performance of *CLUSTAL W*. Similarly, the latest version of *CLUSTAL*, *CLUSTAL Omega* can use searches to the *Pfam* database [34] and align those positions of the input sequences together that match the same position in a *Pfam* domain.

Inspired by this approach, we implemented a version of *DIALIGN* for protein alignment that takes matches to *Pfam* domains as additional input information [35]. More specifically, we construct *blocks* of segments of the input sequences matching to the same segment of a *Pfam* domain. These blocks are then preferentially included into the output MSA. We tested different ways of integrating our "blocks" into our MSA procedure. In a straight forward way, the identified blocks are used as *anchor points* for the subsequent alignment of the input sequences. Here, we also developed an *interactive* approach where the *blocks* that are defined by common *Pfam* matches can be inspected by the user and accepted or rejected based on expert knowledge.

Alternatively, putative homologies identified by matches to the same *Pfam* domain can be used together with similarities found by pairwise *DIALIGN* alignment; these different similarities are finally integrated into one single output MSA using a graph-theoretical algorithm that we recently proposed [23]. We tested these new approaches systematically using *BAliBASE* [36, 37] and *SABmark* [38] as benchmark databases. Using homologies to *Pfam* domains could considerably improve the performance of *DIALIGN* [35].

5 Altavist: Comparing Multiple Alignments

Every molecular biologist knows that the reliability of automated MSA methods is limited. In fact, many biologists argue that the best way of creating multiple alignments is still to have experts creating them *manually*. Such alignments are often superior to computationally computed alignments. For the same reasons, it is

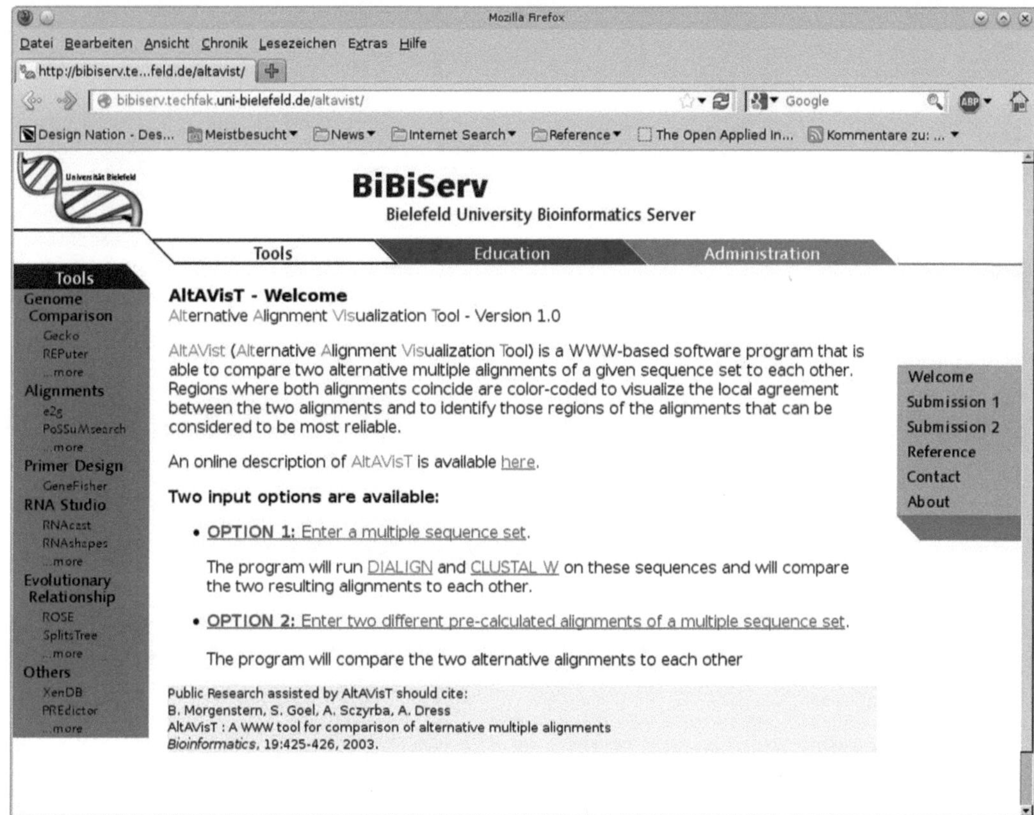

Fig. 1 Comparing alternative alignments of the same sequence set with *ALTAVIST* at *Bielefeld Bioinformaitcs Server (BiBiServ)*. Two ways of using the program are offered. With *OPTION 1*, a set of nucleic acid or protein sequences is uploaded. The server then runs *DIALIGN* and *CLUSTAL W* on the sequences and graphically shows the positions in the two output alignments where these alignments agree. With *OPTION 2*, two pre-calculated MSAs of the same sequence set can be entered and are compared in the same way

common practice to post-process MSAs manually. Various alignment editors are available for this purpose, e.g., the widely used program *JalView* [39].

One way to assess the (local) reliability of automatically calculated multiple alignments is to run different MSA programs on the same set of input sequences and to compare the results. Generally, those parts of the sequences that are aligned in the same way by different MSA programs should be more reliable than regions that are differently aligned. We developed a program called *ALTAVIST (ALTernative Alignment VISualisation Tool)* that compares two MSAs of the same sequence set and highlights those parts of the MSAs where both programs agree [40].

At the bioinformatics server *BiBiServ*, this program can be run in two different ways. (1) A sequence set can be entered, and the MSA programs *CLUSTAL W* and *DIALIGN* are run on these sequences. The program then highlights residues aligned in

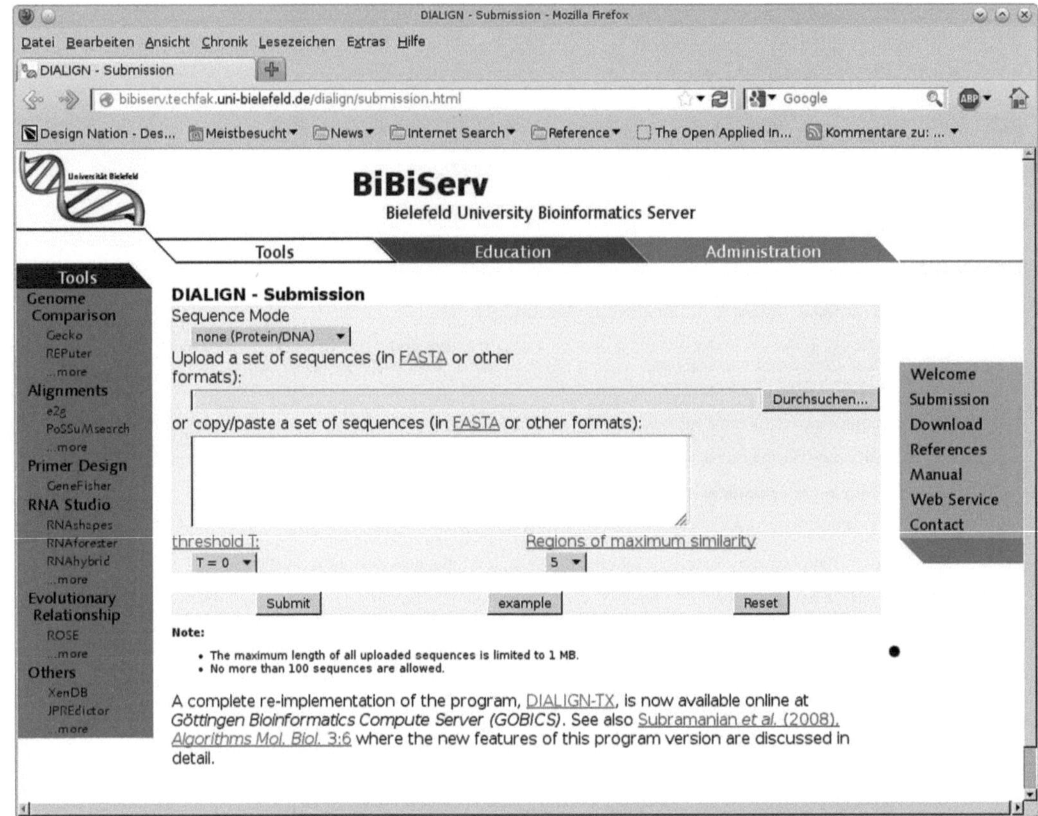

Fig. 2 The standard version of the program, DIALIGN 2.2.1, at *BiBiServ*. Sequences can be uploaded or pasted into a window in FASTA format. Only one parameter needs to be specified, namely a *threshold T* for the quality of local pairwise alignments (fragments) that are incorporated into the final MSA

the same way by both aligners. (2) Alternatively, two MSAs of the same sequence that may have been calculated by arbitrary methods can be uploaded for comparison.

6 Program Availability at GOBICS and BiBiServ

Various versions of DIALIGN are available to the research community through web servers; in addition, the program code is freely available. Since much of the program development has been done at *University of Bielfeld*, Germany, the standard version of *DIALIGN* is available from *Bielefeld University Bioinformatics Server (BiBiServ)* (http://bibiserv.techfak.uni-bielefeld.de/) (Fig. 1). Here, it is also possible to run *DIALIGN 2.2.1* through an easy-to-use web interface. Fig. 2 shows the web interface at BiBiserv.

When I moved to *University of Göttigen*, Germany, in 2002, the main development work on *DIALIGN* continued there, and a

Submit your sequences:

Please upload your file in multiple FastA format

Durchsuchen...

Sequence type Threshold fast_alignment

protein sequence T=0

Your email address (optional)

Reset input | Run DIALIGN

Optional anchor points:

If you want to *upload* your anchor points, please use a file in this format (recommended for large sets of anchor points)

Durchsuchen...

Alternatively, if you have only a few anchor points, you can put them into the table below.

first sequence	second sequence	start position in first sequence	start position in second sequence	length of anchor

Note that each anchor point corresponds to a *pair of equal-length segments* that are to be aligned. So an anchor point is characterized by two sequences, by the start positions of the segments and by the segment length. In addition, each anchor point needs a score that determines its priority in case of conflicting anchor points. If you enter anchor points by filling the above table, no *scores* for anchor points are required. Instead, proposed anchor points are prioritized by the order in which you input them.

Reset input | Run DIALIGN

Fig. 3 Anchored alignment with DIALIGN at *GÖttingen BIoinformatics Compute Server (GOBICS)* (www.gobics.de/anchor/submission.php). Sequences and threshold parameter are specified as with the standard version of the program (Fig. 2). In addition, *anchor points* can be specified by the user. An "anchor point" consists of two equal-length segments of two input sequences that are to be aligned in the output MSA. Thus, each anchor point is characterized by five coordinates: the involved sequences, the starting points in these sequences and its length. In addition, anchor points are given a *scores* to prioritize them in case of *consistency* conflicts, i.e. if not all chosen anchor points can be included in one single output alignment. The coordinates of the selected anchor points can be either uploaded or entered manually through a form. In the latter case no score is required, the priority of anchor points is defined by the order in which they are entered

WWW interfaces for more recent versions of the program were set up at *Göttingen Bioinformatics Compute Server* (GOBICS). In particular, *DIALIGN* with anchor points is available through *GOBICS* at http://dialign.gobics.de/anchor/index.php, as well as *DIALIGN-TX* at http://dialign-tx.gobics.de/submission. The most recent addition to the *DIALIGN* family of MSA programs is the program version using matches to *Pfam* domains which is available at http://dialign-pfam.gobics.de:8080/SequenceAlignment/. Threshold parameters for the *Pfam* search can be adjusted, and the user can interactively view and select/deselect matches to *Pfam* domains that are to be used as anchor points (Figs. 3–5).

University of Göttingen | Faculty of Biology | Inst. of Microbiology and Genetics | Dep. of Bioinformatics

DIALIGN-TX [job submission]

Submit your sequences:

Upload a file containing protein sequences in multiple FastA format.
(Example Multiple Fasta file)

Durchsuchen...

A detailed online help explaining the parameters of DIALIGN-TX is given here.

Recommended values are L=4, T=40.

Length of a low-scoring region (Details)
L = 4

Maximum fragment length that is allowed to contain regions of low quality
T = 40

Sensitivity (-l) = unlimited (0)
☐ fast mode (-F), implies -l0, since it already significantly reduces sensitivity
☐ FASTA output

Run Dialign-TX

Fig. 4 DIALIGN-TX at *GOBICS* (http://dialign-tx.gobics.de/index)

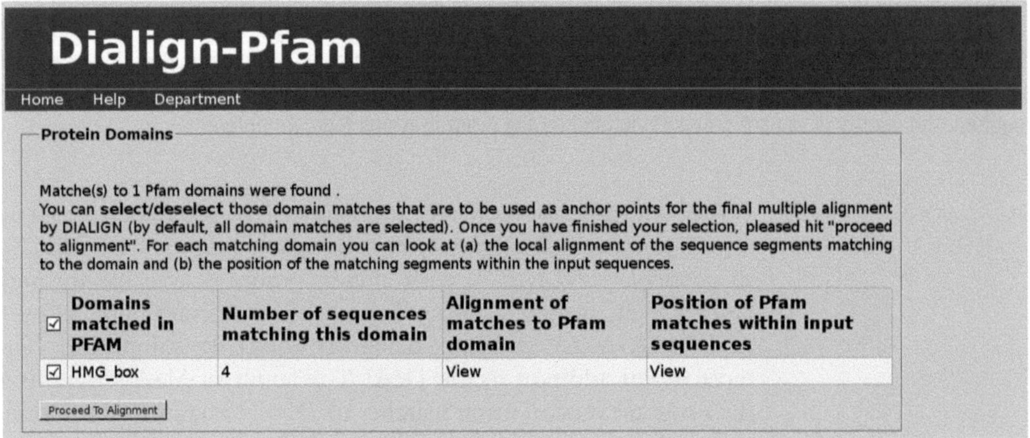

Fig. 5 DIALIGN using *Pfam* matches as anchor points (http://dialign-pfam.gobics.de:8080/SequenceAlignment/). A list of *Pfam* domains matching the input sequences is presented to the user (in this case only a single domain matched). The user can view the alignment of the matching segments of the input sequences and their position within the sequences. Domains that are to be used as anchor points in the final alignment can be selected/deselected by the user

References

1. Feng DF, Doolittle RF (1987) Progressive sequence alignment as a prerequisite to correct phylogenetic trees. J Mol Evol 25:351–360
2. Higgins DG, Sharp PM (1988) CLUSTAL—a package for performing multiple sequence alignment on a microcomputer. Gene 73: 237–244
3. Taylor WR (1988) A flexible method to align large numbers of biological sequences. J Mol Evol 28:161–169
4. Katoh K, Kuma K, Toh H, Miyata T (2005) MAFFT version 5: improvement in accuracy of multiple sequence alignment. Nucleic Acids Res 33:511–518
5. Edgar RC (2004) MUSCLE: multiple sequence alignment with high score accuracy and high throughput. Nucleic Acids Res 32:1792–1797
6. Notredame C, Higgins D, Heringa J (2000) T-Coffee: a novel algorithm for multiple sequence alignment. J Mol Biol 302:205–217
7. Sievers F, Wilm A, Dineen D, Gibson TJ, Karplus K, Li W, Lopez R, McWilliam H, Remmert M, Sding J, Thompson JD, Higgins DG (2011) Fast, scalable generation of high-quality protein multiple sequence alignments using Clustal Omega. Mol Syst Biol 7:539
8. Bailey TL, Elkan C (1994) Fitting a mixture model by expectation maximization to discover motifs in biopolymers. In: Proceedings of the second international conference on intelligent systems for molecular biology, The AAAI Press, Menlo Park, California, pp 28–36
9. Smith RF, Smith TF (1992) Pattern-Induced Multi-sequence Alignment (PIMA) algorithm employing secondary structure-dependent gap penalties for comparative protein modelling. Protein Eng 5:35–41
10. Rigoutsos I, Floratos A (1998) Combinatorial pattern discovery in biological sequences: the Teiresias algorithm. Bioinformatics 14(1): 55–67
11. Morgenstern B, Dress A, Werner T (1996) Multiple DNA and protein sequence alignment based on segment-to-segment comparison. Proc Natl Acad Sci USA 93:12098–12103
12. Loots GG, Locksley RM, Blankespoor CM, Wang ZE, Miller W, Rubin EM, Frazer KA (2000) Identification of a coordinate regulator of interleukins 4, 13, and 5 by cross-species sequence comparisons. Science 288(5463): 136–140
13. Frazer KA, Elnitski L, Church DM, Dubchak I, Hardison RC (2003) Cross-species sequence comparisons: a review of methods and available resources. Genome Res 13:1–12
14. Göttgens B, Gilbert JGR, Barton LM, Grafham D, Rogers J, Bentley DR, Green AR (2001) Long-range comparison of human and mouse SCL loci: localized regions of sensitivity to restriction endonucleases correspond precisely with peaks of conserved noncoding sequences. Genome Res 11:87–97
15. Chapman MA, Charchar FJ, Kinston S, Bird CP, Grafham D, Rogers J, Grützner F, Marshall Graves JA, Green AR, Göttgens B (2003) Comparative and functional analysis of LYL1 loci establish marsupial sequences as a model for phylogenetic footprinting. Genomics 81: 249–259
16. Morgenstern B (2002) A simple and space-efficient fragment-chaining algorithm for alignment of DNA and protein sequences. Appl Math Lett 15:11–16
17. Morgenstern B, Werner N, Prohaska SJ, Steinkamp R, Schneider I, Subramanian AR, Stadler PF, Weyer-Menkhoff J (2005) Multiple sequence alignment with user-defined constraints at GOBICS. Bioinformatics 21: 1271–1273
18. Morgenstern B, Prohaska SJ, Pöhler D, Stadler PF (2006) Multiple sequence alignment with user-defined anchor points. Algorithms Mol Biol 1:6
19. Brudno M, Steinkamp R, Morgenstern B (2004) The CHAOS/DIALIGN WWW server for multiple alignment of genomic sequences. Nucleic Acids Res 32:W41–W44
20. Pöhler D, Werner N, Steinkamp R, Morgenstern B (2005) Multiple alignment of genomic sequences using CHAOS, DIALIGN and ABC. Nucleic Acids Res 33:W532–W534
21. Altschul SF, Gish W, Miller W, Myers EM, Lipman DJ (1990) Basic local alignment search tool. J Mol Biol 215:403–410
22. Stanke M, Tzvetkova A, Morgenstern B (2006) AUGUSTUS+ at EGASP: using EST, protein and genomic alignments for improved gene prediction in the human genome. Genome Biol 7:S11
23. Corel E, Pitschi F, Morgenstern B (2010) A min-cut algorithm for the consistency problem in multiple sequence alignment. Bioinformatics 26:1015–1021
24. Thompson JD, Higgins DG, Gibson TJ (1994) CLUSTAL W: improving the sensitivity of progressive multiple sequence alignment through sequence weighting, position-specific gap penalties and weight matrix choice. Nucleic Acids Res 22:4673–4680

25. Edgar RC (2004) MUSCLE: a multiple sequence alignment method with reduced time and space complexity. BMC Bioinformatics 5:113
26. Do CB, Mahabhashyam MSP, Brudno M, Batzoglou S (2005) ProbCons: probabilistic consistency-based multiple sequence alignment. Genome Res 15:330–340
27. Lenhof H, Morgenstern B, Reinert K (1999) An exact solution for the segment-to-segment multiple sequence alignment problem. Bioinformatics 15:203–210
28. Kececioglu JD, Lenhof H, Mehlhorn K, Mutzel P, Reinert K, Vingron M (2000) A polyhedral approach to sequence alignment problems. Discrete Appl Math 104:143–186
29. Subramanian AR, Weyer-Menkhoff J, Kaufmann M, Morgenstern B (2005) DIALIGN-T: an improved algorithm for segment-based multiple sequence alignment. BMC Bioinformatics 6:66
30. Morgenstern B (2000) A space-efficient algorithm for aligning large genomic sequences. Bioinformatics 16:948–949
31. Subramanian AR, Kaufmann M, Morgenstern B (2008) DIALIGN-TX: greedy and progressive approaches for the segment-based multiple sequence alignment. Algorithms Mol Biol 3:6
32. Clarkson KL (1983) A modification of the greedy algorithm for vertex cover. Inf Process Lett 16:23–25
33. Thompson JD, Plewniak F, Thierry J, Poch O (2000) DbClustal: rapid and reliable global multiple alignments of protein sequences detected by database searches. Nucleic Acids Res 28:2919–2926
34. Finn RD, Mistry J, Tate J, Coggill P, Heger A, Pollington JE, Gavin OL, Gunasekaran P, Ceric G, Forslund K et al (2010) The Pfam protein families database. Nucleic Acids Res 38(Suppl 1):D211–D222
35. Ait LA, Corel E, Morgenstern B (2012) Using protein-domain information for multiple sequence alignment. In: Proceedings of the IEEE 12th international conference on bioinformatics and bioengineering (BIBE 12), Institute of Electrical and Electronics Engineers (IEEE), pp 164–168
36. Thompson JD, Plewniak F, Poch O (1999) BAliBASE: a benchmark alignment database for the evaluation of multiple sequence alignment programs. Bioinformatics 15:87–88
37. Thompson JD, Koehl P, Ripp R, Poch O (2005) BAliBASE 3.0: latest developments of the multiple sequence alignment benchmark. Proteins Struct Funct Bioinformatics 61:127–136
38. Walle IV, Lasters I, Wyns L (2005) SABmark—a benchmark for sequence alignment that covers the entire known fold space. Bioinformatics 21:1267–1268
39. Clamp M, Cuff J, Searle SM, Barton GJ (2004) The Jalview java alignment editor. Bioinformatics 20:426–427
40. Morgenstern B, Goel S, Sczyrba A, Dress A (2003) AltAVisT: a WWW server for comparison of alternative multiple sequence alignments. Bioinformatics 19:425–426

Chapter 13

PicXAA: A Probabilistic Scheme for Finding the Maximum Expected Accuracy Alignment of Multiple Biological Sequences

Sayed Mohammad Ebrahim Sahraeian and Byung-Jun Yoon

Abstract

PicXAA is a probabilistic nonprogressive alignment algorithm that finds protein (or DNA) multiple sequence alignments with maximum expected accuracy. PicXAA greedily builds up the alignment from sequence regions with high local similarity, thereby yielding an accurate global alignment that effectively captures the local similarities across sequences. PicXAA constantly yields accurate alignment results on a wide range of reference sets that have different characteristics, with especially remarkable improvements over other leading algorithms on sequence sets with high local similarities. In this chapter, we describe the overall alignment strategy used in PicXAA and discuss several important considerations for effective deployment of the algorithm.

Key words Multiple sequence alignment, Nonprogressive alignment, Maximum expected accuracy (MEA), Probabilistic consistency transformation, PicXAA

1 Introduction

Multiple sequence alignment (MSA) is an indispensable tool in comparative studies of biological sequences, and it plays a prominent role in many applications such as phylogenetic analysis, structure prediction, function prediction, motif discovery, and modeling sequence homology [1–7]. The mathematically optimal MSA can be found using dynamic programming. However, the dynamic programming approach has a high computational cost that renders it impractical for aligning more than a few sequences. For this reason, the progressive alignment scheme—which successively aligns pairs of sequences (or sequence profiles) along a phylogenetic tree of the given sequences—has gained popularity as a practical alternative [8–16]. In fact, the progressive alignment technique is surprisingly effective for closely related sequences and it yields

David J. Russell (ed.), *Multiple Sequence Alignment Methods*, Methods in Molecular Biology, vol. 1079,
DOI 10.1007/978-1-62703-646-7_13, © Springer Science+Business Media, LLC 2014

accurate alignment results despite the low computational overhead. However, it does not work as well when applied to a set of divergent sequences that share only local similarities. Typically, the progressive scheme tends to propagate early-stage errors throughout the entire alignment process, which can be problematic when we need to align a set of sequences that prominently share local similarities but also possess many differences across sequence regions. In such a case, it may be difficult to build up the MSA through progressive alignment, as it may dilute local similarities and propagate errors that arise in divergent sequence regions. Until now, several techniques have been developed to address the shortcoming of progressive alignment and alleviate these undesirable effects [17–20].

Recently, a novel alignment algorithm called PicXAA [20] has been proposed to address this problem by adopting a computationally efficient non-progressive scheme. Based on the maximum expected accuracy (MEA) principle, PicXAA aims to find the optimal alignment that maximizes the expected number of correctly aligned symbols (i.e., amino acids or nucleotides). Towards this goal, PicXAA first computes the posterior pairwise symbol alignment probability for all pairs of symbol locations for every sequence pair. Next, it updates the estimated probabilities through an improved probabilistic consistency transformation, which aims to refine the symbol alignment probabilities of a given sequence pair by incorporating the information from other sequences. Using an efficient graph-based technique, PicXAA greedily builds up the alignment based on the updated probabilities, starting from confidently alignable regions with high local similarities. Once the initial alignment is constructed, PicXAA goes through an iterative refinement process to further improve the alignment quality in divergent sequence regions that cannot be confidently aligned. In summary, PicXAA can accurately predict the global alignment of multiple biological sequences, in which local homologies are effectively captured. Experimental results confirm that PicXAA consistently yields accurate alignment results in various benchmarks, where the improvements are especially significant on reference sets that consist of sequences with only local similarities [20].

2 Methods

PicXAA [20] aims to find the multiple sequence alignment with the maximum expected accuracy, i.e., the maximum expected number of correctly aligned residue pairs. Through a greedy approach, PicXAA probabilistically builds up the MSA, by starting from high similarity regions and proceeding towards more divergent regions that bear less similarity. In this way, PicXAA effectively avoids the error-propagation problem that many of the current

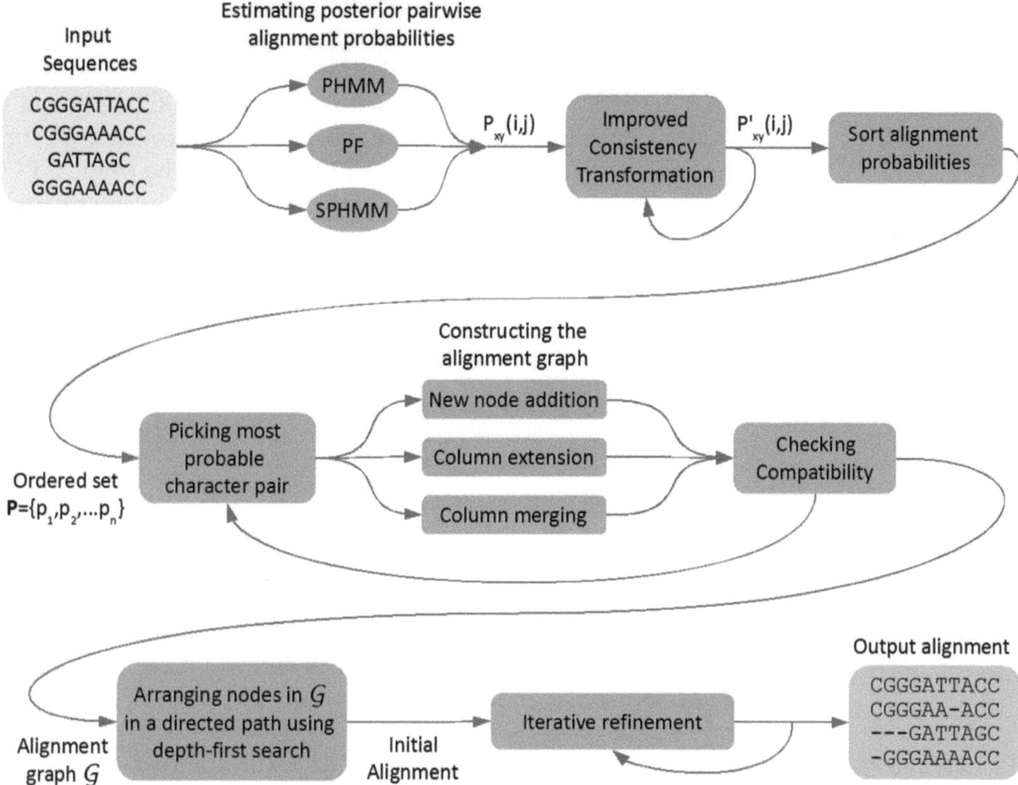

Fig. 1 Diagrammatic overview of the alignment steps in PicXAA

progressive alignment techniques suffer from. In the following, we give a brief summary of the alignment steps that are involved in PicXAA. Fig. 1 provides a diagrammatic overview of the alignment process of PicXAA.

2.1 Improved Probabilistic Consistency Transformation

Suppose we have estimated the posterior pairwise alignment probabilities for all residue pairs in every possible pair of sequences $\mathbf{x}, \mathbf{y}$ in a given sequence set $\mathbf{S}$. We denote this probability as $P(x_i \sim y_j \in a^*|\mathbf{x}, \mathbf{y})$, where $x_i \in \mathbf{x}$ is a residue in sequence $\mathbf{x}$, $y_j \in \mathbf{y}$ is a residue in sequence $\mathbf{y}$, and $x_i \sim y_j \in a^*$ means that the residues x_i and y_j are aligned in the true (unknown) alignment a^*. These probabilities can be computed using various approaches, such as pair-HMMs (hidden Markov model) [10], partition function based methods (*see* **Note 8** for parameters used in this scheme) [21], and structural pair-HMMs [15] (*see* **Notes 1–3**, **6** and 7 for more details on these methods). Given these pairwise alignment probabilities $P(x_i \sim y_j \in a^*|\mathbf{x}, \mathbf{y})$, PicXAA updates them using an improved probabilistic consistency transformation. The probabilistic consistency transformation (PCT) attempts to enhance the reliability of the estimated pairwise residue alignment probabilities by incorporating the information from other sequences in the set **S**. The idea of PCT has been originally proposed in [10] and it has been widely adopted afterwards by many alignment algorithms. It has been shown that such transformation can ultimately lead to a more consistent and accurate MSA. The improved PCT first proposed and adopted in PicXAA [20] improves the original PCT by considering the relative significance of each intermediate sequence $\mathbf{z} \in \mathbf{S} - \{\mathbf{x}, \mathbf{y}\}$ while transforming the pairwise alignment probabilities, originally estimated using pair-HMMs, partition function method, or structural pair-HMMs (*see* ref. 20 for details). The improved transformation is defined as:

$$P'\left(x_i \sim y_j \in a^*|\mathbf{S}\right) \approx \frac{\sum_{\mathbf{z}\in\mathbf{S}} P(x_i \sim z_k \in a^*|\mathbf{x},\mathbf{z})P\left(z_k \sim y_j \in a^*|\mathbf{z},\mathbf{y}\right)P(\mathbf{x}\diamondsuit\mathbf{z})P(\mathbf{z}\diamondsuit\mathbf{y})}{\sum_{\mathbf{z}\in\mathbf{S}} P(\mathbf{x}\diamondsuit\mathbf{z})P(\mathbf{z}\diamondsuit\mathbf{y})}$$

where $P(\mathbf{x}\diamondsuit\mathbf{z})$ is the probability that sequences **x** and **z** are homologous to each other. This probability $P(\mathbf{x}\diamondsuit\mathbf{z})$ is estimated by computing the average residue alignment probability in the optimal pairwise alignment between **x** and **z**. This transformation can be applied for more than one round of iterations (*see* **Note 8**).

2.2 Construction of the Alignment Graph

To find the alignment that maximizes the number of correctly aligned residues and effectively captures the local similarities between the given sequences, PicXAA constructs the MSA by adding one aligned residue pair at a time, starting from the most confidently alignable regions (i.e., residue pairs with high alignment probabilities) and progressing towards less confident regions (i.e., residue pairs with relatively low alignment probabilities) (*see* **Note 5**).

During this process, PicXAA preserves the internal consistency of the alignment by avoiding any conflicts between the current alignment and the potential residue pair to be added to the alignment. In order to verify this compatibility in an efficient manner, PicXAA adopts a graph-based strategy for building up the alignment. In this approach, the MSA is represented as a directed acyclic graph $\mathcal{G}$, where the nodes in $\mathcal{G}$ correspond to the columns in the alignment and the directed edges between nodes reflect the relative order of the corresponding columns in the final sequence alignment. To construct the alignment graph $\mathcal{G}$, PicXAA first sorts all possible residue pairs for all pairs of sequences in **S** according to the consistency transformed posterior alignment probabilities, in a descending order, to get an ordered set **P**. Starting from the most probable residue pair, we successively add residue pairs in **P** to the alignment graph $\mathcal{G}$ one pair at a time, provided that the pair being added to the alignment is compatible with the current alignment graph. This compatibility can be easily verified by finding out whether the graph remains acyclic after adding the new residue pair. To further improve the overall speed of this graph-based alignment process, the alignment graph is pruned after each update by removing any redundant edges, which makes the compatibility verification step more efficient.

2.3 Mapping the Graph to a Multiple Sequence Alignment

Except for possibly a few nodes that may not have any priority to each other in $\mathcal{G}$, there is a one-to-one relationship between the final alignment graph $\mathcal{G}$ and the multiple sequence alignment. To find the final MSA, we only have to arrange the columns (represented as nodes in $\mathcal{G}$) such that the relative order of the corresponding nodes in this linear arrangement does not conflict with that in the final alignment graph $\mathcal{G}$. This can be easily achieved by using a depth-first-search algorithm to arrange the nodes in a linear directed path **P**, according to their topological ordering.

2.4 Improving the Alignment Quality in Low Confidence Regions

The alignment quality of the regions that mainly consist of residue pairs with low alignment probabilities can be further improved by performing selective profile-profile alignments. Rather than taking a random split and realignment strategy as in [21], which may break the confidently aligned residue pairs that have high alignment probabilities, PicXAA adopts an iterative refinement technique, which first aligns each sequence with a set of highly similar sequences in **S**, and then aligns the resulting sequence profile with the profile that consists of the remaining sequences (*see* **Note 8**). In this way, PicXAA takes advantage of both the intra-family similarity as well as the inter-family similarity, thereby improving the overall quality of the MSA in low similarity regions without disrupting the residue alignments in high confidence regions (*see* **Note 4**).

2.5 Other Relevant Versions of PicXAA

A similar approach can be also used for the structural alignment of noncoding RNAs (ncRNAs). Recently, PicXAA-R [23] has extended the basic idea of PicXAA by additionally incorporating RNA folding information to predict accurate multiple RNA sequence alignments. There is also a Web-based platform called PicXAA-Web [24], which is designed to integrate PicXAA and PicXAA-R in a user-friendly Web environment for accurate alignment and analysis of multiple protein, DNA, and RNA sequences. PicXAA-Web can be freely accessed at: http://gsp.tamu.edu/picxaa

3 Notes

1. Generally, PicXAA can be used with any estimation scheme for computing the pairwise residue alignment probabilities. Currently, PicXAA allows the user to choose from three different methods for computing the alignment probabilities: (a) the pair-HMM approach implemented in ref. 10, (b) the structural pair-HMM approach used in ref. 15, and (c) the partition

function-based method adopted in ref. 21. These schemes are respectively called PicXAA-PHMM, PicXAA-SPHMM, and PicXAA-PF. Detailed description of each posterior probability computation scheme can be found in ref. 20.

2. PicXAA-PF and PicXAA-PHMM have comparable computational cost, which is considerably lower than that of PicXAA-SPHMM. The increased computational cost of PicXAA-SPHMM mainly arises from its computationally intensive probability estimation step that uses a complicated structural pair-HMM.
3. PicXAA-PF and PicXAA-PHMM can be used for aligning both protein sequences as well as nucleotide sequences, while PicXAA-SPHMM can be only used for multiple protein sequence alignment.
4. Although the main focus of PicXAA lies in effectively capturing the local similarities across sequences while predicting the global alignment of multiple sequences, it consistently yields accurate alignment results for various reference sets with diverse characteristics. In fact, PicXAA can accurately predict the alignment of sequences that belong to closely related sequence families (thus bearing strong global similarities) as well as those that belong to distant families (thus sharing only local similarities).
5. For distantly related sequences that share local similarities that are limited to relatively short subsequences, PicXAA has a clear advantage over other progressive alignment techniques in terms of alignment accuracy. This is a direct effect of the probabilistic greedy alignment approach adopted by PicXAA, which first builds up the MSA from sequence regions that can be aligned with high confidence.
6. Typically, PicXAA-PF outperforms PicXAA-PHMM on many datasets, while PicXAA-PHMM yields better alignment results for locally similar sequences.
7. Incorporating structural similarities can be advantageous for aligning protein sequences that share many structural similarities in addition to sequence similarities. PicXAA-SPHMM uses the SPHMM implemented in [15] to estimate the pairwise residue alignment probabilities by incorporating such structural information. As a result, PicXAA-SPHMM often yields improved alignment results for structurally similar proteins, but at the price of increased computational overhead.
8. Parameters used in PicXAA:
   (a) *Number of iterations for the probabilistic consistency transformation (PCT)*: In general, increasing this parameter will improve the consistency of the predicted alignment while

reducing the specificity of the predicted result. The default value of this parameter is two.

(b) *Number of iterations for the refinement step*: This is the number of times the refinement steps are applied to the sequence set. Experiments show that, typically, 100 iterations are sufficient to obtain an accurate and consistent multiple sequence alignment.

(c) *Scoring matrix for PicXAA-PF*: This parameter specifies the scoring matrix that will be used for computing the posterior pairwise residue alignment probabilities in the PicXAA-PF scheme. The default matrix will be the Gonnet 160 scoring matrix [22] for protein sequences, and the identity nucleotide scoring matrix for DNA sequences.

(d) *Gap open and gap extension penalties for PicXAA-PF*: These parameters control the affine gap penalties that will be used to compute the posterior pairwise residue alignment probabilities in the PicXAA-PF scheme. In general, higher gap penalty results in higher alignment probability for mismatching (i.e., nonidentical) residues. For protein sequences, the default gap open and gap extension penalties are −22 and −1, respectively (for the Gonnet 160 scoring matrix). For nucleotide sequences, the default gap open and gap extension penalties are −4 and −0.25, respectively.

Acknowledgment

This work was supported in part by the National Science Foundation through NSF Award CCF-1149544.

References

1. Phillips A, Janies D, Wheeler W (2000) Multiple sequence alignment in phylogenetic analysis. Mol Phylogenet Evol 16:317–330
2. Wong KM, Suchard MA, Huelsenbeck JP (2008) Alignment uncertainty and genomic analysis. Science 319:473–476
3. Cuff JA, Barton GJ (2000) Application of multiple sequence alignment profiles to improve protein secondary structure prediction. Proteins 40:502–511
4. Kemena C, Notredame C (2009) Upcoming challenges for multiple sequence alignment methods in the high-throughput era. Bioinformatics 25:2455–2465
5. Edgar R, Batzoglou S (2006) Multiple sequence alignment. Curr Opin Struct Biol 16:368–373
6. Pei J (2008) Multiple protein sequence alignment. Curr Opin Struct Biol 18:382–386
7. Kumar S, Filipski A (2007) Multiple sequence alignment: in pursuit of homologous DNA positions. Genome Res 17:127–135
8. Thompson JD, Higgins DG, Gibson TJ (1994) CLUSTAL W: improving the sensitivity of progressive multiple sequence alignment through sequence weighting, position-specific gap penalties and weight matrix choice. Nucleic Acids Res 22:4673–4680
9. Notredame C, Higgins DG, Heringa J (2000) T-Coffee: a novel method for fast and accurate multiple sequence alignment. J Mol Biol 302:205–217
10. Do CB, Mahabhashyam MS, Brudno M, Batzoglou S (2005) ProbCons: probabilistic

consistency-based multiple sequence alignment. Genome Res 15:330–340
11. Edgar RC (2004) MUSCLE: multiple sequence alignment with high accuracy and high throughput. Nucleic Acids Res 32:1792–1797
12. Katoh K, Misawa K, Kuma K, Miyata T (2002) MAFFT: a novel method for rapid multiple sequence alignment based on fast Fourier transform. Nucleic Acids Res 30:3059–3066
13. Katoh K, Kuma K, Toh H, Miyata T (2005) MAFFT version 5: improvement in accuracy of multiple sequence alignment. Nucleic Acids Res 33:511–518
14. Katoh K, Toh H (2008) Recent developments in the MAFFT multiple sequence alignment program. Brief Bioinform 9:286–298
15. Pei J, Grishin NV (2006) MUMMALS: multiple sequence alignment improved by using hidden Markov models with local structural information. Nucleic Acids Res 34:4364–4374
16. Paten B, Herrero J, Beal K, Birney E (2009) Sequence progressive alignment, a framework for practical large-scale probabilistic consistency alignment. Bioinformatics 25:295–301
17. Subramanian AR, Kaufmann M, Morgenstern B (2008) DIALIGN-TX: greedy and progressive approaches for segment-based multiple sequence alignment. Algorithms Mol Biol 3:6
18. Schwartz AS, Pachter L (2007) Multiple alignment by sequence annealing. Bioinformatics 23:e24–e29
19. Bradley RK, Roberts A, Smoot M, Juvekar S, Do J, Dewey C, Holmes I, Pachter L (2009) Fast statistical alignment. PLoS Comput Biol 5:e1000392
20. Sahraeian SM, Yoon BJ (2010) PicXAA: greedy probabilistic construction of maximum expected accuracy alignment of multiple Sequences. Nucleic Acids Res 38:4917–4928
21. Roshan U, Livesay DR (2006) Probalign: multiple sequence alignment using partition function posterior probabilities. Bioinformatics 22:2715–2721
22. Gonnet GH, Cohen MA, Benner SA (1992) Exhaustive matching of the entire protein sequence database. Science 256:1443–1445
23. Sahraeian SM, Yoon BJ (2010) PicXAA-R: efficient structural alignment of multiple RNA sequences using a greedy approach. BMC Bioinformatics 11(Suppl 1):S38
24. Sahraeian SM, Yoon BJ (2011) PicXAA-Web: a web-based platform for non-progressive maximum expected accuracy alignment of multiple biological sequences. Nucleic Acids Res 39: W8–W12

Chapter 14

Multiple Protein Sequence Alignment with MSAProbs

Yongchao Liu and Bertil Schmidt

Abstract

Multiple sequence alignment (MSA) generally constitutes the foundation of many bioinformatics studies involving functional, structural, and evolutionary relationship analysis between sequences. As a result of the exponential computational complexity of the exact approach to producing optimal multiple alignments, the majority of state-of-the-art MSA algorithms are designed based on the progressive alignment heuristic. In this chapter, we outline MSAProbs, a parallelized MSA algorithm for protein sequences based on progressive alignment. To achieve high alignment accuracy, this algorithm employs a hybrid combination of a pair hidden Markov model and a partition function to calculate posterior probabilities. Furthermore, we provide some practical advice on the usage of the algorithm.

Key words Multiple sequence alignment, Progressive alignment, Hidden Markov models, Partition function, Consistency-based scheme

1 Introduction

Multiple sequence alignment (MSA) is fundamental to many bioinformatics analysis studies that involve analyzing functional, structural, and evolutionary relationships between sequences. The exact approach to producing optimal MSAs relies on exhaustive dynamic programming. However, this approach has an exponential computational complexity and thus prohibits its use for large-scale data analysis. Therefore, many heuristics have been proposed to accelerate the computation of MSAs, among which the progressive alignment heuristic [1] is most widely used. However, the MSAs produced by these heuristics are generally suboptimal and may not meet the requirements of biologists. To further improve alignment accuracy, many modern progressive alignment-based MSA algorithms have fused other techniques into progressive alignment, such as introducing iterative refinement or consistency-based schemes.

David J. Russell (ed.), *Multiple Sequence Alignment Methods*, Methods in Molecular Biology, vol. 1079,
DOI 10.1007/978-1-62703-646-7_14,

In this chapter, we outline MSAProbs [2], a progressive alignment-based multiple protein sequence alignment algorithm, and give some practical guidance on how to use this software. MSAProbs employs a hybrid combination of a pair hidden Markov model (pair-HMM) [3] and a partition function [4] to calculate pairwise posterior probabilities for sequence pairs. Furthermore, weighted probabilistic consistency transformation, weighted profile–profile alignment, and random-split iterative refinements are incorporated to progressive alignment to further improve alignment accuracy. In addition, this algorithm has been parallelized using multi-threading to leverage the compute power of multi-core CPUs. Evaluated using some popular benchmarks such as BAliBASE [5] and PREFAB [6], MSAProbs has been demonstrated to be one of the most accurate MSA algorithms in some recent studies [7–9]. While yielding high alignment accuracy, our algorithm has also demonstrated competitive execution speed compared to other top performing MSA algorithms.

2 Materials

2.1 Hardware

Standard personal computers and workstations based on multi-core CPUs.

2.2 Software

1. **Program name**: MSAProbs.
2. **Home page**: http://msaprobs.sourceforge.net.
3. **Operating system**: Linux or Windows.
4. **Programming language**: C++.
5. **Parallelization**: Multi-threaded using OpenMP.

3 Methods

3.1 Program Workflow

MSAProbs is basically designed based on the progressive alignment pipeline that typically consists of three stages, namely, pairwise sequence distance matrix computation, guide-tree construction, and profile–profile progressive alignment. However, in MSAProbs, some extensions to the basic pipeline have been introduced.

MSAProbs works in five major stages: (1) calculating all pairwise posterior probability matrices using both a pair-HMM and a partition function; (2) calculating a pairwise sequence distance matrix from pairwise posterior probability matrices; (3) constructing a guide tree from the pairwise sequence distance matrix; (4) performing a weighted probabilistic consistency transformation of all pairwise posterior probability matrices; and (5) computing a profile–profile progressive global alignment along the guide tree using the transformed posterior probability matrices. In addition, an optional iterative

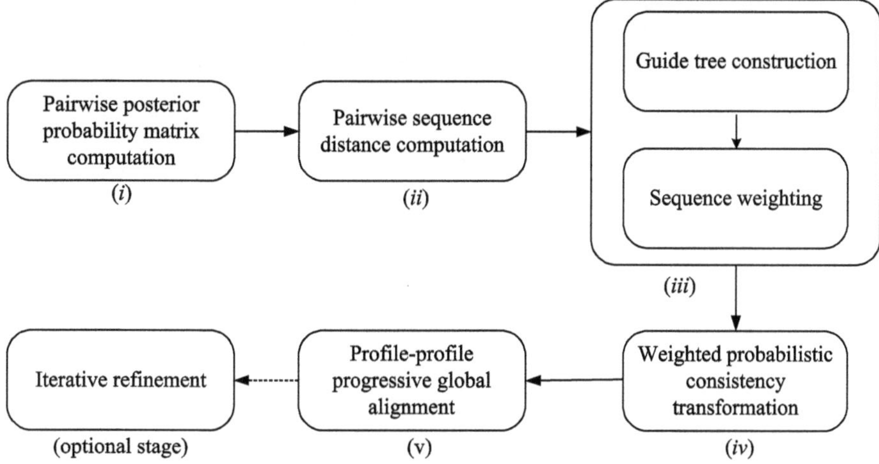

Fig. 1 Diagram of the program workflow

refinement can be performed as a post-processing step of stage (5) to further improve alignment accuracy. Figure 1 shows the diagram of the program workflow.

3.2 Pair Hidden Markov Model

Given two sequences X and Υ, define X_i to denote the ith residue in X and Υ_j to denote the jth residue in Υ. Assuming A to be the space of all possible global alignments of X and Υ and $a^* \in A$ be the "*true*" alignment of the two sequences, the posterior probability that X_i is aligned to Υ_j (denoted as $X_i \sim \Upsilon_j$) in a^* is defined as

$$P(X_i \sim \Upsilon_j \in a^*|X, \Upsilon) = \sum_{a \in A} P(a|X, \Upsilon)\delta\{X_i \sim \Upsilon_j \in a\}, \quad (1)$$

where $1 \leq i \leq |X|$ and $1 \leq j \leq |\Upsilon|$. The indicator function $\delta\{cond\}$ returns 1 if the condition *cond* is true and 0, otherwise. $P(a|X, \Upsilon)$ represents the probability that a is the *true* alignment a^*. Thus, $P(X_i \sim \Upsilon_j \in a^*|X,\Upsilon)$, i.e., $P(X_i \sim \Upsilon_j)$ for short, can be considered as the probability that X_i is aligned to Υ_j in the true alignment a^*. The posterior probability matrix $P_{X\Upsilon}$ of X and Υ is a two-dimensional table of size $|X| \times |\Upsilon|$, consisting of all values $P(X_i \sim \Upsilon_j)$ for $1 \leq i \leq |X|$ and $1 \leq j \leq |\Upsilon|$.

Figure 2 shows the used pair-HMM model to specify the probability distribution over all alignments A of a sequence pair. This pair-HMM model has three states: M, I, and D. At state M, one residue is emitted for each of the sequences X and Υ, meaning that the two residues are aligned together. At state I, it only emits one residue for sequence X, meaning that this residue from X is aligned to a gap. Similarly, state D only emits one residue for sequence Υ, meaning that this residue from Υ is aligned to a gap. To compute the posterior probabilities, we used both the forward and backward algorithms as described in [3].

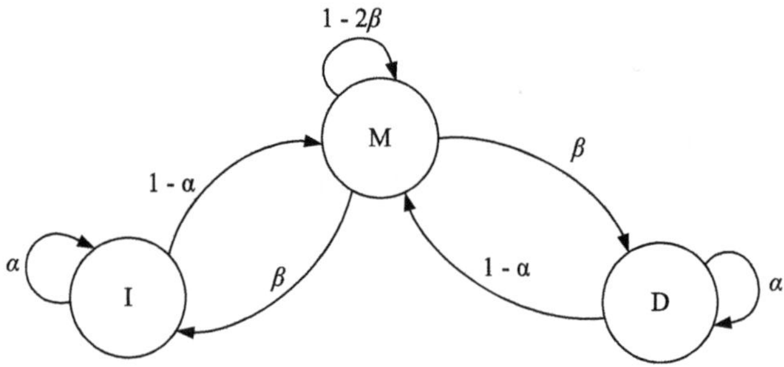

Fig. 2 Basic pair-HMM model for sequence alignment of two sequences

3.3 Partition Function

The partition function of alignments calculates the pairwise posterior probabilities by generating suboptimal alignments using dynamic programming. For all global alignments of protein sequences X and Υ ending at position (i, j), define $Z(i, j)$ to denote the partition function, $Z_M(i, j)$ to denote the partition function with X_i aligned to Υ_j, $Z_E(i, j)$ to denote the partition function with Υ_j aligned to a gap, and $Z_F(i, j)$ to denote the partition function with X_i aligned to a gap. The partition function can then be defined recursively as

$$\begin{aligned} Z_M(i,j) &= Z(i-1,j-1)e^{\beta s(X_i,\Upsilon_j)} \\ Z_E(i,j) &= Z_M(i,j-1)e^{\beta\rho} + Z_E(i,j-1)e^{\beta\sigma} \\ Z_F(i,j) &= Z_M(i-1,j)e^{\beta\rho} + Z_F(i-1,j)e^{\beta\sigma} \\ Z(i,j) &= Z_M(i,j) + Z_E(i,j) + Z_F(i,j), \end{aligned} \tag{2}$$

where $s(X_i, \Upsilon_j)$ is the substitution score between residues X_i and Υ_j, ρ ($\rho \leq 0$) is the gap open penalty, σ ($\sigma \leq 0$) is the gap extension penalty, and β measures the deviation between suboptimal and optimal alignments.

Once having the alignment partition function matrices constructed, the posterior probability $P(X_i \sim \Upsilon_j)$ can be computed as

$$P(X_i \sim \Upsilon_j) = \frac{Z_M(i-1,j-1)Z'_M(i+1,j+1)}{Z}e^{\beta s(X_i,\Upsilon_j)}, \tag{3}$$

where $Z'_M(i, j)$ represents the partition function of all the reverse alignments starting from position $(|X|, |\Upsilon|)$ and ending at (i, j) with X_i aligned to Υ_j, for $1 \leq i \leq |X|$ and $1 \leq j \leq |\Upsilon|$. As mentioned above, a pair-HMM and a partition function are combined together to compute the pairwise posterior probability matrix. After computing the probability matrix $P^a{}_{X\Upsilon}$ using the pair-HMM and matrix $P^b{}_{X\Upsilon}$ using the partition function, the final probability matrix $P_{X\Upsilon}$ is calculated by combining these two matrices as the root mean square of the corresponding values in $P^a{}_{X\Upsilon}$ and $P^b{}_{X\Upsilon}$:

$$P_{X\Upsilon}(X_i \sim \Upsilon_j) = \sqrt{\frac{P^a_{X\Upsilon}(X_i \sim \Upsilon_j)^2 + P^b_{X\Upsilon}(X_i \sim \Upsilon_j)^2}{2}}. \tag{4}$$

3.4 Consistency-Based Schemes

Several conventional consistency-based schemes have been proposed in the literature [10–13]. These schemes are primarily designed to overcome the drawback of progressive alignment through evaluating the accuracy consistency between the resulting MSA and pairwise alignments. ProbCons [14] introduced a new probabilistic consistency model, which is used to reestimate more accurate pairwise posterior probabilities by a three-way alignment criterion. Different from the conventional schemes, this probabilistic consistency model does not iteratively evaluate the accuracy consistency between the resulting MSA and pairwise alignments. Since MSAProbs relies on pairwise posterior probabilities to compute MSA, it employs a similar probabilistic consistency model to ProbCons with the difference that sequence weights are introduced to the model to avoid a biased sampling of sequences. The transformations are further performed for a fixed number of iterations to refine the posterior probabilities. By default, our program uses two iterations in order to offer a good trade-off between alignment accuracy and speed.

3.5 Program Usage

Before showing how to align sequences using MSAProbs, we list the major command line parameters accepted by the software.

- **-num_threads** ***n***: Specifies the number of threads to be used for parallel execution. Otherwise, the program will automatically detect the number of available CPU cores in the computer and then use the same number of threads with the number of CPU cores.
- *-c* ***n***: Specifies the number of iterations performed for weighted probabilistic consistency transformation (default = 2).
- **-ir** ***n:*** Specifies the number of iterations performed for random-split iterative refinements (default = 10).
- **-o** ***file***: Specifies the alignment output file name *file*. By default, the resulting alignments will be output to STDOUT.
- **-clustalw**: Indicates that the resulting alignments will be displayed in CLUSTAL format rather than the default FASTA format.
- **-a**: Indicates that the aligned sequences will be displayed in alignment order rather than the input order. Otherwise, they will be displayed in the same order as in the input file.

Unlike other MSA algorithms requiring users to tune a number of parameters in order to yield higher alignment accuracy, MSAProbs attempts to unload this tedious burden from users by fully automatic execution while achieving high alignment accuracy. However, we still provide two parameters to allow users to have some control over the procedures relating to alignment accuracy.

```
>1aab_
---GKGDPKKPRGKMSSYAFFVQTSREEHKKKHPDASVNFSEFSKKCSERWKTMSAKEKG
KFEDMAKADKARYEREMKTY----------IPPKGE
>1j46_A
---M---QDRVKRPMNAFIVWSRDQRRKMALENPRM--RNSEISKQLGYQWKMLTEAEKW
PFFQEAQKLQAMHREKYPNYKYRP---RRKAKMLPK
>1k99_A
MKKLKKHPDFPKKPLTPYFRFFMEKRAKYAKLHPEM--SNLDLTKILSKKYKELPEKKKM
KYIQDFQREKQEFERNLARFREDH---PDLIQNAKK
>2lef_A
---M-----HIKKPLNAFMLYMKEMRANVVAESTLK--ESAAINQILGRRWHALSREEQA
KYYELARKERQLHMQLYPGWSARDNYGKKKKRKREK
```

Fig. 3 Example MSA produced by MSAProbs in multi-FASTA format

```
1aab_      ---GKGDPKKPRGKMSSYAFFVQTSREEHKKKHPDASVNFSEFSKKCSERWKTMSAKEKG
1j46_A     ---M---QDRVKRPMNAFIVWSRDQRRKMALENPRM--RNSEISKQLGYQWKMLTEAEKW
1k99_A     MKKLKKHPDFPKKPLTPYFRFFMEKRAKYAKLHPEM--SNLDLTKILSKKYKELPEKKKM
2lef_A     ---M-----HIKKPLNAFMLYMKEMRANVVAESTLK--ESAAINQILGRRWHALSREEQA
               * . .  :  :..:  :    * :     .        :.:  . ::: :.  ::

1aab_      KFEDMAKADKARYEREMKTY----------IPPKGE
1j46_A     PFFQEAQKLQAMHREKYPNYKYRP---RRKAKMLPK
1k99_A     KYIQDFQREKQEFERNLARFREDH---PDLIQNAKK
2lef_A     KYYELARKERQLHMQLYPGWSARDNYGKKKKRKREK
            : :  :  :  . .    :               :
```

Fig. 4 Example MSA produced by MSAProbs in CLUSTAL format

MSAProbs takes as input protein sequences in multi-FASTA format. The resulting alignments can be displayed in either multi-FASTA format or CLUSTAL format. Typical usages of the program (on Linux) can be as follows:

- **msaprobs infile >outfile**
- **msaprobs infile -o outfile**
- **msaprobs -num_threads 4 infile1 infile2**

In the following, we show how to use MSAProbs to construct MSA by taking the alignment of four sequences from BAliBASE 3.0 [5] as an example. The default settings are used to produce the multiple alignments. Figures 3 and 4 illustrate the resulting alignments in multi-FASTA format and CLUSTAL format, respectively, where the used command lines are "*msaprobs infile*" and "*msaprobs infile –clustalw*," respectively.

3.6 Parallel Scalability

As mentioned above, we have parallelized MSAProbs using multithreading to accelerate the construction of MSA on multi-core CPUs. In the following, we have used a protein sequence dataset,

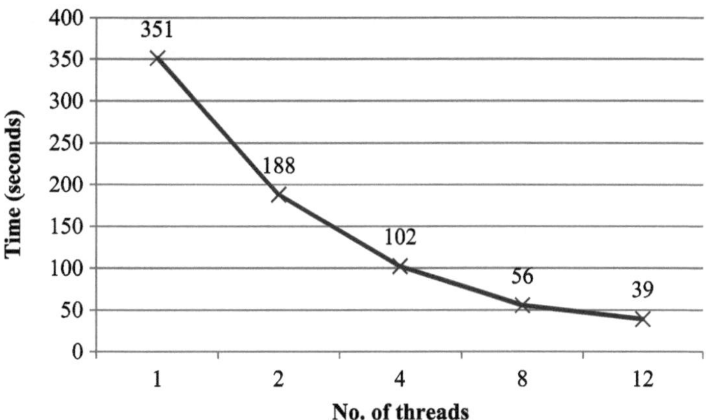

Fig. 5 Runtimes in terms of different number of threads

which comprises 100 sequences and has an average length of 408, to evaluate the parallel scalability of MSAProbs. All tests in this evaluation are conducted in a workstation with two Intel six-core 2.67 GHz CPUs and 96 GB memory, running the Linux operating system (Ubuntu 12.10). Figure 5 shows the runtimes (measured in wall clock time) in terms of different number of threads. MSAProbs achieves a speedup of about 1.9 using 2 threads, about 3.4 using 4 threads, about 6.3 using 8 threads, and about 8.9 using 12 threads.

4 Practical Issues

1. In terms of speed, the most time-consuming parts are the pairwise posterior probability matrix computation and the weighted probabilistic consistency transformation. The time complexity for the former is $O(N^2L^2)$ and $O(N^2L^3)$ for the latter, where N is the number of sequences and L is the average sequence length. Even though we have proposed some optimizations to improve the speed, the inherent high computational complexity still results in slow speed. In this case, we recommend the use of multiple threads on multi-core CPUs to accelerate the execution, considering the relatively good parallel scalability of our program.
2. In terms of memory overhead, the memory space complexity of our program is $O(|X||Y|)$ for sequence pair X and Y. In addition, we must store the pairwise posterior probabilities for each sequence pair in the dataset. Hence, for a dataset of N sequences and with average sequence length L, the memory space complexity can be calculated as $O(N^2L^2)$. Fortunately, the posterior probability matrices tend to be sparse with most entries near zero. Hence, we can significantly reduce the memory overhead by storing all pairwise posterior probability matrices in sparse matrix format. However, even after using

sparse matrices, the memory footprint is still large for large-scale datasets. Hence, we recommend the use of a computer with a large amount of shared memory to compute MSA of large-scale data. In the future, we plan to design an out-of-core pairwise probability matrix computation by storing all matrices on disk.

3. In our program, the profile–profile progressive alignment stage has not yet been parallelized. Hence, for large-scale data, this stage might become a parallel scalability bottleneck when using multiple threads. In MSA-CUDA [15], a dynamic scheduling parallelization has been proposed to parallelize the profile–profile progressive alignment stage of ClustalW [16] on graphics processing units. This method is also suitable for parallelization based on multi-threading on CPUs. Hence, our future work also includes the parallelization of this stage.

References

1. Feng DF, Doolittle RF (1987) Progressive sequence alignment as a prerequisite to correct phylogenetic trees. J Mol Evol 25:351–361
2. Liu Y, Schmidt B, Maskell DL (2010) MSAProbs: multiple sequence alignment based on pair hidden Markov models and partition function posterior probabilities. Bioinformatics 26:1958–1964
3. Durbin R, Eddy SR, Krogh A, Mitchison G (1998) Biological sequence analysis: probabilistic models of proteins and nucleic acids. Cambridge University Press, Cambridge
4. Miyazawa S (1995) A reliable sequence alignment method based on probabilities of residue correspondences. Protein Eng 8:999–1009
5. Thompson JD, Koehl P, Ripp R, Poch O (2005) BAliBASE 3.0: latest developments of the multiple sequence alignment benchmark. Proteins 61:127–136
6. Edgar RC (2004) MUSCLE: multiple sequence alignment with high accuracy and high throughput. Nucleic Acids Res 32:1792–1797
7. Sievers F, Wilm A, Dineen D et al (2011) Fast, scalable generation of high-quality protein multiple sequence alignments using Clustal Omega. Mol Syst Biol 7:539
8. Chang JM, Di Tommaso P, Taly JF et al (2012) Accurate multiple sequence alignment of transmembrane proteins with PSI-Coffee. BMC Bioinformatics 13:S1
9. Deng X, Cheng J (2011) MSACompro: protein multiple sequence alignment using predicted secondary structure, solvent accessibility, and residue–residue contacts. BMC Bioinformatics 12:472
10. Vingron M, Argos P (1989) A fast and sensitive multiple sequence alignment algorithm. Comput Appl Biosci 5:115–121
11. Gotoh O (1990) Consistency of optimal sequence alignments. Bull Math Biol 52: 509–525
12. Notredame C, Holm L, Higgins DG (1998) COFFEE: an objective function for multiple sequence alignments. Bioinformatics 14: 407–422
13. Notredame C, Higgins DG, Heringa J (2000) T-coffee: a novel method for fast and accurate multiple sequence alignment. J Mol Biol 302: 205–217
14. Do CB, Mahabhashyam MS, Brudno M et al (2005) ProbCons: probabilistic consistency-based multiple sequence alignment. Genome Res 15:330–340
15. Liu Y, Schmidt B, Maskell DL (2009) MSA-CUDA: multiple sequence alignment on graphics processing units with CUDA. 20th IEEE international conference on application-specific systems, architectures and processors, pp 121–128
16. Thompson JD, Higgins DG, Gibson TJ (1994) CLUSTAL W: improving the sensitivity of progressive multiple sequence alignment through sequence weighting, position-specific gap penalties and weight matrix choice. Nucleic Acids Res 22:4673–4680

Chapter 15

Large-Scale Multiple Sequence Alignment and Tree Estimation Using SATé

Kevin Liu and Tandy Warnow

Abstract

SATé is a method for estimating multiple sequence alignments and trees that has been shown to produce highly accurate results for datasets with large numbers of sequences. Running SATé using its default settings is very simple, but improved accuracy can be obtained by modifying its algorithmic parameters. We provide a detailed introduction to the algorithmic approach used by SATé, and instructions for running a SATé analysis using the GUI under default settings. We also provide a discussion of how to modify these settings to obtain improved results, and how to use SATé in a phylogenetic analysis pipeline.

Key words Multiple sequence alignment, Maximum likelihood, Phylogenetics, SATé, Species tree estimation, Gene tree estimation, Phylogenomics

1 Introduction

A typical phylogenetic study estimates a multiple sequence alignment (MSA) from biomolecular sequence data, and then infers a phylogeny using the MSA [1]. While much has been established about the relative performance of phylogeny estimation methods and the importance of picking a highly accurate estimation method, only in recent years has there been substantial study of the impact of the alignment method on the final phylogenetic estimation. It is now understood that the accuracy of the inferred phylogeny depends on the accuracy of the multiple sequence alignments estimated in the preceding phase [2–9], and that inaccurate multiple sequence alignments tend to produce inaccurate trees. While datasets with low enough rates of evolution can be aligned well using existing fast alignment methods (such as ClustalW [10], Muscle [11, 12], and MAFFT [13]), alignments of datasets that evolve more quickly are substantially harder to estimate, and standard methods typically produce poor alignments on these datasets [3, 4, 14]. Furthermore, many of the highly accurate alignment methods cannot be run on datasets with many sequences, due to

David J. Russell (ed.), *Multiple Sequence Alignment Methods*, Methods in Molecular Biology, vol. 1079,
DOI 10.1007/978-1-62703-646-7_15, © Springer Science+Business Media, LLC 2014

computational requirements (running time or memory), and these issues are all particularly challenging for very large datasets with upwards of 5,000 sequences [3, 14]. Thus, the standard two-phase approach to phylogeny estimation is limited by the alignment estimation step, at least for large datasets.

To improve the quality of both the inferred MSA and phylogeny in comparison to the traditional "two-phase" approach (first align, then estimate the tree), methods for simultaneous inference of alignments and phylogenies have been proposed [15–21]. Some of these methods are extensions of maximum parsimony to minimize the total "treelength," taking both number of substitutions and number of gap events into account; while POY [15] is the most popular of these methods, the accuracy of trees estimated using POY is substantially debated [16, 17]. Co-estimation methods based on statistical models of evolution that include gap events as well as substitutions have also been developed [18–21], of which BAli-Phy [18] is probably the most used and most scalable. Statistical co-estimation methods have the potential to be much more accurate than two-phase methods, because the standard treatment of gaps as missing data in phylogenetic analysis can be statistically inconsistent, even given the true alignment [22]. Statistical co-estimation methods have the potential to be much more accurate than two-phase methods, because the standard treatment of gaps as missing data in phylogenetic analysis can be statistically inconsistent, even given the true alignment [74]. However, not even BAli-Phy is able to run on datasets with more than about 200 sequences.

SATé (an acronym for "Simultaneous Alignment and Tree estimation") was developed [23] to address the need for highly accurate alignment estimation on datasets with more than a few hundred sequences. SATé uses an iterative technique in which each iteration computes an alignment (using a divide-and-conquer strategy that estimates alignments on subsets and then merges the subset alignments) and then computes a tree for that alignment using maximum likelihood heuristics. SATé produces highly accurate alignments and trees in no more than 24 h, even on datasets with 1,000 sequences. A modification to the divide-and-conquer strategy led to a substantially improved version, called SATé-II, that achieves even better accuracy in less time [24]. Finally, while both versions of SATé were studied using RAxML [25] for the maximum likelihood tree estimator, the use of FastTree [26] led to additional speed-ups and comparable alignment accuracy (unpublished data). We also extended SATé by adding other techniques for estimating alignments on subsets and/or merging subset alignments [27, 28]. This new version of SATé is available in the public distribution, and is able to analyze datasets that were too large for the original version. Most importantly, on large datasets (with 500 or more sequences), especially those that evolve quickly, SATé can provide much more accurate alignments and trees than other methods. SATé has been used to analyze protein as well as nucleotide datasets for many different types of organisms (birds, plants, bacteria, etc.). Many of these analyses have been on small datasets, with less than 100 sequences; however, SATé has also been used to analyze almost 28,000 rRNA sequences, spanning the domains of Archaea, Bacteria, and Eukaryota [24].

Although the first publication [23] of SATé has been cited over 100 times, the current implementation in the public distribution (available from the University of Kansas Web site at http://phylo.bio.ku.edu/software/sate/sate.html) is based on the second publication [24]. The focus of this chapter, therefore, is on the new implementation of SATé. We limit our discussion to the GUI usage, but readers interested in command-line usage can obtain additional information from the tutorial available online from the Kansas Web site (*see* **Note 1**), or from the SATé user group (*see* **Note 2**). SATé is under active development, with extensions to handling fragmentary data (as created by next generation sequencing technologies), improved analysis of protein sequences, etc., and users may wish to contact the UT-Austin SATé group for information about these plans, or to suggest new developments (*see* **Note 3**). Finally, phylogenetic estimation is a large and complex research discipline, and we direct the interested reader to [29] for a more in-depth discussion.

2 SATé Design Goals and Limitations

SATé was designed to enable fast and accurate estimation of alignments and trees for nucleotide datasets with hundreds to thousands of sequences [23, 24]. Its design, which is based on divide-and-conquer, improves accuracy on those datasets for which the best alignment methods cannot run due to computational requirements (either memory or time). Therefore, SATé is not designed to improve accuracy on those datasets that are small enough to be handled well by standard methods. In addition, although SATé is designed for large datasets, the largest dataset ever analyzed by SATé is the 16S.B.ALL dataset with 27,643 rRNA sequences with 6,857 sites [23], and we do not know how well it will scale for very large datasets with many tens of thousands of sequences.

Some datasets fall clearly outside of the design goals of SATé. SATé is also not designed for alignment estimation of datasets that are extremely long (hundreds of thousands of nucleotides) or that evolve with rearrangements rather than just indels (insertions and deletions) and substitutions; thus, whole genome alignment [30] is not part of SATé's capabilities. SATé has also not been designed for datasets with substantial missing data or fragmentary data from short read sequencing projects. Phylogeny estimation for highly fragmentary data can be obtained through methods based on "phylogenetic placement" [31–33]. Multiple sequence alignment estimation for highly fragmentary data can also be addressed through these phylogenetic placement methods, but has not been sufficiently studied in this context.

3 Algorithm

SATé uses iteration to produce improved alignments and trees, so that each iteration uses the results from the previous iteration to start its analysis, and then reestimates a multiple sequence alignment and tree for the dataset. Empirically, our studies have shown that the iterations quickly converge to good alignments and trees, with the biggest improvement occurring in the first iteration. Therefore, even a single iteration can result in an improved tree and alignment, and several iterations provide increased accuracy.

The studies in [3, 14] showed that the most accurate alignment methods (such as MAFFT, in its most accurate setting) could not be run on datasets with thousands of sequences, and that all methods have reduced accuracy for large enough rates of evolution and numbers of sequences. SATé overcomes this barrier using divide-and-conquer. It divides an input sequence dataset into subsets that are small enough that highly accurate alignment methods can be run on them, thus producing "subset alignments". SATé then merges the subset alignments together to produce an alignment of the full dataset, on which a tree can then be estimated using maximum likelihood methods. By repeating this process several times, the alignments and trees become increasingly accurate.

By design, SATé has several algorithmic parameters that determine how it runs. These have default values, but can also be reset by the user. Understanding the algorithmic parameters is helpful to obtaining improved accuracy for dataset analyses. Here we describe the algorithmic structure, and point out the algorithmic parameters that the user can set.

The input to SATé is a set of unaligned sequences. However, the user can also provide an initial alignment and/or tree, which can then be used by SATé to begin the iterative process. If none are provided, then SATé will estimate its own alignment and tree for the input sequences.

The main analysis then proceeds by repeating the following steps in an iterative fashion. The tree T from the previous iteration is used to guide the divide-and-conquer strategy for the current iteration. The tree itself is estimated using maximum likelihood (either using RAxML or FastTree) on the alignment on the sequences, and so has branch "lengths" (indicating substitution parameters for the Markov model of evolution). A branch e is selected in the tree T (either the "centroid" branch, which divides the tree into two subtrees with roughly equal numbers of taxa, or the "longest" branch). When this branch is removed from *T*, it divides the leaves of *T* into two subtrees. This decomposition is repeated until every subtree has a small enough number of leaves, as determined by the "maximum subproblem" size provided by the user (this is one of the algorithmic parameters). Once every subtree is small enough, the decomposition ceases, and each of these subtrees defines a "subproblem" of sequences (associated to the leaves of the subtree). The sequences in each subproblem are realigned using a multiple sequence alignment method selected by the user (the "aligner"), and the resulting subset alignments are then merged into an alignment on the full set of sequences. This merger step is handled by repeatedly applying an alignment "merger" method (also specified by the user) in the reverse order of the decomposition. Finally, a phylogenetic tree is estimated using either RAxML or FastTree.

Each iteration of SATé produces an alignment and tree, and thus each SATé analysis produces a sequence of alignment/tree pairs (one pair per iteration). Each alignment/tree pair has a maximum likelihood (ML) score as well, which can help the user to select a tree and alignment from the sequence of alignment/tree pairs. SATé terminates the iterative process based on a user-specified termination condition, which can be either elapsed wall-clock time, or a maximum number of iterations, or a lack of improvement in ML score. The final alignment/tree pair output by SATé is chosen from among the sequence of alignment/tree pairs generated during the course of analysis, and can be the pair with the best maximum likelihood score or the final pair produced by SATé.

4 Algorithmic Parameters and Software Settings

The SATé algorithm specifies several algorithmic parameters, and can be adapted to the needs of a particular dataset analysis by changing these parameters. However, it can also be run in default mode, so that the user does not need to set any parameters.

The software implementation of the SATé algorithm provides user-selectable settings for each of the algorithmic parameters. Table 1 describes the relationship between the algorithmic parameters and software settings; additional discussion of these parameters (and guidance on how to set these parameters for improved performance) is provided in the text.

After loading the input files in the SATé program, the software provides the option to automatically select all software settings based upon the properties of the input dataset. The following sections cover this usage scenario first, and we recommend the automatically selected settings unless more advanced analyses are required. Advanced usage scenarios involving changes to the automatically selected software settings are discussed later in this chapter.

Table 1
Relationship between algorithmic parameters and software settings

Algorithmic parameter	Software setting	Software setting choices	Description
Subproblem alignment method	"Aligner" dropbox	MAFFT, ClustalW, Prank [26], Opal [27]	This determines the method used to align the subsets of the sequences, and the default is MAFFT
Alignment merge method	"Merger" dropbox	Muscle, Opal	This determines how subset alignments are merged together. Muscle is the default, but Opal should be used if the dataset is small enough
ML-based phylogenetic estimation method	"Tree Estimator" dropbox	FastTree, RAxML	This determines how phylogenetic trees are estimated in each iteration; the default is FastTree, due to its improved speed relative to RAxML
Substitution model	"Model" dropbox	Many models, depending on type of data and ML tree estimation method	The model selected determines the parameters optimized by the ML tree estimation method. See [29]
Maximum subproblem size	"Max. Subproblem" dialog	Percentage (1–50 %), size (1–200)	This determines the maximum size subset given to the "aligner" method
Decomposition edge	"Decomposition" dropbox	Centroid, longest	This determines the edge used to decompose the dataset into subsets. The default setting is the centroid edge
Termination condition	"Apply Stop Rule" dropbox	After Last Improvement, After Launch	This determines when the stopping rule is evaluated. We recommend using "After Last Improvement" unless your dataset is very large
Termination condition	"Stopping Rule" dialog - "Blind Mode Enabled" checkbox	Checked/unchecked	This determines which tree (best ML or current tree) is used in the subsequent iteration
Termination condition	"Stopping Rule" dialog - "Time Limit (hr)" dialog/"Iteration Limit" dialog	0.01–72 h ("Time Limit (hr)" dialog) 1+ ("Iteration Limit" dialog)	This determines whether time or number of iterations is used to define when SATé stops
Final tree/alignment pair output by SATé	"Return" dropbox	Best, Final	This determines which tree and alignment pair (Best ML or last pair computed) is output

(continued)

Table 1 (continued)

Algorithmic parameter	Software setting	Software setting choices	Description
Parallelization	"CPU(s) Available"	1–16	This determines whether SATé will be run in parallel mode
Multi-gene analysis	"Multi-Locus Data" checkbox/"Sequence files" button	Checked/unchecked Folder dialog box	This enables a multi-gene analysis. See the "Advanced Analysis" section
Miscellaneous algorithmic modifications	"Extra RAxML Search" checkbox	Checked/unchecked	Checking this makes SATé perform a RAxML analysis of the final alignment
Miscellaneous algorithmic modifications	"Two-Phase (not SATe)" checkbox	Checked/unchecked	Check to run a two-phase analysis (first align and then compute an ML tree)

Choosing one of the settings in the "Quick Set" dropbox will automatically configure the software settings to perform one of the SATé-II analyses described in ref. 23. Subsequent modifications to software settings will cause the "Quick Set" dropbox to display the "(Custom)" choice

5 Additional Guidelines for Selecting Algorithmic Parameters

"Aligner" method. The choice of method to align the subsets has a large impact on the resultant alignment and tree. The default is MAFFT, due to its high accuracy on both simulated and biological data on both nucleotides and amino acid datasets [2, 3, 13, 14, 23, 24]. However, Prank has also been used in studies [24], and has the advantage over MAFFT and other standard alignment methods of not "over-aligning" as much. Because Prank is slower than MAFFT, the use of Prank to align subsets should be accompanied by a reduction in the maximum subset size so that the runs can complete. Finally, Opal and ClustalW are also enabled. Opal presents memory challenges on large datasets, and is not recommended unless the dataset is small enough. ClustalW is fast and can be used on any dataset size, but may not provide the same accuracy as MAFFT.

"Merger" method. Only Muscle and Opal are enabled for merging alignments. Muscle is the current default, because it has low memory requirements while Opal has high memory requirements. However, we strongly recommend Opal because it generally produces more accurate alignments. Therefore, we recommend using Opal unless you do not have sufficient memory for your dataset analysis. However, this is unlikely to be a problem except for very large datasets (with more than 10,000 sequences), if you have a reasonable amount of memory on your laptop or desktop machine.

"Tree Estimator" method. Only RAxML and FastTree are enabled for estimating trees from alignments, and FastTree is the default. Both are heuristics for maximum likelihood, which is a computationally hard problem. FastTree is much faster than RAxML, and generally produces trees of very similar accuracy [34]. Furthermore, in our unpublished studies, the use of FastTree instead of RAxML within SATé produces alignments of comparable accuracy and only a small decrease in accuracy for the trees. Because of its great speed advantage, however, we recommend the use of FastTree. If FastTree is used, a final RAxML run can be applied to the output alignment in order to obtain a RAxML tree (and thus potentially improved accuracy).

Substitution model. This refers to the statistical model [29] used by the maximum likelihood method (RAxML or FastTree) to estimate trees from alignments. The choice of statistical model depends on whether your data are nucleotide or amino-acid sequences, and also on whether you are using RAxML or FastTree as the tree estimator, since these enable somewhat different models. For nucleotide data, the default using RAxML is GTRCAT, while the default using FastTree is GTR + G20. GTR stands for the General Time Reversible (GTR) model, which is the most general substitution model available within SATé. G20 and CAT refer to how the model handles the Gamma rates-across-sites model; G20 is the GAMMA distribution approximated by 20 rate categories, while CAT [35] is a heuristic approximation to the GAMMA rate-variation model. Alternative settings for RAxML include GTRGAMMA (GTR + GAMMA) and GTRGAMMAI (GTR + Gamma + Invariable). Alternative settings for FastTree include JC (the Jukes-Cantor model) [36] instead of GTR, but this simplified model is not recommended except under very unusual circumstances where the data seem to fit the Jukes-Cantor model best (unlikely for most data). Note that the GAMMA setting is usually used in phylogenetic analyses, but the CAT setting improves speed at a potential loss of phylogenetic accuracy. For amino-acid datasets, the choice of substitution model is more complicated; see the section below on Amino-Acid Datasets for more information.

Maximum subproblem size. This is the maximum allowed size of the subsets of sequences, and so determines how many times the decomposition strategy is applied. The default depends on the dataset size (and will be set by SATé after you input your data). However, the main issue in setting the maximum subproblem size is the method used to align subsets. When MAFFT is the aligner method, then keeping the maximum subproblem size to at most 200 allows the most accurate version of MAFFT (L-INS-i) to be used to align the subsets, and this results in the best accuracy. If you wish to use Prank instead of MAFFT to align subsets, the maximum subproblem size should be reduced substantially, because Prank is computationally more expensive. Similarly, the use of Opal to align subsets will require a reduction in subproblem size because of Opal's memory requirements (and hence increased running time). Less is known about how to set the maximum subproblem size when ClustalW is used for aligning subsets, but the default settings are probably fine.

Decomposition edge. This algorithmic parameter determines how the dataset is decomposed—through the centroid edge (which produces a roughly equal decomposition into two datasets) or the longest branch. The default is the centroid edge, and this produces results of similar accuracy to the longest edge, while being much faster.

Stopping rule. There are various settings that determine the stopping condition and when it is evaluated. You can set the stopping rule to be defined by the number of iterations or time, and at least one of these must be specified (selecting both means that either can trigger the stopping rule). You can begin this stopping rule immediately ("After Launch") or only after the ML score stops improving ("After Last Improvement"). We recommend using "After Last Improvement" unless your dataset is so large that you need to limit the number of iterations. "Blind mode" means that the previous iteration's tree will always be used as the tree in the beginning of the next iteration, regardless of its ML score. Disabling "blind mode" means that the best-scoring tree so far will be used in the beginning of the next iteration, which can cause the iterative search to become stuck in local optima. We recommend enabling "blind mode".

Final tree/alignment. This determines which tree is the output for the SATé run. The "Best" setting returns the best-scoring alignment/tree pair encountered during the SATé analysis. The "Final" setting returns the final alignment/tree pair from SATé.

CPUs available. This is the number of CPUs in your machine that SATé should use, and using multiple CPUs can speed up the analysis. However, do not set this number to more CPUs than you have! *See* **Note 4**.

Extra RAxML search. Checking this box makes SATé perform a RAxML analysis of the final alignment. If time is not highly constrained (and your dataset is not too large), checking this box is recommended if you have used FastTree for the ML tree estimator. However, when you use RAxML for the ML tree estimator, it automatically computes a RAxML tree on the final alignment, and so it does not make any sense to check this box.

Two-phase (not SATé*).* Check this box to run a two-phase analysis (first align and then compute an ML tree) using the settings in the "External Tools" window, instead of running a SATé analysis. This may not produce alignments and trees as accurate as those produced by SATé, but should be faster.

6 Advanced Topics

Amino-acid datasets. The analysis of amino-acid datasets presents some additional challenges and opportunities. Compared to nucleotide sequences, the selection of the substitution model is more complicated, since the models are not "nested". The best model for your data needs to be selected using a statistical test [37, 38], however, JTT [39] and WAG [40] models are often used for amino acid datasets and are reasonable defaults. The models available for use for amino-acid analyses are displayed within SATé after you check the box indicating that your data are proteins, and depend upon the ML method you have selected (RAxML or FastTree). RAxML enables many more models than FastTree, and so may be preferable. The other amino-acid models available in SATé when used with RAxML are DAYHOFF [41], DCMUT [42], MTREV [43], RTREV [44], CPREV [45], VT [46], BLOSUM62 [47], MTMAM [48], and LG [49], *each* in combination with a rates-across-sites model. To set base frequencies for these amino-acid models to empirical base frequencies, add an "F" suffix to the name of the model; see the RAxML documentation for details (available from http://sco.h-its.org/exelixis/oldPage/RAxML-Manual.7.0.4.pdf). SATé has been used to analyze protein datasets [50, 51], but we have not studied SATé as a protein aligner nearly as thoroughly as we have studied it as a nucleotide sequence aligner; therefore, the default settings for the algorithmic parameters may not be optimized well. Finally, amino-acid alignment estimation in particular can be enhanced with structural (secondary or tertiary) information about the proteins, information that the aligner methods (MAFFT, ClustalW, Prank, and Opal) used by SATé do not use. Therefore, there is the potential for improved accuracy to be obtained through the use of a different set of protein alignment methods, including methods such as SATCHMO-JS [52] that employ Hidden Markov Models to take advantage of particular properties of protein alignments.

Large datasets. We now present guidelines for the analysis of datasets with 1,000 or more sequences. However, because SATé has not been tested on datasets with more than 28,000 sequences, our recommendation on very large datasets should be taken as our best guess, at this time, for how to handle such datasets. We strongly recommend the use of FastTree rather than RAxML for ML tree estimation in each iteration: FastTree is much faster than RAxML, and our preliminary studies (unpublished) suggest that using FastTree instead of RAxML produces the same quality alignments in a fraction of the time. However, switching to FastTree can reduce the tree accuracy slightly, and so the user may wish to use

RAxML on the final alignment returned by SATé. For very large datasets, the final RAxML analysis could take a long time, and so an alternative is to run SATé using FastTree and without any final RAxML run, save the resultant alignment and tree, and then run RAxML on the final alignment. We recommend using MAFFT for aligning subsets, using a maximum alignment subset size of 200, and the centroid edge decomposition. We recommend using Opal to merge subset alignments instead of Muscle, unless the dataset is so large (in number of sequences and/or sequence length) that the memory requirements for using Opal exceed what you have available on your machine. Opal should never be used as the subset alignment technique on extremely large datasets (its memory requirements will slow down the analysis dramatically). Prank is too slow to use on even moderately large datasets, and therefore Prank should not be used as the subset aligner. The use of ClustalW for the subset aligner will not cause running time issues, but there is little evidence that ClustalW is likely to produce more accurate alignments than MAFFT; therefore, it is not recommended as a subset aligner.

For very large datasets, providing an initial alignment (and possibly initial tree) to SATé can speed up and potentially improve the analysis. If you run SATé without providing it an initial alignment and/or tree, this initial alignment will be estimated using MAFFT, which is run in its less accurate setting (in extreme cases, this will be MAFFT-PartTree [53]) on very large datasets. However, faster and potentially more accurate estimations of initial alignments might be achievable using other methods, such as Clustal-Omega [54] for amino-acid sequences or MAFFT-profile [55] for nucleotide sequences. Once the initial alignment is provided, SATé will use FastTree to estimate the initial tree on the alignment. Because SATé is quite robust to its initial tree [23, 24], this means that the initial alignment need not be particularly accurate. The analysis of very large datasets presents both memory and running time challenges; *see* **Notes 5–7** for advice on how to handle problems that may arise.

Small datasets. Using SATé to estimate trees and alignments on very small datasets (with less than 200 sequences) may not result in improved accuracy, since these datasets can be analyzed well using methods such as MAFFT; however, datasets of this size have been analyzed using SATé (see, for example, [50, 51, 56–58]).The main recommendation we make for the analysis of small datasets is to use 50 % as the maximum subproblem size, rather than a smaller percentage, and to otherwise use the standard defaults. In addition, for small enough datasets, phylogeny estimation methods that are generally too computationally intensive to use on even moderately large datasets (such as MrBayes [59]) can be used to estimate a tree on the resultant SATé alignment.

Exploring the solution space. One of the appealing aspects of SATé is that it provide opportunities for exploration of the set of alignments and trees that are returned during the SATé run, which can allow you to explore how alignments impact the tree estimation, among other things. This is particularly useful on small datasets because each iteration can be done quickly, and so many iterations can be run on small datasets. To enable this exploration, we recommend setting the stopping rule to an iteration limit, and setting that limit to a large number (how large, of course, depends upon how much time you wish to devote). There are many methods for exploring sets of trees [60–64], each aimed at extracting different types of information. Similar analyses for exploring sets of alignments are not yet in standard use, but pairs of alignments are often compared to determine common homologies [65].

Multi-locus datasets. Often the objective is the estimation of a species tree from a set of different genes, each of which involves an alignment and tree estimation. You have several options for how to do a multi-locus analysis, depending on whether you are concerned about the potential for gene trees to be different from the species tree. That is, true gene trees can differ from the true species tree due to biological processes such as incomplete lineage sorting, gene duplication and loss, and horizontal gene transfer [66]. Therefore, the choice of how to estimate the species tree from a set of estimated gene trees can take some care. If you have concerns about potential conflict between gene trees, you can run SATé on each marker separately, thus producing independently estimated gene trees and alignments for each gene, and these estimated trees and alignments can then be used to estimate a species tree using techniques that are specifically designed to combine estimated gene trees into a species tree. See [67–74] and references therein for an introduction to methods that can estimate phylogenetic trees and networks in the presence of these processes that cause gene tree incongruence. If you are not concerned about potential gene tree conflict, we recommend using SATé in its default setting for multi-locus datasets. This analysis operates by concatenating the datasets together, and then uses the standard iterative divide-and-conquer strategy to produce alignments of each locus and a tree on the entire dataset.

General advice. We recommend that you back up your files (*see* **Note 8**) for all SATé analyses. This is generally a good practice, but especially for large dataset analyses or when you wish to explore the solution space, which can take a substantial amount of time to run. Some analyses may benefit from the use of archival systems (*see* **Note 9**), especially if your analyses involve very large datasets that you plan to explore in multiple ways.

7 Materials

The following sections pertain to Apple computers running recent versions of the Mac OS X operating system (including versions 10.4, 10.5, 10.6, and 10.7). For alternative operating systems and hardware, please consult the relevant software documentation.

7.1 Software

1. File format conversion software. SATé utilizes FASTA-formatted sequence files and Newick-formatted tree files. Many software packages and Web portals provide a format conversion capability. For example, the European Molecular Biology Open Software Suite (http://emboss.sourceforge.net/) [75] contains a module to convert sequence formats (see instructions for the seqret command at http://emboss.sourceforge.net/docs/themes/SequenceFormats.html). In addition, the savetrees command in PAUP* [76] will output a tree in Newick format when the "FORMAT = PHYLIP" option is specified.
2. SATé software. Available from http://www.cs.utexas.edu/~phylo/software/SATé/ and http://phylo.bio.ku.edu/software/SATé/SATé.html. We used version 2.2.5, although the methodology in this chapter is compatible with any recent version of the SATé software.

7.2 Hardware

1. We recommend using an Apple computer with a recent Intel processor and at least 1 GB of available memory. Large-scale analyses are primarily constrained by memory requirements (*see* **Note 6**). Clock speed and related CPU features primarily affect running time.
2. While the SATé installation requires less than 100 MB of disk space, much more disk space is required for a large-scale analysis, especially if the analysis runs for many iterations over a long period of time (*see* **Note 7**).

8 Methods

8.1 Install Software

1. Download the latest SATé software package from http://phylo.bio.ku.edu/software/SATé/SATé.html. *See* **Note 10** in the event of installation problems.
2. Open the downloaded package file and view its contents (Fig. 1).
3. Create a new folder for the new SATé installation in a separate location on the hard drive (*see* **Note 11**).
4. Drag and drop the package contents to the new folder.
5. In the new folder, double-click the SATé icon to start the SATé program.

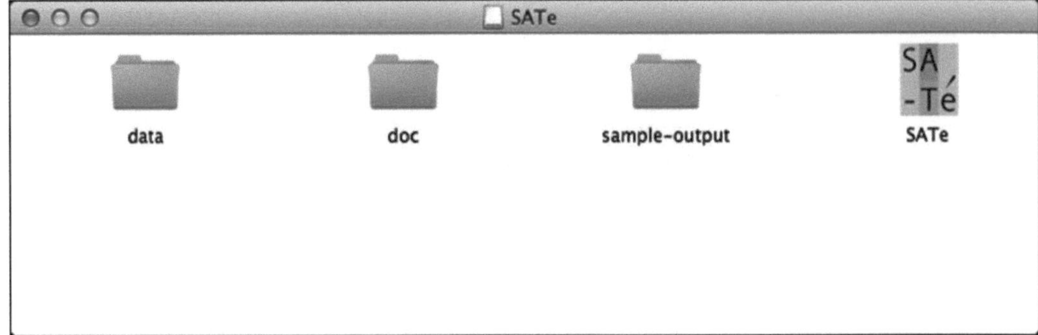

Fig. 1 Contents of the SATé software package. The SATé application is represented by the rightmost icon. The "doc" folder contains software documentation. The "data" folder contains example input files, and the corresponding output files are contained in the "sample-output" folder

8.2 Preparing Input File and Output Folder

1. If the input sequence file is not in FASTA format, use a third-party program to convert the input file to FASTA format. *See* **Note 12**.
2. We recommend that a new output folder be created for each SATé analysis.

8.3 Basic Analysis: Nucleotide Datasets

1. After starting the SATé program, the main analysis window appears (Fig. 2). The default settings correspond to the "SATé-II fast" analysis described in [24], which are appropriate for a wide range of phylogenetic studies. Make sure that the "SATé-II-fast" option is selected in the "Quick Set" drop-down menu.
2. Click the "Sequence file" button to load the input file. Locate and select the FASTA-formatted input file in the dialog box.
3. A dialog box appears with a query about automatic customization of some analysis settings based on the input file (Fig. 3). Click "OK" to enable the automatic customization. Customized settings will be reflected in the "SATé Settings" box of the application.
4. In the "Decomposition" drop-down menu, select "Centroid". If this changes the current defaults, the "Quick Set" menu will change to the "(Custom)" option.
5. In the "Job Name" field, provide a unique name for the analysis.
6. Click the "Output Dir." button. Locate and select the desired output folder in the dialog box. If output files for a job with the same name exists in the output folder, the output files of the current analysis will contain an additional integer to prevent file collisions.

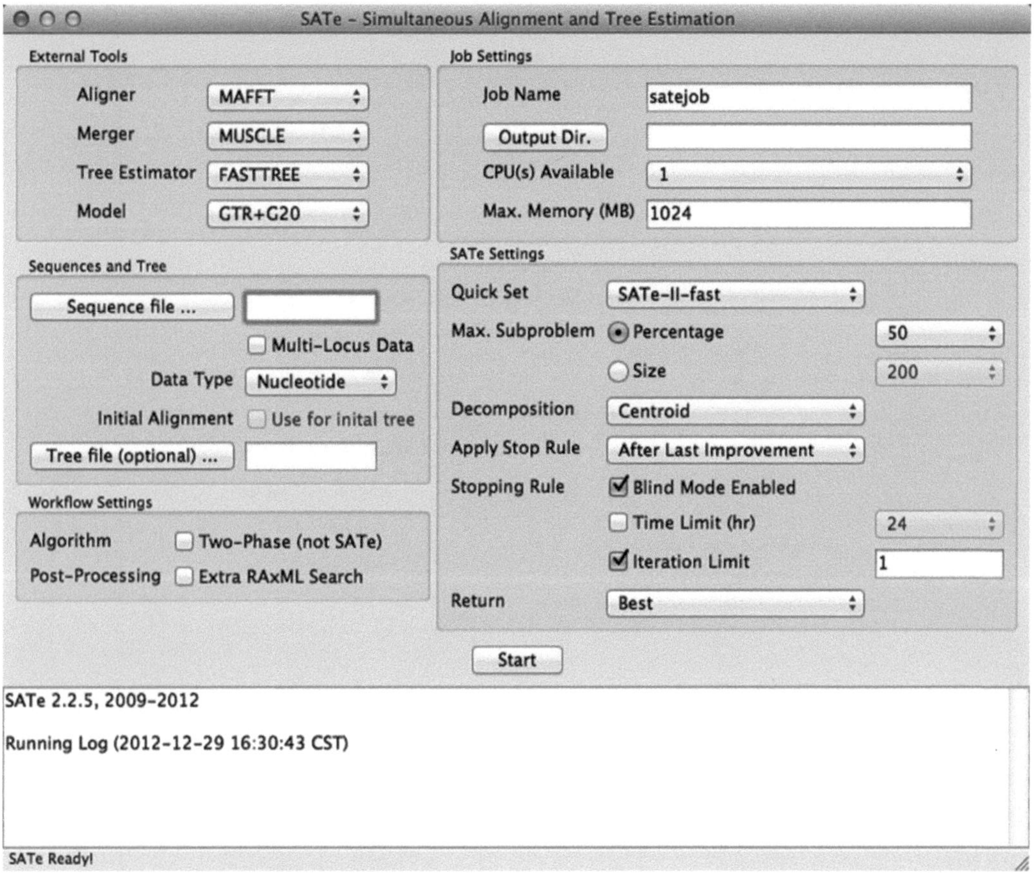

Fig. 2 The main SATé application window

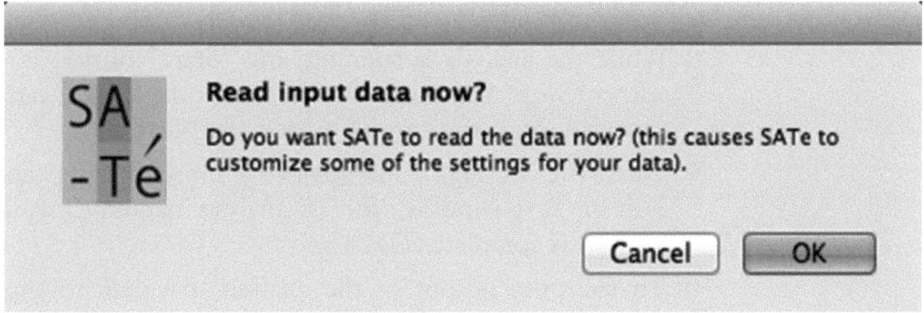

Fig. 3 After pressing the "Sequence file" button and selecting an input file to read, SATé responds with a prompt about automatic configuration. Selecting "OK" will enable automatic configuration of analysis settings based on the input file

7. Press the "Start" button to begin the SATé analysis. As the analysis proceeds, the bottom text window shows progress updates. The time duration required for an analysis depends on many factors, especially dataset size and complexity (*see* **Notes 6** and 7).

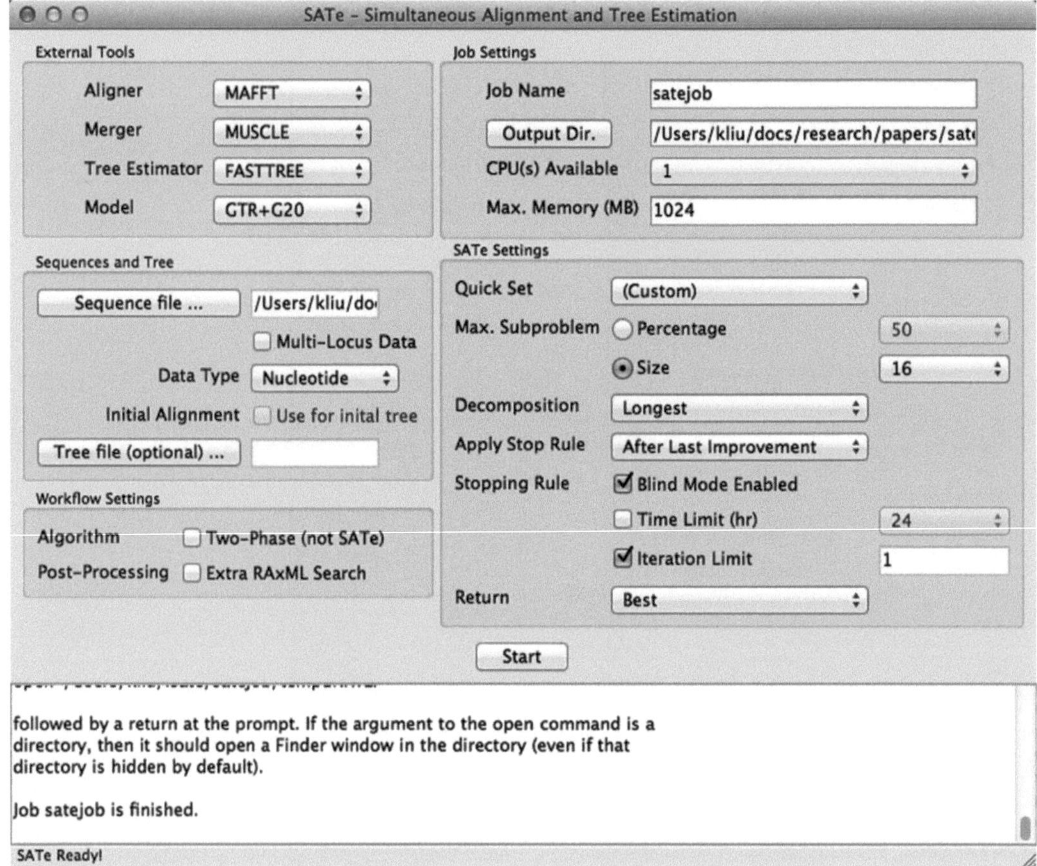

Fig. 4 The conclusion of a typical SATé analysis

8. While the analysis is running, the "Start" button is replaced with a "Stop" button. In the event that the analysis needs to be canceled, press the "Stop" button (*see* **Note 6**).
9. Once the message "Job myjob is finished." appears in the bottom text window (for an analysis named "myjob"), the analysis is complete (Fig. 4).
10. To view the output of the analysis, navigate to the output folder. The output files are described in Table 2.

8.4 Basic Analysis: Amino Acid Datasets

1. *See* **Note 3**. Follow the steps listed in the Subheading 8.3, but be sure to pick "Protein" for Data Type (in the Sequences and Tree dialog box). The automatic customization step will configure the settings appropriately (Fig. 5).

Table 2
Output files from a SATé analysis

Output file name	Description
myjob.marker001.sequence.aln	SATé alignment
myjob.tre	SATé tree
myjob.score.txt	ML score for the SATé alignment/tree pair
myjob.out.txt	Diagnostic messages
myjob.err.txt	Error messages. If this file is not empty, check your settings and retry the analysis
myjob_temp_iteration_initialsearch_seq_alignment.txt	Starting alignment
myjob_iteration_initialsearch_tree.tre	Starting tree
myjob_temp_iteration_0_seq_alignment.txt, myjob_temp_iteration_1_seq_alignment.txt, etc.	Intermediate alignments
myjob_iteration_0_tree.tre, myjob_iteration_1_tree.tre, etc.	Intermediate trees
myjob_temp_name_translation.txt	Taxa in intermediate trees and alignments are renamed according to this translation table. The temporary substitute name for a taxon is shown on one line, followed by its original name, and then a blank line

The analysis used a job name of "myjob" and the input file was named "sequence.fasta"

8.5 Advanced Analysis: More Iterations Desired

1. If additional time and computational resources are available, an extended, more thorough SATé analysis can be run. To do this, run the following steps.
2. In the "Quick Set" drop-down menu, select the "SATé-II-ML" option.
3. In the "Decomposition" drop-down menu, select the "Centroid" option.
4. Proceed with **steps 7** through **10** from the "Basic Analysis" sections 8.3 and 8.4, but be sure to pick a large enough number of iterations.

8.6 Advanced Analysis: Providing an Initial Alignment (and/or Initial Tree)

1. Providing a precomputed alignment and/or tree to SATé can save substantial time. To begin, follow **steps 1** through **6** in the "Basic Analysis" sections 8.3 and 8.4. During **step 2**, provide a FASTA-formatted file with aligned sequences. The "Initial Alignment" dialog will have the "Use for initial tree" checkbox enabled. If a user-specified starting tree is available, click the "Tree file (optional)" button and provide the Newick-formatted starting tree file name.
2. Proceed with **steps 7** through **10** from the "Basic Analysis" section.

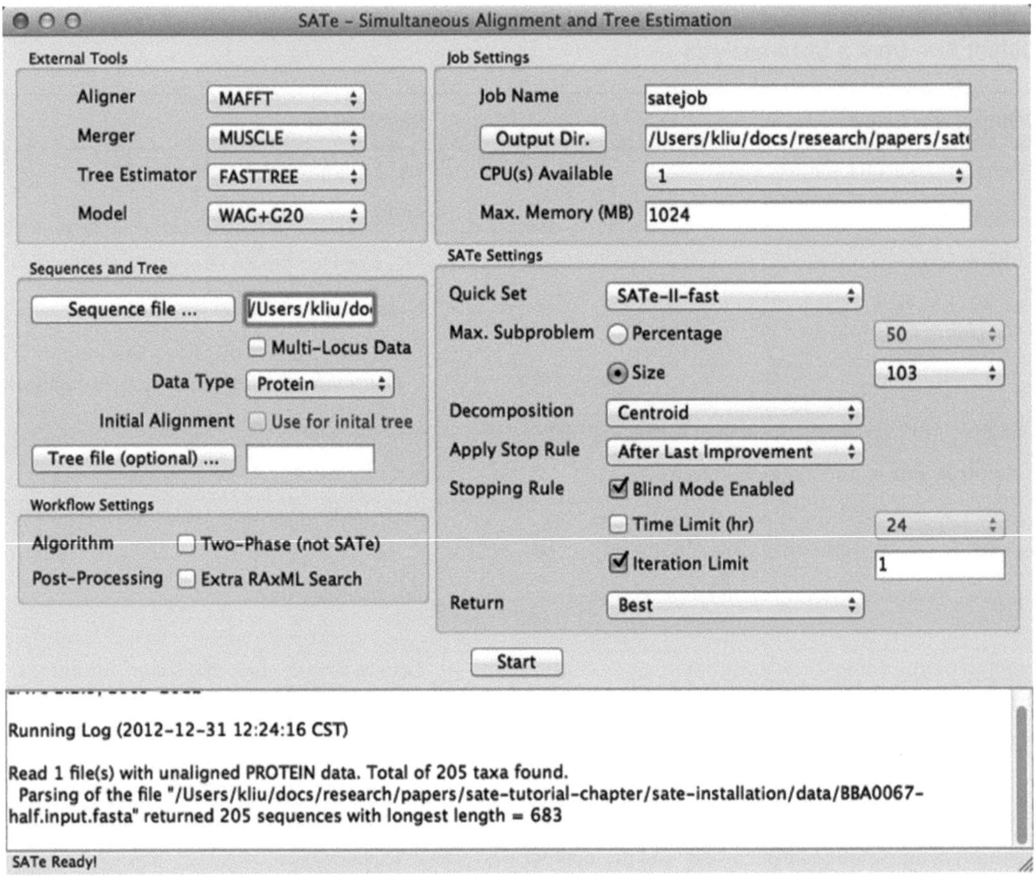

Fig. 5 The start of a SATé analysis of an amino acid dataset

8.7 Advanced Analysis: Very Large Datasets (More Than 10,000 Sequences)

1. Very large datasets with tens of thousands of sequences or more pose a special computational challenge. Changing software settings is recommended in this instance, although the optimal settings for a particular dataset depend upon many factors. Thus, while we provide specific suggestions for this case, experimenting with software settings is also advisable, with the caveats described in **Notes 5** through 7. See the discussion above (for "large dataset analyses") for some explanations for why we make the following recommendations.
2. If an alignment and tree are already available, we recommend providing them to SATé. This recommendation is strongly recommended for very large datasets (with 10,000 sequences or more), but beneficial for all analyses.
3. Otherwise, we recommend computing an initial alignment using either MAFFT's PartTree algorithm or Clustal Omega; these tools are not available within the GUI usage of SATé, and so this will need to be done offline. For an input file named

"sequence.fasta", the PartTree algorithm can be invoked using the following command: mafft -parttree -retree 2 -partsize 1000 sequence.fasta > startingAlignment.fasta. The command to run Clustal Omega is: clustalo –auto –dealign -i sequence.fasta > startingAlignment.fasta. Once you have the alignment, you can provide this to SATé as the initial alignment (see above).

4. In the "External Tools" window, choose the following software settings: "MAFFT" for the "Aligner" dropbox, "Muscle" for the "Merger" dropbox, and "FastTree" for the "Tree Estimator" dropbox. For nucleotide analyses, select "GTR + CAT" for the "Model" dropbox, and for protein analyses, select JTT + CAT.
5. In the "Sequences and Tree" window, provide your initial alignment (if available), and click on "initial alignment (use for initial tree)". Follow from **step 3** in Subheading 8.6.
6. In Workflow Settings, do not select "Extra RAxML Search", unless your dataset is not particularly big–the final RAxML search could be the most computationally intensive part of your analysis, and may not provide substantial benefits.
7. In the "Job Settings" window, make sure you provide the number of CPU(s) available (this will have a large impact on the running time, if more than 1 CPU can be used in the analysis). Also make sure that the "Max. Memory (MB)" dialog specifies the correct amount of available memory, since memory limitations are often a problem that cause running times to increase. *See* **Note 7**.
8. In the "SATé settings" window, you can use Quick Set to select " SATé-II-fast"; this will set all the settings appropriately. Alternatively, you can modify the settings as follows. Select the "Size" radio button in the "Max. Subproblem" field and a size of 200 in the dropdown menu. Set the decomposition to "centroid" (because using "Longest" will not only slow down the analysis, but also should only be run with Opal, and Opal should not be run with large datasets). Set the "Apply Stop Rule" to either "After Launch" (for very large datasets) or to "After Last Improvement". Do not select "Blind Mode Enabled" if your dataset is very large. It is also probably not a good idea to use a time limit for the stopping rule if your dataset is very large, since it is possible for a single iteration to not complete in the time you pick. Therefore, we recommend instead picking an iteration limit. The number of iterations you pick should depend on your dataset, but for very large datasets, it may be best to have a small number (say, 2) of iterations. If these complete quickly, you can always use the output alignment and tree to initialize another SATé run! We recommend setting "Return" to "Best".

8.8 Advanced Analysis: Multi-gene Datasets

1. Prepare your dataset by creating a new folder and saving the sequence data for each gene (or marker) in a separate FASTA-formatted file in the new folder. Each FASTA-formatted file name must end with the suffix .fasta or .fas. Make sure that the set of taxon names are identical across all of the FASTA files.
2. Begin by following **step 1** from the "Basic Analysis" section.
3. Click the "Multi-Locus Data" checkbox in the "Sequences and Tree" pane. Notice that the "Sequence file" dialog changes into the "Sequence files" dialog. Click the "Sequence files" button and choose the folder containing the input files.
4. Now run the analysis by following **steps 3** through **9** in the "Basic Analysis" section.
5. After the analysis finishes, the output files will be saved to the output directory. The file names and descriptions will match Table 2, with one exception. For an analysis with job name "myjob" and input files named "geneA.fasta", "geneB.fasta", "geneC.fasta", and so on, SATé saves the output alignments in files named myjob.marker001.geneA.aln, myjob.marker002.geneB.fasta.aln, myjob.marker003.geneC.fasta.aln, and so on.

9 Summary and Related Work

SATé is a method for large-scale alignment and tree estimation that has been shown to give very good results on both biological and simulated datasets of both nucleotide and amino-acid datasets. However, the reasons for its good performance are subtle: for example, it is not the case that allowing the alignment to change arbitrarily and seeking the alignment with the best maximum likelihood score (treating gaps as missing data) will lead to good trees [75]. Instead, the benefits to using SATé come because alignment methods with great accuracy but poor scalability can be used to estimate alignments on small subsets of the sequence dataset, and the resultant subset alignments can then be merged into an alignment on the full dataset. This design strategy means that SATé can continue to improve in accuracy as new alignment methods are developed. Similarly, as better tree estimation methods are developed (including ones that might use gap events in a more informative manner), SATé can continue to improve in accuracy and/or scalability though the incorporation of these improved methods.

Alternative approaches to large-scale phylogeny estimation that do not require the estimation of a multiple sequence alignment have also been developed; of these, DACTAL [14] has been shown to give results that are almost as accurate as SATé, while being able to run on very large datasets. However, DACTAL is not completely alignment-free; instead, it computes alignments and trees on small subsets (carefully selected from the taxon set), and combines these smaller trees into a tree on the full set of taxa using SuperFine [77–79]. By combining this divide-and-conquer strategy with iteration, it quickly produces highly accurate trees. Truly alignment-free estimation has also been considered [80, 81], with some methods have strong theoretical guarantees [82]. Certainly the benefits of not requiring a full multiple sequence alignment are significant, especially in terms of running time. Ongoing research will show whether methods that do not require full sequence alignments are able to produce trees of comparable accuracy to the best of the tree estimation methods that do (at some point) estimate an alignment on the entire dataset.

This tutorial is limited to the GUI usage of SATé; readers interested in using the command line version are directed to the online tutorial [83]. Datasets and software to study alignment and phylogeny estimation methods are available through the SATé group Webpages at UT-Austin [84]. For additional discussion on methods for phylogenetic analysis, including data selection, see [85, 86].

10 Notes

1. This chapter is adapted from part of the SATé tutorial materials available at http://phylo.bio.ku.edu/software/sate/sate_tutorial.pdf.
2. A SATé user group provides announcements about SATé development, user support, and a general discussion area for all matters related to SATé. See the "SATé User Group" section of the SATé Webpage (http://phylo.bio.ku.edu/software/SATé/SATé.html).
3. SATé is under active development. See the SATé Webpage at UT-Austin, http://www.cs.utexas.edu/~phylo/software/sate/ for discussion about new and experimental features.
4. Do not set the "CPU(s) Available" setting to greater than the number of physical computing cores available on the computer. Doing so can overload computational resources and slow down the SATé analysis. To find out the number of computing cores on your computer, open the System Information utility in OS X (Apple icon in top left > About This Mac > More Info > System Report). The main screen of this utility will show the number of cores ("Total Number of Cores" field in the "Hardware Overview" panel).
5. MSA and biomolecular sequence files can require a significant amount of disk space. Since intermediate results—including intermediate MSA file—are stored during a SATé analysis,

make sure that the output folder contains enough free space for the analysis. A general rule of thumb is to provide one to two orders of magnitude more free space in the output folder than the size of the input file. If an analysis uses up available disk space, retry the analysis on a computer with more available disk space.

6. While SATé was designed with scalability in mind, SATé analyses of extremely large datasets (for example, datasets with 100,000 sequences or more) may overburden some desktop computers. If your computational resources are exceeded and your computer becomes unresponsive, first try to click the "Stop" button while the analysis is running. WARNING: the following two steps may lead to data loss, and should only be used as a last resort. If the situation still is not resolved, next try to quit the SATé application by either clicking the close button on the SATé application window or pressing COMMAND-Q. As a last resort if the previous steps did not work, force-quit the SATé application by pressing COMMAND-OPTION-ESC, choosing the SATé application, and pushing the "Force Quit" button. For more details about force-quitting an application, see http://support.apple.com/kb/HT3411.

7. If an analysis was canceled due to memory or time limitations, retry the analysis under one or more of the following conditions. Increase the "Max. Memory (MB)" setting in the "Jobs Settings" window up to 90 % of physical memory. Use a computer with a more powerful CPU and/or additional physical memory.

8. Always backup data and analysis files frequently. We suggest using the Time Machine feature in Mac OS X to schedule regular and frequent backups. Revert to backup files in the event of accidental modification to or loss of files.

9. For more powerful archival and other capabilities, use a version control system to manage storage for computational analyses and experiments. We particularly recommend Git (http://git-scm.com/).

10. In the event of installation issues, first try to update your system using the "Software Update" feature in Mac OS X and retry the installation.

11. Make sure that the new installation folder for the SATé application is not contained within the downloaded package. The SATé application will not run correctly within the downloaded package and must be installed to a separate location.

12. Make sure that your input file is compatible with your operating system. This situation can arise if your input file was created on a computer running a different operating system than the operating system on the computer running the SATé application. Incompatibility can prevent the SATé application from

reading the input file properly. For example, the line break character(s) differ across popular operating systems. To convert line breaks from a non-Mac format to a Mac format, try external utilities like TextWrangler's "Translate Line Breaks" command (http://www.barebones.com/products/textwrangler/).

Acknowledgments

This work was supported by a training fellowship to KL from the Keck Center of the Gulf Coast Consortia, on the NLM Training Program in Biomedical Informatics, National Library of Medicine (NLM) T15LM007093. This work was also partially supported by NSF grant DEB 0733029 to TW. This material was based on work supported by the National Science Foundation, while TW was working at the Foundation. Any opinion, finding, and conclusions or recommendations expressed in this material are those of the authors and do not necessarily reflect the views of the National Science Foundation.

References

1. Kemena C, Notredame C (2009) Upcoming challenges for multiple sequence alignment methods in the high-throughput era. Bioinformatics 25:2455–2465
2. Nelesen S, Liu K, Zhao D et al (2008) The effect of the guide tree on multiple sequence alignments and subsequent phylogenetic analyses. Pac Symp Biocomput 2008:25–36
3. Liu K, Linder CR, Warnow T (2010) Multiple sequence alignment: a major challenge to large-scale phylogenetics. PLoS Curr 2, RRN1198
4. Wang L-S, Leebens-Mack J, Wall PK, Beckman K, de Pamphilis CW, Warnow T (2011) The impact of multiple protein sequence alignment on phylogenetic estimation. IEEE Trans Comput Biol Bioinform 8:1108–1119
5. Cantarel BL, Morrison HG, Pearson W (2006) Exploring the relationship between sequence similarity and accurate phylogenetic trees. Mol Biol Evol 11:2090–100
6. Löytynoja A, Goldman N (2008) Phylogeny-aware gap placement prevents errors in sequence alignment and evolutionary analysis. Science 320(5883):1632–5
7. Hall BG (2005) Comparison of the accuracies of several phylogenetic methods using protein and DNA sequences. Mol Biol Evol 22(3): 792–802
8. Morrison DA, Ellis JT (1997) Effects of nucleotide sequence alignment on phylogeny estimation: a case study of 18S rDNAs of *Apicomplexa*. Mol Biol Evol 14(4):428–41
9. Ogden TH, Rosenberg MS (2006) Multiple sequence alignment accuracy and phylogenetic inference. Syst Biol 55(2):314–28
10. Larkin MA, Blackshields G, Brown NP et al (2007) ClustalW and ClustalX version 2.0. Bioinformatics 23:2947–2948
11. Edgar RC (2004) MUSCLE: a multiple sequence alignment method with reduced time and space complexity. BMC Bioinformatics 5:113
12. Edgar RC (2004) MUSCLE: a multiple sequence alignment with high accuracy and high throughput. Nucleic Acids Res 32: 1792–1797
13. Katoh K, Toh H (2008) Recent developments in the MAFFT multiple sequence alignment program. Brief Bioinformatics 9:286–298
14. Nelesen S, Liu K, Wang L-S et al (2012) DACTAL: fast and accurate estimations of trees without computing full sequence alignments. Bioinformatics 28:i274–i282
15. Varón A, Vinh LS, Wheeler WC (2010) POY version 4: phylogenetic analysis using dynamic homologies. Cladistics 26:72–85
16. Liu K, Nelesen S, Raghavan S, Linder CR, Warnow T (2009) Barking up the wrong tree-length: the impact of gap penalty on alignment and tree accuracy. IEEE/ACM Trans Comput Biol Bioinform 6(1):7–21

17. Liu K, Warnow T (2012) Treelength optimization for phylogeny estimation. PLoS One 7(3): e33104
18. Suchard MA, Redelings BD (2006) BAli-Phy: simultaneous Bayesian inference of alignment and phylogeny. Bioinformatics 22:2047–2048
19. Fleissner R, Metzler D, von Haeseler A (2005) Simultaneous statistical multiple alignment and phylogeny reconstruction. Syst Biol 54:548–561
20. Novák A, Miklós I, Lyngsoe R et al (2008) StatAlign: an extendable software package for joint Bayesian estimation of alignments and evolutionary trees. Bioinformatics 24:2403–2404
21. Lunter G, Miklós I, Drummond A et al (2005) Bayesian coestimation of phylogeny and sequence alignment. BMC Bioinformatics 6:83
22. Warnow T (2012) Standard maximum likelihood analyses of alignments with gaps can be statistically inconsistent. PLoS Curr 4: RRN1308. doi:10.1371/currents.RRN1308
23. Liu K, Raghavan S, Nelesen S et al (2009) Rapid and accurate large-scale coestimation of sequence alignments and phylogenetic trees. Science 324:1561–1564
24. Liu K, Warnow T, Holder MT et al (2012) SATé-II: very fast and accurate simultaneous estimation of multiple sequence alignments and phylogenetic trees. Syst Biol 61(1):90–106
25. Stamatakis A (2006) RAxML-VI-HPC: maximum likelihood-based phylogenetic analyses with thousands of taxa and mixed models. Bioinformatics 22:2688–2690
26. Price M, Dehal P, Arkin A (2010) FastTree 2 – approximately maximum-likelihood trees for large alignments. PLoS One 5:e9490
27. Löytynoja A, Goldman N (2005) An algorithm for progressive multiple alignment of sequences with insertions. Proc Natl Acad Sci U S A 102:10557–10562
28. Wheeler T, Kececioglu J (2007) Multiple alignment by aligning alignments. Bioinformatics 23:i559–i568
29. Felsenstein J (2004) Inferring phylogenies. Sinauer, Sunderland, MA
30. Dewey CN (2012) Whole-genome alignment. Methods Mol Biol 855:237–257
31. Mirarab S, Nguyen N-P, Warnow T (2012) SEPP: SATé-enabled phylogenetic placement. Pac Symp Biocomput 2012:247–58
32. Matsen F, Kodner R, Armbrust EV (2010) pplacer: linear time maximum-likelihood and Bayesian phylogenetic placement of sequences onto a fixed reference tree. BMC Bioinformatics 11:538
33. Berger SA, Krompass D, Stamatakis A (2011) Performance, accuracy, and web server for evolutionary placement of short sequence reads under maximum likelihood. Syst Biol 60:291–302
34. Liu K, Linder CR, Warnow T (2011) RAxML and FastTree: comparing two methods for large-scale maximum likelihood phylogeny estimation. PLoS One 6(11):e27731. doi:10.1371/journal.pone.0027731
35. Stamatakis A (2006) Phylogenetic models of rate heterogeneity: a high performance computing perspective. Proc IPDPS, Rhodes, Greece, 2006
36. Jukes TH, Cantor CR (1969) Evolution of protein molecules. Mammalian protein metabolism. Academic, New York, pp 21–132
37. Posada D, Buckley T (2004) Model selection and model averaging in phylogenetics: advantages of Akaike Information criterion and Bayesian approaches over likelihood ratio tests. Syst Biol 53(5):793–808
38. Abascal F, Zardoya R, Posada D (2005) ProtTest: selection of best-fit models of protein evolution. Bioinformatics 21(9):2104–2105
39. Jones DT, Taylor WR, Thornton JM (1992) The rapid generation of mutation data matrices from protein sequences. Comput Appl Biosci 8:275–282
40. Whelan S, Goldman N (2001) A general empirical model of protein evolution derived from multiple protein families using a maximum-likelihood approach. Mol Biol Evol 18:691–699
41. Dayhoff M, Schwartz R, Orcutt B (1978) A model of evolutionary change in proteins. Atlas Protein Sequence Struct 5:345–352
42. Kosiol C, Goldman N (2005) Different versions of the Dayhoff rate matrix. Mol Biol Evol 22:193–199
43. Adachi J (1996) Model of amino acid substitution in proteins encoded by mitochondrial DNA. J Mol Evol 42:459–468
44. Dimmic M, Rest J, Mindell D, Goldstein R (2002) rtREV: an amino acid substitution matrix for inference of retrovirus and reverse transcriptase phylogeny. J Mol Evol 55:65–73
45. Adachi J, Waddell P, Martin W, Hasegawa M (2000) Plastid genome phylogeny and a model of amino acid substitution for proteins encoded by chloroplast DNA. J Mol Evol 50:348–358
46. Mueller T, Vingron M (2000) Modeling amino acid replacement. J Comput Biol 7:761–776
47. Henikoff S, Henikoff J (1992) Amino acid substitution matrices from protein blocks. Proc Natl Acad Sci U S A 89:10915–10919
48. Yang Z (1998) Synonymous and nonsynonymous rate variation in nuclear genes of mammals. J Mol Evol 46:409–418

49. Le S, Gascuel O (2008) An improved general amino acid replacement matrix. Mol Biol Evol 25(7):1307–1320
50. Bodaker I, Suzuki MT, Oren A, Béjà O (2012) Dead Sea rhodopsins revisited. Environ Microbiol Rep 4(6):617–621
51. Andam C, Harlow T, Papke RT, Gogarten JP (2012) Ancient origin of the divergent forms of leucyl-tRNA synthetases in the Halobacteriales. BMC Evol Biol 12(1):85
52. Hagopian R, Davidson JR, Datta RS et al (2010) SATCHMO-JS: a webserver for simultaneous protein multiple sequence alignment and phylogenetic tree construction. Nucleic Acids Res 38(suppl 2):W29–W34
53. Katoh K, Toh H (2007) PartTree: an algorithm to build an approximate tree from a large number of unaligned sequences. Bioinformatics 23:372–374
54. Sievers F, Wilm A, Dineen D et al (2011) Fast, scalable generation of high-quality protein multiple sequence alignments using Clustal Omega. Mol Syst Biol 7:539
55. Katoh K, Frith MC (2012) Adding unaligned sequences into an existing alignment using MAFFT and LAST. Bioinformatics, 28 (23):3144–3146. doi:10.1093/bioinformatics/bts578
56. Wang N, Braun EL, Kimball RT (2012) Testing hypotheses about the sister group of the Passeriformes using an independent 30-locus data set. Mol Biol Evol 29(2):737–750
57. Xiang C-L, Gitzendanner MA, Soltis DE et al (2012) Phylogenetic placement of the enigmatic and critically endangered genus *Saniculiphyllum* (Saxifragaceae) inferred from combined analysis of plastid and nuclear DNA sequences. Mol Phylogenet Evol 64:357–367
58. Andam C, Harlow T, Thane R et al (2012) Ancient origin of the divergent forms of leucyl-tRNA synthetases in the Halobacteriales. Evol Biol 12:85
59. Huelsenbeck JP, Ronquist R (2001) MrBayes: Bayesian inference of phylogeny. Bioinformatics 17:754–755
60. Stockham C, Wang L-S, Warnow T (2002) Postprocessing of phylogenetic analysis using clustering. Bioinformatics 18(Suppl 1):i285–i293
61. Amenta N, Klinger J (2002). Case study: visualizing sets of evolutionary trees. In: Proceedings IEEE symposium on information visualization, pp 71–74
62. Bryant D (2003) A classification of consensus methods for phylogenetics. DIMACS series in discrete mathematics and theoretical computer science 51:163–184
63. Kannan S, Warnow T, Yooseph S (1998) Computing the local consensus of trees. SIAM J Comput 27(6):1695–1724
64. Phillips C, Warnow T (1996) The asymmetric median tree – a new model for building consensus trees. Discrete Appl Math 71(1–3): 311–335
65. Mirarab S, Warnow T (2011) FAST-SP: linear time calculation of alignment accuracy. Bioinformatics 27(23):3250–3258
66. Maddison W (1997) Gene trees in species trees. Syst Biol 46(3):523–536
67. Boussau B, Szöllősi G, Duret L et al (2013) Genome-scale coestimation of species and gene trees. Genome Res 23(2):323–30
68. Yu Y, Degnan JH, Nakhleh L (2012) The probability of a gene tree topology within a phylogenetic network with applications to hybridization detection. PLoS Genet 8(4):e1002660
69. Degnan JH, Rosenberg NA (2009) Gene tree discordance, phylogenetic inference and the multispecies coalescent. Trends Ecol Evol 24(6):332–340
70. Chaudhary R, Bansal MS, Wehe A et al (2010) iGTP: a software package for large-scale gene tree parsimony analysis. BMC Bioinformatics 11:547
71. Bansal MS, Alm EJ, Kellis M (2012) Efficient algorithms for the reconciliation problem with gene duplication, horizontal transfer, and loss. Bioinformatics 28(12):i283–i291
72. Yang J, Warnow T (2011) Fast and accurate methods for phylogenomic analyses. RECOMB comparative genomics, 2011. BMC Bioinformatics 12(Suppl 9):S4
73. Bayzid MS, Warnow T (2012) Finding optimal species trees from incomplete gene trees under incomplete lineage sorting. J Comput Biol 19(6):591–605
74. Bayzid MS, Warnow T (2013) Naive binning improves phylogenomic analyses. Bioinformatics first published online July 9, 2013 doi:10.1093/bioinformatics/btt394
75. Rice P, Longden I, Bleasby A (2000) EMBOSS: The European Molecular Biology Open Software Suite. Trends Genet 16:276–277
76. Swofford DL (2003) PAUP*: phylogenetic analysis using parsimony (*and other methods), Version 4
77. Swenson MS, Suri R, Linder CR et al (2012) SuperFine: fast and accurate supertree estimation. Syst Biol 61(2):214–227
78. Neves DT, Warnow TJ, Sobral L et al (2012) Parallelizing SuperFine. 27th Symp Appl Comp 1361–1367. doi: 10.1145/2245276.2231992

79. Nguyen N, Mirarab S, Warnow T (2012) MRL and SuperFine + MRL: new supertree methods. Algorithms Mol Biol 7:3

80. Daskalakis C, Roch S (2010) Alignment-free phylogenetic reconstruction. Proc Res Comp Molec Biol (RECOMB), Lecture Notes Computer Science 6044: 123–137

81. Chan CX, Ragan RA (2013) Next-generation phylogenomics. Biol Direct 8:30. doi: 10.1186/1745-6150-8-3

82. Vinga S, Almeida J (2003) Alignment-free sequence comparison – a review. Bioinformatics 19(4):513–523

83. Holder M, Warnow T, Mirarab S et al (2012) Online tutorial for SATé. http://phylo.bio.ku.edu/software/sate/sate_tutorial.pdf

84. Linder CR, Suri R, Liu K et al (2010) Benchmark datasets and software for developing and testing methods for large-scale multiple sequence alignment and phylogenetic inference. PLoS Curr 2:RRN1195. doi:10.1371/currents.RRN1195

85. Linder CR, Warnow T (2005) Overview of phylogeny reconstruction. In: Aluru S (ed) Handbook of Computational Biology. CRC computer and information science series. Chapman & Hall, Boca Raton, FL

86. Warnow T (2013) Large-scale multiple sequence alignment and phylogeny estimation, Chapter 6, in “Models and Algorithms for Genome Evolution”, edited by Cedric Chauve, Nadia El-Mabrouk and Eric Tannier. Springer, Series on “Computational Biology”

Chapter 16

PRALINE: A Versatile Multiple Sequence Alignment Toolkit

Punto Bawono and Jaap Heringa

Abstract

Profile ALIgNmEnt (PRALINE) is a versatile multiple sequence alignment toolkit. In its main alignment protocol, PRALINE follows the global progressive alignment algorithm. It provides various alignment optimization strategies to address the different situations that call for protein multiple sequence alignment: global profile preprocessing, homology-extended alignment, secondary structure-guided alignment, and transmembrane aware alignment. A number of combinations of these strategies are enabled as well.

PRALINE is accessible via the online server http://www.ibi.vu.nl/programs/PRALINEwww/. The server facilitates extensive visualization possibilities aiding the interpretation of alignments generated, which can be written out in pdf format for publication purposes. PRALINE also allows the sequences in the alignment to be represented in a dendrogram to show their mutual relationships according to the alignment. The chapter ends with a discussion of various issues occurring in multiple sequence alignment.

Key words Multiple sequence alignment, Progressive alignment, Sequence preprocessing, Homology-extended MSA, Secondary structure-guided MSA, Transmembrane-aware protein alignment

1 Introduction

Multiple sequence alignments (MSAs) are pervasive in biology. They are often used to elucidate conserved and variable regions in protein or DNA sequences, which can reveal crucial information regarding the functional and evolutionary relationships between the aligned sequences. One of the initial breakthroughs in the field of MSA, which addressed the computational burden associated with MSA, was the invention of the progressive alignment strategy [1].This strategy builds up an MSA by first constructing an approximate phylogenetic tree (*guide tree*) for the query sequences [1, 2]. In many methods the guide tree is constructed from the scores of all-against-all pairwise alignments of the query proteins. The sequences are then progressively aligned according to the order specified by the tree. However, an MSA produced using this method might contain errors due to the so-called greediness of this algorithm; i.e., alignments affected are not reconsidered anymore and any match error occurring in the process will be

David J. Russell (ed.), *Multiple Sequence Alignment Methods*, Methods in Molecular Biology, vol. 1079,
DOI 10.1007/978-1-62703-646-7_16,

propagated into subsequent alignment steps ("Once a gap, always a gap") [3]. Several methods exist that try to alleviate the greediness of the progressive alignment, for example by implementing an iterative alignment protocol, as first proposed by Hogeweg and Hesper [2].

Profile ALIgNmEnt (PRALINE) adopts a global progressive alignment algorithm that reevaluates at each alignment step which sequence or sequence block pairs to align. This means that unlike many other progressive MSA methods [2, 4–6], PRALINE determines at each step during progressive alignment which alignment between any alignment block or hitherto unaligned sequence will be optimal such that a tree reflecting the order in which sequences are aligned is produced on the fly without the use of a precalculated guide tree.

In order to minimize the effects of the greediness of the progressive alignment protocol and to improve alignment quality, PRALINE includes a number of alignment strategies to improve the basic progressive protocol: global profile preprocessing, homology-extended alignment, secondary structure-guided alignment, and transmembrane (TM)-aware alignment. It also allows combinations of different strategies to cater for the various needs researchers might have, for example combining profile preprocessing with secondary structure-guided alignment or with TM-aware alignment.

PRALINE employs various profile preprocessing protocols to address the problems caused by the greediness of progressive alignment method. These protocols can be categorized into three types: global, local, and homology-extended profile preprocessing [7, 8]. The main principle behind these profile preprocessing techniques is avoiding early error in progressive alignment by projecting information from other sequences onto each input sequence prior to progressive alignment. This is done by converting each input sequence into a pre-profile, which is abstracted from a master–slave sequence alignment of the sequence considered with the other input sequences. In the global preprocessing strategy, sequences are stacked upon the key sequence, i.e., the sequence considered, by means of global alignment, while in the local preprocessing protocol, local alignments are used to enrich the information of the key sequence. The homology-extended multiple alignment strategy is an extension of the local preprocessing method. In this method, information to enrich the input sequences is not gleaned from other input sequences, but from putatively homologous sequences residing in sequence databases. It has been shown in previous studies that the addition of homology information has distinctly positive effects on alignment quality, particularly in cases of distantly related protein sets [8–11].

PRALINE provides the option to allow the incorporation of secondary structure and/or transmembrane information to guide the alignment and further optimize its quality. Here the rationale is to integrate predicted structural information into the alignment, following the principle that protein structural aspects tend to be more conserved than the associated sequences during evolution. PRALINE incorporates secondary structure and/or TM information by using specific residue exchange matrices during alignment.

PRALINE is available as an online server (URL: http://www.ibi.vu.nl/programs/PRALINEwww/), which is also equipped with a SOAP service, allowing the users easy access to the Web service from within their own programs or scripts.

2 Method

2.1 The "Core" MSA Protocol in PRALINE

PRALINE employs a profile-based progressive alignment strategy. As stated above, after initial all-against-all pairwise alignment, the highest scoring sequence pair is joined into the first sequence block. Then, this sequence block is aligned with all the remaining single sequences, after which the highest scoring pair is selected. Note that at this stage, the highest scoring alignment can be between the sequence block and a single sequence, while at a later stage also alignment of sequence blocks may occur. Alignment proceeds until all sequences have been aligned in a single MSA. By following this protocol, PRALINE does not utilize a precomputed guide tree in its alignment protocol, but calculates the guide tree on the fly by utilizing the information afforded by pre-aligned blocks at each stage, such that the tree reflecting the progressive alignment steps becomes available at the end. Since successive profile scores during the PRALINE progressive protocol descend uniformly, they can be used to construct a dendrogram reflecting the alignment order.

Alignment in PRALINE is carried out using the dynamic programming technique [7]. The following simple profile-scoring scheme is used to score a pair of profile positions (columns) x and y:

$$\text{Score}\,(x, y) = \sum_{\text{i}}^{20} \sum_{\text{j}}^{20} \alpha_{\text{i}} \beta_{\text{j}} \log\left(\frac{P_{\text{ij}}}{P_{\text{i}} P_{\text{j}}}\right), \tag{1}$$

where α_{i} and β_{j} are the frequencies with which amino acids i and j appear in columns x and y, respectively, and M (i, j) is the exchange value for amino acids i and j according to substitution matrix M (e.g., BLOSUM62 [12] or PAM250 [13]).

PRALINE adopts a semi-global alignment strategy, which means that it aligns sequences over their whole length, but without penalizing the so-called end gaps, i.e., gaps occurring N- or C-terminally to any of the sequences. Global alignment strategy is known to be optimal for sequences of high-to-medium sequence similarity. Since interesting biological alignments can have sequences that diverged considerably beyond the level that can

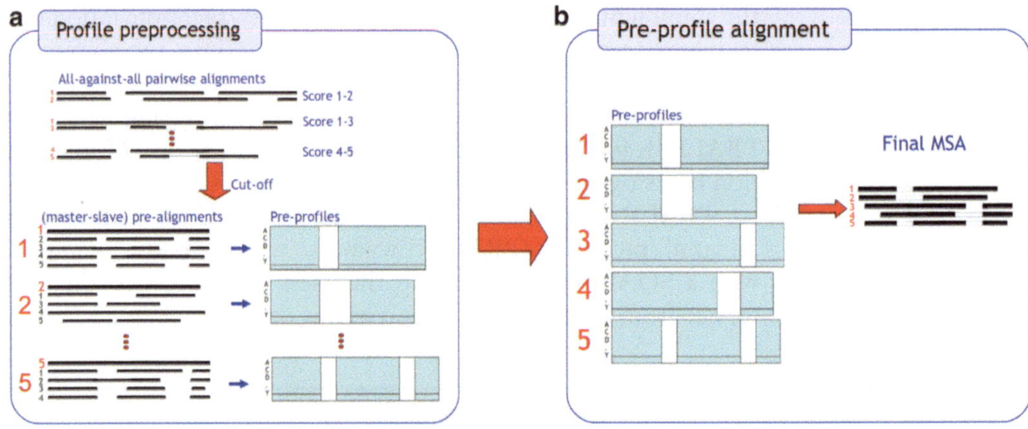

Fig. 1 Schematic overview of the profile preprocessing (**a**) and the pre-profile alignment (**b**) routines. For details, see text. Adapted from ref. 8

be recognized by global alignment, PRALINE offers a number of strategies to address evolutionary divergent alignment situations.

2.2 Global Pre-profile Preprocessing

Pre-profile processing is an optimization method aimed at minimizing error propagation during progressive alignment by including prior knowledge about the other sequences during alignment [7]. In this method each of the input sequences is represented as a preprocessed profile (pre-profile) instead of a single sequence. For each input sequence a master–slave alignment is constructed by stacking other input sequences whose pairwise global alignment score against the master sequence is higher than a user-specified threshold (Fig. 1). The user can determine whether to include distant sequences in the pre-profile or not to use an alignment score threshold value. Although distant sequences might contribute significant information, there is the chance that they contribute noise due to the fact that alignment error is known to increase super-linearly with sequence distance [14].

PRALINE allows the alignment score threshold value to be specified as a factor relating to the sequence lengths: $S \geq tL$, where L is the length of the shortest sequence in the alignment and t is the alignment score threshold. This means that the alignment score S should be at least as high as the threshold score multiplied by L in order to become included in the pre-profile such that the average score over L positions is at least t. Using a score threshold which is linearly related to alignment length is in agreement with observations made for global alignments of random sequences [8, 15].

The pre-profiles in PRALINE further incorporate position-specific gap penalties, enabling increased matching of distant sequences and likely placement of gaps outside ungapped core regions in the pre-profiles during progressive alignment.

The preprocessing strategy can be further optimized by means of an iterative protocol. Each iteration is based upon the consistency of a preceding MSA. Consistency is defined here as the agreement between matched amino acids in the MSA and those in associated pairwise alignments. PRALINE calculates a consistency score for each amino acid in the MSA. These are then used as position-specific weight in subsequent alignment. The effect of this is that alignments in next iterations tend to maintain consistently aligned regions, while less consistent regions are more likely to become aligned differently. Iterations are terminated when convergence or limit cycle is reached. The latter means that a given MSA has been encountered during iteration earlier than the preceding round. The user must specify the maximum number of iterations for cases where convergence or limit cycle is not reached.

2.3 Homology-Extended Alignment

Protein sequences accumulate varying degrees of mutation during evolution. This situation has an important bearing on the quality of alignment methods which use generic amino acid scoring matrices since these matrices are mostly derived from a specific set of carefully curated alignments. Such generalization implies a standardized evolutionary model, which might lead to inconsistencies in the alignments. Although the quality of alignments of closely related proteins is hardly influenced by this issue, alignments of distant protein sequences (<30 % sequence identity) are much more sensitive to this issue. This is because evolutionary traces become largely obfuscated in divergent cases [16].

Two main approaches have led to improvements in distant protein alignment. In the first approach, the generic substitution matrix is readjusted to the evolutionary relation observed in the input sequence set [17]. The second approach attempts to identify the distant relation between the sequences through the incorporation of additional structural or homologous sequence information. The homology-extended alignment strategy in PRALINE attempts to address the problem of distant protein sequence alignment by enriching the information content for each of the input sequence with the help of homologous sequences collected using PSI-BLAST. In this alignment strategy, a PSI-BLAST search is performed for each input sequence against a particular sequence database; the default is the nonredundant (NR) database. The user can set the initial *E*-value threshold and the number of PSI-BLAST iterations. In order to filter for redundant sequences, all PSI-BLAST hits with 100 % sequence identity are not taken into account. In cases where no hits are found or only redundant hits are found, the PSI-BLAST search is rerun using an *E*-value threshold which is ten times higher, i.e., ten times less stringent, than the previous one. This process is reiterated until each input sequence has at least one homologues sequence. The final local PSI-BLAST

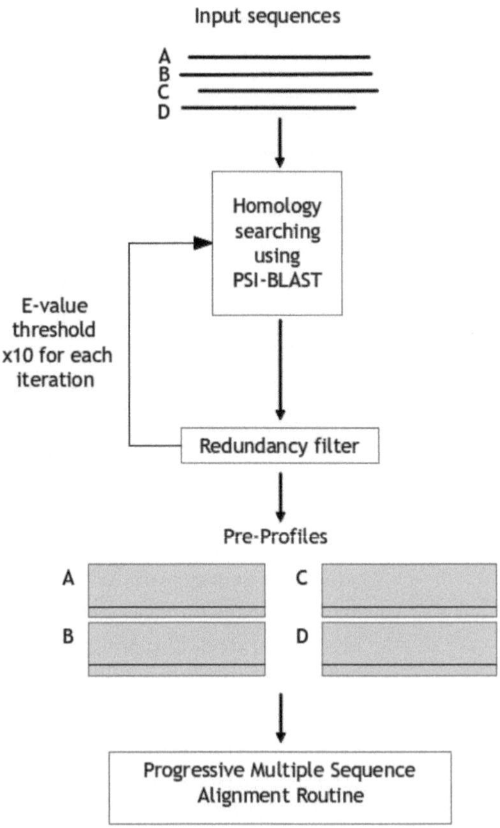

Fig. 2 Schematic overview of the homology-extended alignment strategy. For details, see text. Adapted from ref. 18

alignments for each sequence are converted to a pre-profile and then become progressively aligned using the core MSA method described in Subheading 1 (Fig. 2).

The main advantage of this approach is that it uses a much greater amount of position-specific information in the profile scoring scheme due to the homology-extended profiles based upon potentially large numbers of putatively homologous sequences. This greatly helps the alignment of sequences, particularly those in the twilight zone (<30 % sequence identity) or even in more divergent cases [18].

In addition to global profile preprocessing, PRALINE also incorporates local profile preprocessing. However, since this implementation is essentially equivalent to the homology-extended strategy based upon PSI-BLAST searches described here, but is less successful in general due to the inclusion of input sequences only, it is not enabled as an option in the PRALINE Web server.

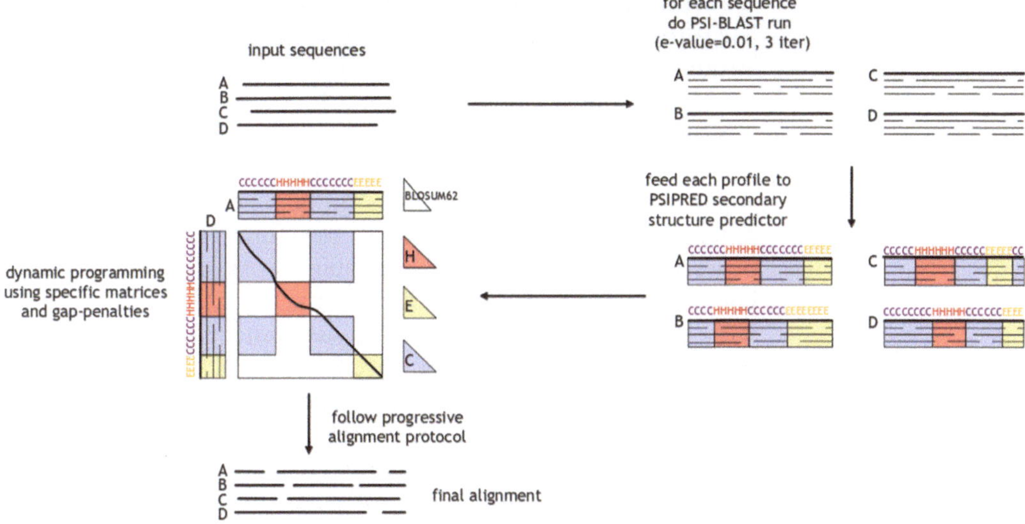

Fig. 3 Schematic overview of the secondary structure-guided alignment strategy in PRALINE (Pirovano, Simossis, and Heringa, unpublished). For details, see text

2.4 Secondary Structure-Guided Alignment

It is well known that the secondary structure elements of proteins are much more conserved than their amino acid sequence during evolution [16, 19–21]. Therefore, secondary structure information can be used to guide the alignment process, particularly in the case of distantly related proteins [7, 8, 18, 22–27].

The secondary structure-guided alignment strategy in PRALINE works by combining secondary structure prediction with a secondary structure-based scoring scheme (Fig. 3). When using predicted secondary structure, however, the gain in information might be overshadowed by prediction error. Fortunately, during earlier tests with the PRALINE secondary structure-guided strategy, it turned out that the inclusion of secondary structure information improves alignment whenever a prediction accuracy of 65 % or more is achieved (Simossis and Heringa, unpublished), and this is easily attained by modern prediction methods.

PRALINE starts the strategy by predicting the secondary structure elements of each sequence using a secondary structure prediction tool. PRALINE provides the user with the choice of four different secondary structure predictors: PSIPRED [28], SSPRO 4.0 [29], PORTER [30], and YASPIN [31]. Each of these predictors has its own strengths and weaknesses, the choice of which is therefore left to the user's discretion. The secondary structure prediction methods perform a PSI-BLAST search for each input sequence and then perform the secondary structure prediction using the position-dependent scoring matrix (PSSM) produced by PSI-BLAST, thereby making use of the amino acid conservation as observed in the putative homologous sequences. If an input sequence has an associated 3D structure deposited in the PDB [32], then the secondary structure elements of this protein are assigned using DSSP [33] and so do not need to be predicted. In PRALINE, the DSSP information is found based on the (FASTA) sequence definition line.[1]

After the secondary structure delineation step, PRALINE applies its secondary structure scoring scheme, which is a soft scheme to align the secondary structure elements by using observed residue mutation probabilities as observed in alpha helix, beta strand, or coil conformations (Fig. 3). Residue positions with identical secondary structure assignments are scored using Lüthy helix-, strand-, and coil-specific matrices [34], while residue positions with nonidentical secondary structure assignments are scored using the generic scoring matrix (e.g., BLOSUM62 [12]). Since specific exchange values are used to discriminate the matching of the secondary structures, different structures can become matched (e.g., a helix with a coil structure). This means that the method can reasonably deal with errors in the annotation of secondary structure elements.

2.5 Transmembrane-Aware Protein Alignment

The TM regions of membrane-bound proteins show a different hydrophobicity pattern compared to globular soluble proteins [35]. This is because they are immersed in a largely hydrophobic environment as opposed to the more hydrophilic nature of the cytosol. Conventional scoring matrices which are tailored for soluble proteins are therefore not optimally suited for aligning membrane-bound proteins.

PRALINE is not the first alignment method that combines information from different substitution matrices in order to improve the quality of TM protein alignment. One of the earlier methods that attempted a similar approach was STAM [36]. However, this method incorporates the TM information in a "hard" way. First, TM regions are aligned separately, thereby anchoring the alignment, after which the intervening stretches are aligned. This means that the method is crucially dependent upon the quality of the annotation of TM regions, and would have difficulty, for example, in aligning 7-TM sequences if for some of the sequences less than seven TM regions would have been predicted.

In PRALINE, TM information is taken into account in a more flexible way, which consists of three steps [37]. Firstly, the TM topology for each input sequence is predicted using a TM prediction tool. The user can select one out of three predictors:

[1]PRALINE finds the PDB identifier of a protein by extracting it from the fasta definition line of that protein. For example, these description lines are fine: ">102L_A," ">102L|A," and ">102LA". For any other description line, PDB identifier is not extracted. No description may follow the sequence identifier. Thus ">pdb|102L|A", ">gi|157829524|pdb|102L|A", and also ">102L_A " (note the trailing space) are skipped.

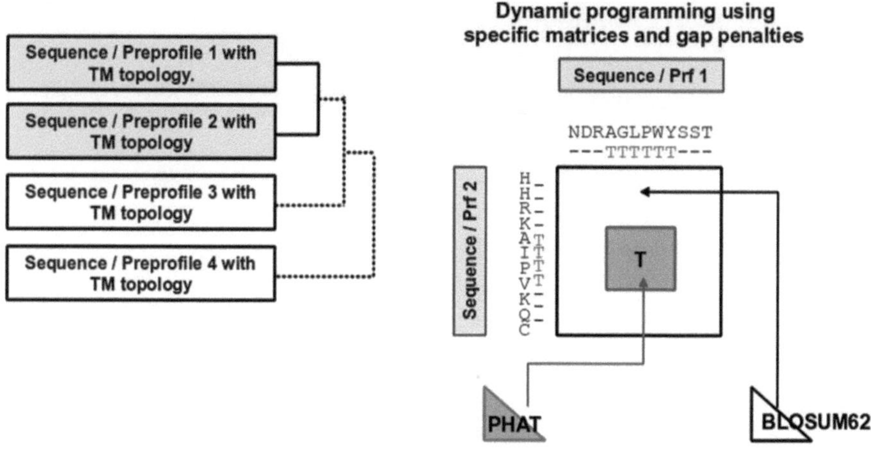

Fig. 4 Schematic overview of the TM-aware strategy in PRALINE. For details, see text. Adapted from ref. 37

PHOBIUS [38], TMHMM [39], or HMMTOP [40]. Secondly, TM-specific substitution scores from the PHAT [41] matrix are used to align residues that are predicted to be members of a TM segment (Fig. 4). The remaining soluble fragments are aligned using the generic BLOSUM62 matrix.

A tree-based consistency iteration scheme is then performed to enhance the MSA quality, which is similar to the tree-dependent partitioning method proposed by Hirosawa et al. [42] and its implementation in the MUSCLE alignment tool [43, 44]. In this scheme each edge of the guide tree is used to divide the alignments into two sub-alignments, which are then successively realigned. A new alignment is selected only if the alignment score is higher than the current score. The alignment score in the TM-aware alignment strategy is calculated as the sum of the substitution values of the BLOSUM and PHAT matrices (depending on the TM topology of the alignment positions). One iterative cycle in this tree-based consistency strategy is completed when each edge of the guide tree is visited once. The maximum number of iteration cycles has been set to 20 [37].

2.6 The PRALINE Online Server

The PRALINE server is accessible via the Web site of the IBIVU center at VU University Amsterdam (URL: http://www.ibi.vu.nl/programs/PRALINEwww/). The server is aimed to assist both specialist and nonspecialist users. It provides the user with extensive online documentation for each of the different parameters PRALINE may be run with, and also provides a "sample output" page which contains examples of the possible outputs of the PRALINE server using the various alignment strategies described above. PRALINE accepts sequences in FASTA [45] format as input. For each alignment job, the maximum number of sequences that can be

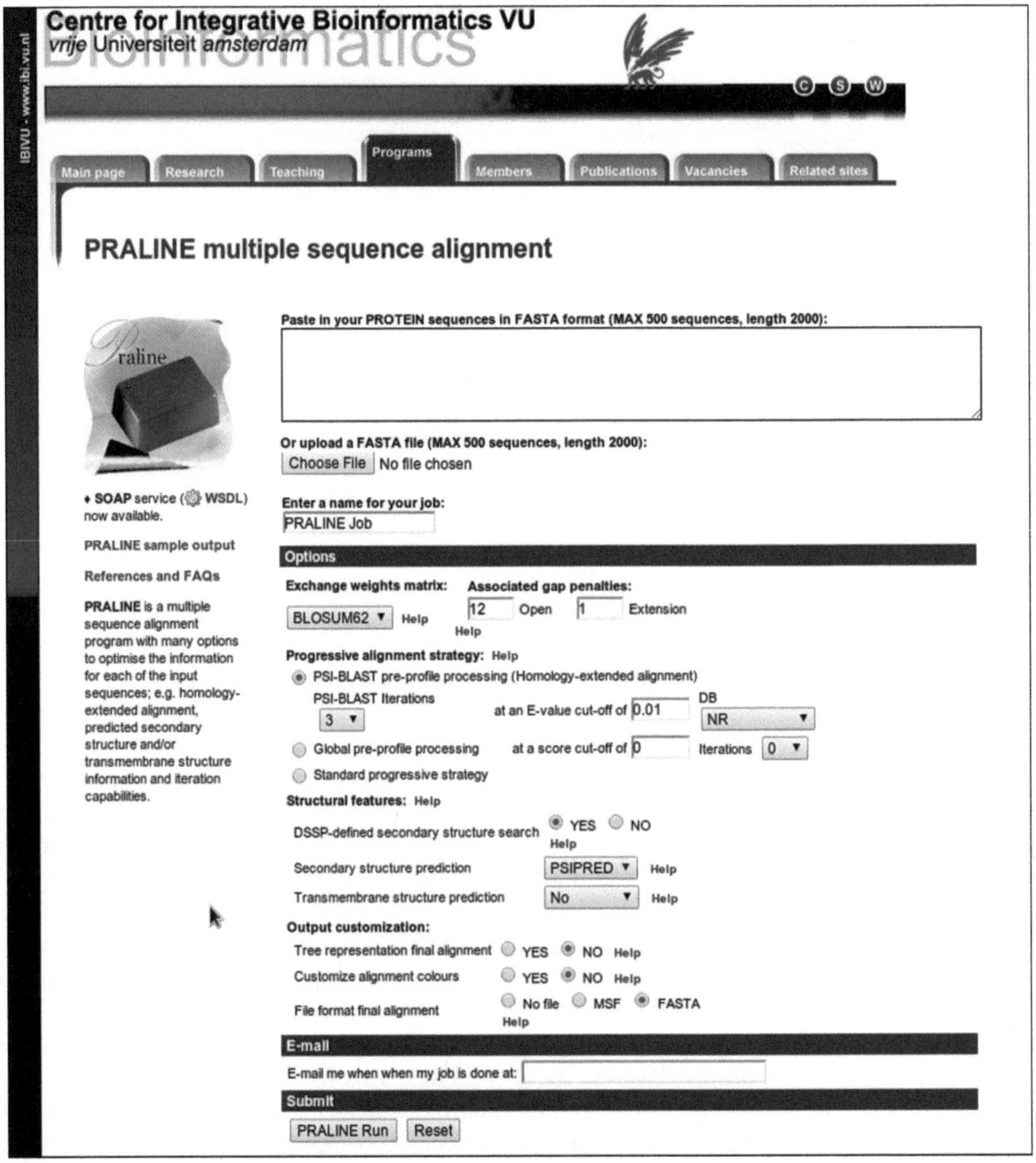

Fig. 5 The user interface of PRALINE server

submitted is 500 with a maximum length of 2,000 residues for each sequence. This is to limit the server load and is not due to any limitation of the PRALINE algorithm itself.

On the main page (Fig. 5), the user can manually set the gap opening and gap extension penalties, choose the appropriate substitution matrix, and set the parameters for various alignment strategies available in PRALINE. The default setting is 12 for gap open penalty, 1 for gap extension penalty, and BLOSUM62 as the amino acid substitution matrix. Other amino acid substitution matrices

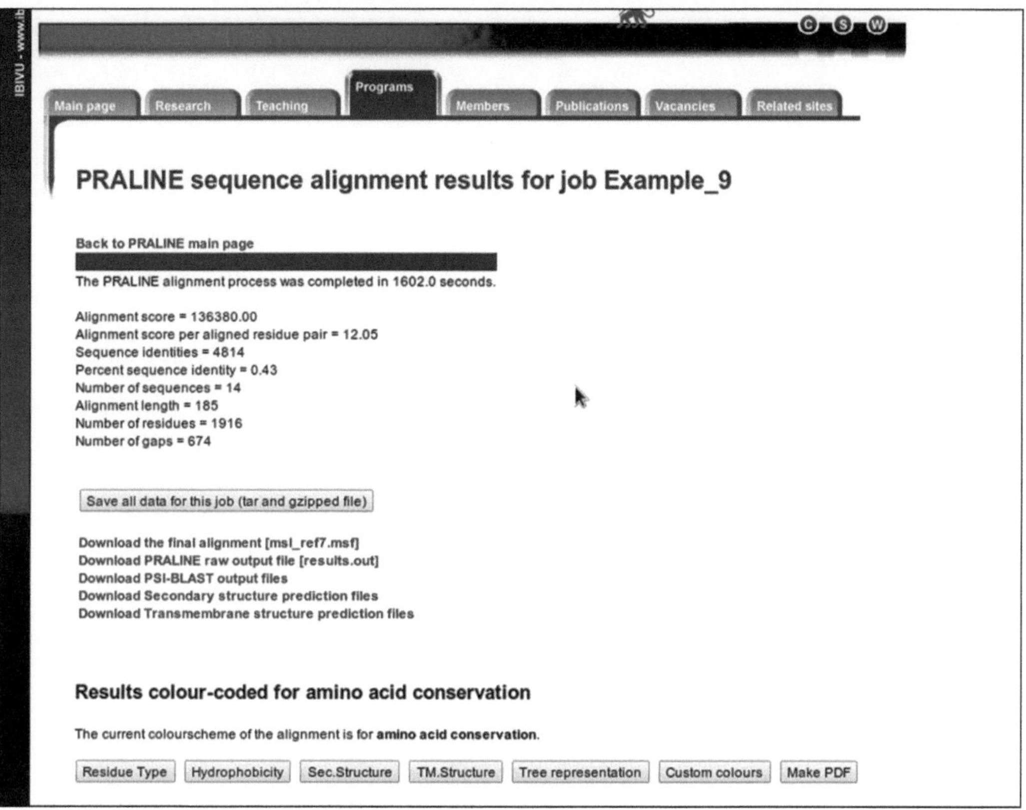

Fig. 6 PRALINE server output page header

available to the user are PAM250 [13], BLOSUM62 and BLOSUM50 [12], and GON120 and GON250 [46].

Once a job is submitted to the PRALINE server, the user is presented with a holding page that refreshes automatically. This holding page shows which alignment steps are being performed by the PRALINE server. Due to longer running times needed for certain alignment strategies (e.g., homology-extended alignment), the PRALINE server also provides the user with the possibility to get an e-mail notification once the job is finished; this notification e-mail contains a link to the outputs and some alignment statistics.

The output page presents general information about the alignment (alignment score, alignment length, number of gaps, etc.) (Fig. 6). It also contains information such as PSI-BLAST output, secondary structure predictions, or TM predictions depending on the alignment strategy selected by the user. On this page the user can also select various predefined color schemes to visualize the alignment according to residue type, hydrophobicity, secondary structure (if applicable), or TM structure (if applicable). Each color scheme comes with a concise explanation as to how to interpret the different colors. Apart from the predefined color schemes, the users can also define their own color scheme using a custom

A/A	Gray	Blue	Yellow	Orange	Red	Green	Pink	No colour
Alanin [A]	○	○	○	○	○	○	○	◉
Arginine [R]	○	○	○	○	○	○	○	◉
Aspartic Acid [D]	○	○	○	○	○	○	○	◉
Asparagine [N]	○	○	○	○	○	○	○	◉
Cysteine [C]	○	○	○	○	○	○	○	◉
Glutamic Acid [E]	○	○	○	○	○	○	○	◉
Glutamine [Q]	○	○	○	○	○	○	○	◉
Glycine [G]	○	○	○	○	○	○	○	◉
Histidine [H]	○	○	○	○	○	○	○	◉
Isoleucine [I]	○	○	○	○	○	○	○	◉
Leucine [L]	○	○	○	○	○	○	○	◉
Lysine [K]	○	○	○	○	○	○	○	◉
Methionine [M]	○	○	○	○	○	○	○	◉
Phenylalanine [F]	○	○	○	○	○	○	○	◉
Proline [P]	○	○	○	○	○	○	○	◉
Serine [S]	○	○	○	○	○	○	○	◉
Threonine [T]	○	○	○	○	○	○	○	◉
Tryptophan [W]	○	○	○	○	○	○	○	◉
Tyrosine [Y]	○	○	○	○	○	○	○	◉
Valine [V]	○	○	○	○	○	○	○	◉

Fig. 7 PRALINE user-defined amino acid color table

color scheme table (Fig. 7). Finally, PRALINE includes the option to generate a tree based upon the MSA. However, the user should note that trees generated by PRALINE are not phylogenetic trees, but simply show the relationships between the sequences as determined by the alignment scores (Fig. 8).

The following output (Figs. 6, 8, and 9) is taken from an alignment of 14 proteins belonging to the MscL family of large-conductance mechanosensitive channels compiled together in the BaliBASE 3.0 benchmarking database [47]. The alignment was performed using the homology-extended strategy with both integrated transmembrane and secondary structure information from the predictions of PHOBIUS and PSIPRED, respectively. The alignment shown in Fig. 9 is colored using the "Residue Type" coloring scheme. The alignment shows conserved elements as well as regions with extensive gaps. The associated tree (Fig. 8) clearly shows that the *1msla* sequence (bottom sequence in the alignment) is an outlier, missing elements at both the N- and C-termini.

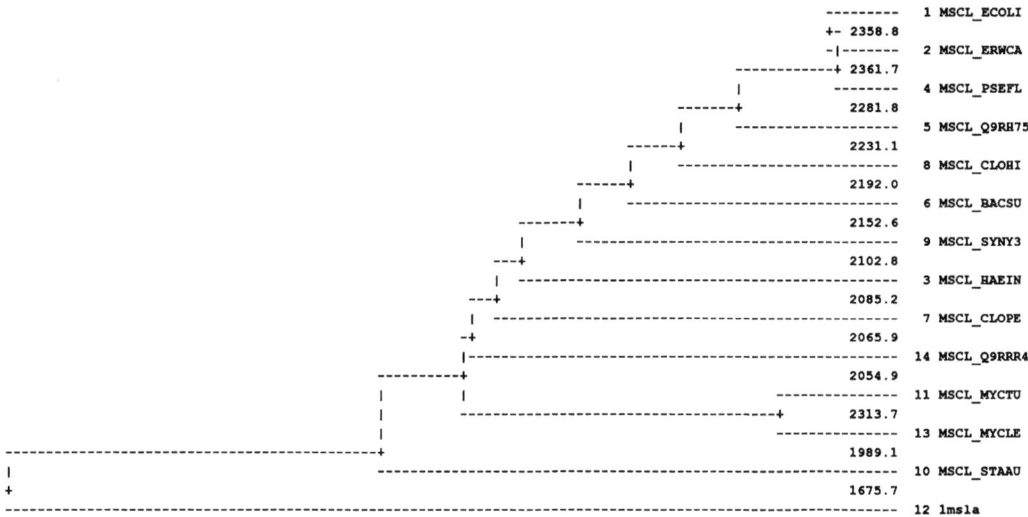

Fig. 8 Tree representation of alignment shown in Fig. 9

2.7 Practical Issues

1. *Aligning distantly related protein sequences.* Although state-of-the-art alignment methods are able to make very accurate MSAs, inaccurate MSA can arise due to divergent evolution. It has been shown that the accuracy of alignment methods decreases dramatically when the sequence identity between the aligned sequences is lower than 30 % [16]. Given this limitation, it is advisable to compile a number of MSAs using different amino acid substitution matrices (e.g., PAM and BLOSUM matrices). It is helpful to know that higher PAM numbers and low BLOSUM numbers (e.g., PAM250 or BLOSUM45) correspond to exchange matrices that are suited for the alignment of more divergent sequences, respectively, whereas matrices with lower PAM and higher BLOSUM numbers are more suitable for more closely related protein sequences. It is also important to try different gap penalties when aligning distant protein sequences. Gap penalties play an important role in the dynamic programming algorithm; therefore they can have considerable influence on the alignment quality. The higher the gap penalties, the stricter the insertion of gaps into the alignment and consequently the fewer gaps inserted. Gap regions in an MSA often correspond to loop regions in the associated tertiary structure, which are more likely to be altered by divergent evolution. Therefore, it can be useful to lower the gap penalty values when aligning divergent proteins, although care should be taken not to deviate too much from the recommended settings. Excessive gap penalty values will enforce a gap-less alignment, whereas low gap penalties will lead to alignments with very many gaps, allowing (near) identical amino acids to be matched. In both cases the resulting alignment will be biologically inaccurate.

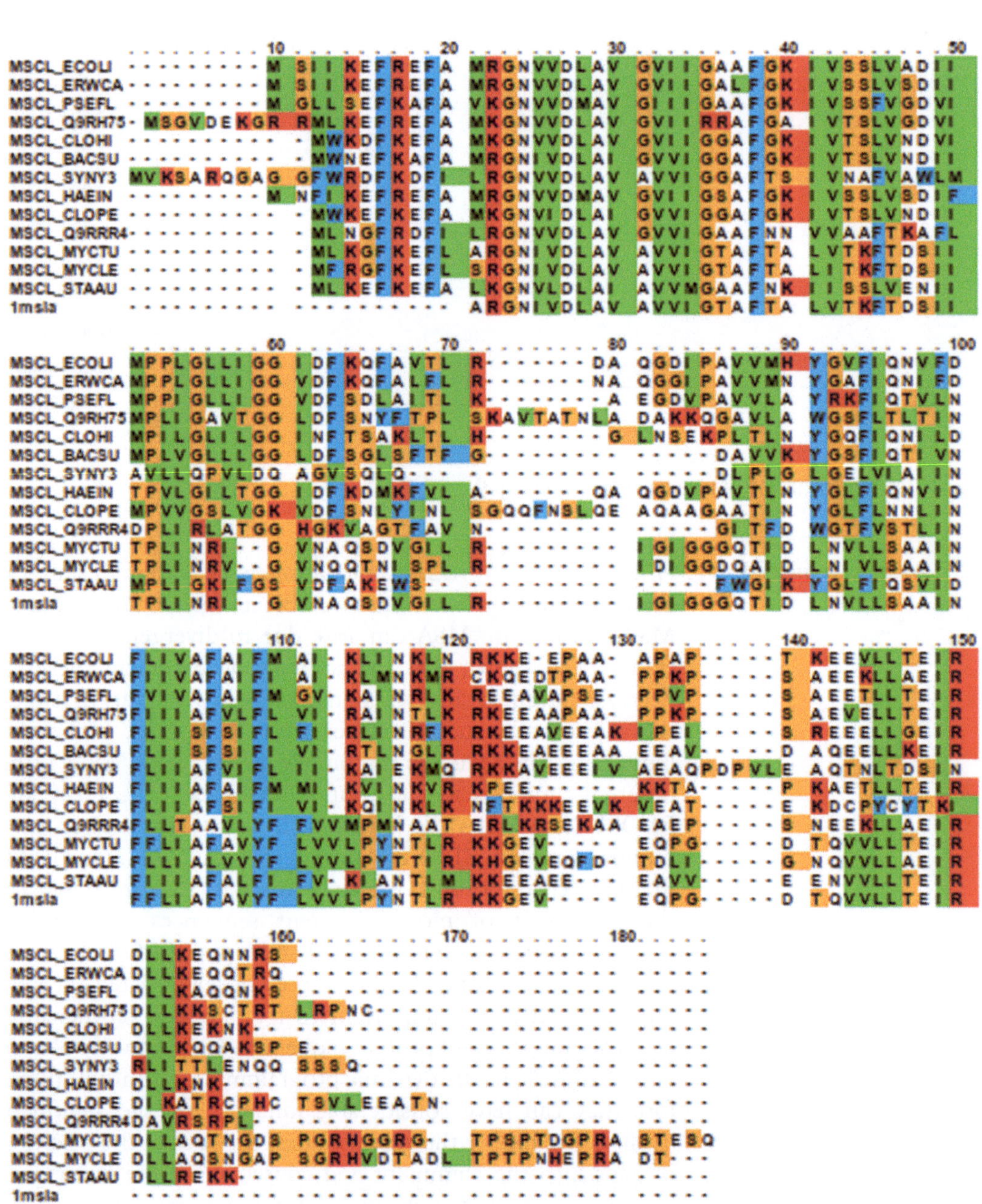

Fig. 9 MSA of 14 proteins belonging to the MscL family of large-conductance mechanosensitive channels

Although the recommended combinations of exchange matrices and gap penalties have been described in the literature, there is no formal theory yet as to how gap penalties should be chosen given a particular residue exchange matrix. Therefore, the opening and extending gap penalties are set empirically: for example, penalties of 11 (open) and 1 (extend) are recommended for BLOSUM62, whereas the suggested values for PAM250 are 10 (open) and 1 (extend).

2. *Multi-domain proteins.* Proteins with multiple domains can be a particular challenge for multiple alignment methods. Whenever there has been an evolutionary change in the domain order of the query protein sequences, or if some domains have been inserted or deleted across the sequences, this leads to serious problems for global alignment methods. Global alignment methods are not suited to deal with permuted domain orders and normally exploit gap penalty regimes that make it difficult to insert long gaps corresponding to the length of one or more protein domains. Therefore, it is advisable to align multi-domain proteins using local multiple alignment methods. MSA tools that are (partly) based on local alignment method (for example T-COFFEE [6]) are good alternatives for this kind of situation.

3. *Repeats in protein sequences.* The occurrence of repeats in many sequences can significantly reduce the accuracy of MSA methods, mostly because the methods are not able to deal with different repeat copy numbers. Sammeth and Heringa have developed an MSA method that is able to perform global MSA on protein sequences under the constraints of a given repeat analysis [48]. This method requires the specification of the individual repeats, which can be obtained by running one of the available repeat detection algorithms, after which a repeat-aware MSA is produced. Although the alignment result can be markedly improved by this method, it is sensitive to the accuracy of the repeat information provided.

4. *Preconceived knowledge.* In a number of cases, there is already some preconceived knowledge about the final alignment. For example, consider a protein family containing a disulfide bond between two specific cysteine (Cys) residues. Given the structural importance of a disulfide bond, Cys residues that form disulfide bonds are generally conserved, so it is important that the final MSA matches such Cys residues correctly. However, depending on conservation patterns and overall evolutionary distances of the sequences, it is sometimes necessary for the alignment method to have special guidance in order to match the Cys residues correctly. The main hurdle in this type of alignment is in marking the positions of amino acids that have to be correctly aligned and assigning specific parameters for their consistency. The following suggestions are therefore offered for (partially) resolving this type of problem:

   (a) *Chopping alignments.* Instead of aligning whole sequences, one can decide to chop the alignment in different parts.

For example, this could be done if the sequences have some known domains with known boundaries. An added advantage in such cases is that no undesirable overlaps will occur between these pre-marked regions if aligned separately. Finally, the whole alignment can be built by concatenating the aligned blocks. It should be stressed that each of the separate alignment operations is likely to follow a different evolutionary scenario, as for example the guide tree or the additionally homologous background sequences in the homology-extended strategy in PRALINE can well be different in each case. It is entirely possible, however, that these different scenarios reflect true evolutionary differences, such as unequal rates of evolution of the constituent domains.

(b) *Altering amino acid exchange weights.* Multiple alignment programs make use of amino acid substitution matrices in order to score alignments. Therefore, it is possible to change individual amino acid exchange values in a substitution matrix. Referring to the disulfide bond example mentioned above, one could decide to up-weight the substitution score for a cysteine self-conservation. As a result, the alignment will obtain a higher score when cysteines are matched, and as a consequence the method will attempt to create an alignment where this is the case. However, some protein families have a number of known pairs of Cys residues that form disulfide bonds, where mixing up of the Cys residues involved in different disulfide bridges might happen in that Cys residues involved in different disulfide bonds become aligned at a given single position. To avoid such incorrect matches in the alignment, one can add a few extra amino acid designators in the amino acid exchange matrix that can be used to identify Cys residue pairs in a given bond (for example J, O, or U). The exchange scores involving these "alternative" Cys residues should be identical to those for the original Cys, except for the cross-scores between the alternative letters for Cys that should be given low (or extremely negative) values to avoid cross alignment. It must be stressed that such alterations are heuristics that may compromise the evolutionary model underlying a given residue exchange matrix.

References

1. Sankoff D, Cedergren RJ (1983) Simultaneous comparison of three or more sequences related by a tree, time warps, string edits and macromolecules. The theory and practice of sequence comparison. Addison-Wesley, Reading, MA, pp 253–263
2. Hogeweg P, Hesper B (1984) The alignment of sets of sequences and the construction of phyletic trees: an integrated method. J Mol Evol 20:175–186
3. Feng DF, Doolittle RF (1987) Progressive sequence alignment as a prerequisite to correct phylogenetic trees. J Mol Evol 25:351–360
4. Thompson JD, Higgins DG, Gibson TJ (1994) CLUSTAL W: improving the sensitivity of progressive multiple sequence alignment

through sequence weighting, position-specific gap penalties and weight matrix choice. Nucleic Acids Res 22:4673–4680

5. Gotoh O (1996) Significant improvement in accuracy of multiple protein sequence alignments by iterative refinement as assessed by reference to structural alignments. J Mol Biol 264:823–838
6. Notredame C, Higgins DG, Heringa J (2000) T-Coffee: a novel method for fast and accurate multiple sequence alignment. J Mol Biol 302:205–217
7. Heringa J (1999) Two strategies for sequence comparison: profile-preprocessed and secondary structure-induced multiple alignment. Comput Chem 23:341–364
8. Heringa J (2002) Local weighting schemes for protein multiple sequence alignment. Comput Chem 26:459–477
9. Katoh K, Kuma K, Toh H et al (2005) MAFFT version 5: improvement in accuracy of multiple sequence alignment. Nucleic Acids Res 33: 511–518
10. Edgar RC, Sjölander K (2004) A comparison of scoring functions for protein sequence profile alignment. Bioinformatics 20: 1301–1308
11. Wang G, Dunbrack RL Jr (2004) Scoring profile-to-profile sequence alignments. Protein Sci 13:1612–1626
12. Henikoff S, Henikoff JG (1992) Amino acid substitution matrices from protein blocks. Proc Natl Acad Sci U S A 89:10915–10919
13. Dayhoff MO, Barker WC, Hunt LT (1983) Establishing homologies in protein sequences. Methods Enzymol 91:524–545
14. Vogt G, Etzold T, Argos P (1995) An assessment of amino acid exchange matrices in aligning protein sequences: the twilight zone revisited. J Mol Biol 249:816–831
15. Yona G, Brenner SE (2000) Comparison of protein sequences and practical database searching. In: Higgins D, Taylor W (eds) Bioinformatics: sequence, structure, and databanks. A practical approach. Oxford University Press, New York, pp 167–190
16. Rost B (1999) Twilight zone of protein sequence alignments. Protein Eng 12:85–94
17. Yu Y-K, Wootton JC, Altschul SF (2003) The compositional adjustment of amino acid substitution matrices. Proc Natl Acad Sci 100: 15688–15693
18. Simossis VA, Kleinjung J, Heringa J (2005) Homology-extended sequence alignment. Nucleic Acids Res 33:816–824
19. Sander C, Schneider R (1991) Database of homology-derived protein structures and the structural meaning of sequence alignment. Proteins 9:56–68
20. Chothia C, Lesk AM (1986) The relation between the divergence of sequence and structure in proteins. EMBO J 5:823–826
21. Simossis VA, Heringa J (2004) The influence of gapped positions in multiple sequence alignments on secondary structure prediction methods. Comput Biol Chem 28:351–366
22. Heringa J (2000) Computational methods for protein secondary structure prediction using multiple sequence alignments. Curr Protein Pept Sci 1:273–301
23. Chung R, Yona G (2004) Protein family comparison using statistical models and predicted structural information. BMC Bioinformatics 5:183
24. Ginalski K, Pas J, Wyrwicz LS et al (2003) ORFeus: Detection of distant homology using sequence profiles and predicted secondary structure. Nucleic Acids Res 31:3804–3807
25. Söding J (2005) Protein homology detection by HMM-HMM comparison. Bioinformatics 21:951–960
26. von Ohsen N, Sommer I, Zimmer R et al (2004) Arby: automatic protein structure prediction using profile-profile alignment and confidence measures. Bioinformatics 20:2228–2235
27. Ginalski K, von Grotthuss M, Grishin NV et al (2004) Detecting distant homology with Meta-BASIC. Nucleic Acids Res 32:W576–W581
28. Jones DT (1999) Protein secondary structure prediction based on position-specific scoring matrices. J Mol Biol 292:195–202
29. Pollastri G, Przybylski D, Rost B et al (2002) Improving the prediction of protein secondary structure in three and eight classes using recurrent neural networks and profiles. Proteins 47:228–235
30. Pollastri G, McLysaght A (2005) Porter: a new, accurate server for protein secondary structure prediction. Bioinformatics 21:1719–1720
31. Lin K, Simossis VA, Taylor WR et al (2005) A simple and fast secondary structure prediction method using hidden neural networks. Bioinformatics 21:152–159
32. Berman HM, Westbrook J, Feng Z et al (2000) The protein data bank. Nucleic Acids Res 28:235–242
33. Kabsch W, Sander C (1983) Dictionary of protein secondary structure: pattern recognition of hydrogen-bonded and geometrical features. Biopolymers 22:2577–2637

34. Lüthy R, McLachlan AD, Eisenberg D (1991) Secondary structure-based profiles: use of structure-conserving scoring tables in searching protein sequence databases for structural similarities. Proteins 10:229–239
35. Jones DT, Taylor WR, Thornton JM (1994) A mutation data matrix for transmembrane proteins. FEBS Lett 339:269–275
36. Shafrir Y, Guy HR (2004) STAM: simple transmembrane alignment method. Bioinformatics 20:758–769
37. Pirovano W, Feenstra KA, Heringa J (2008) PRALINETM: a strategy for improved multiple alignment of transmembrane proteins. Bioinformatics 24:492–497
38. Käll L, Krogh A, Sonnhammer ELL (2004) A combined transmembrane topology and signal peptide prediction method. J Mol Biol 338: 1027–1036
39. Krogh A, Larsson B, von Heijne G et al (2001) Predicting transmembrane protein topology with a hidden Markov model: application to complete genomes. J Mol Biol 305:567–580
40. Tusnády GE, Simon I (2001) The HMMTOP transmembrane topology prediction server. Bioinformatics 17:849–850
41. Ng PC, Henikoff JG, Henikoff S (2000) PHAT: a transmembrane-specific substitution matrix. Bioinformatics 16:760–766
42. Hirosawa M, Totoki Y, Hoshida M et al (1995) Comprehensive study on iterative algorithms of multiple sequence alignment. Comput Appl Biosci 11:13–18
43. Edgar RC (2004) MUSCLE: a multiple sequence alignment method with reduced time and space complexity. BMC Bioinformatics 5:113
44. Edgar RC (2004) MUSCLE: multiple sequence alignment with high accuracy and high throughput. Nucleic Acids Res 32:1792–1797
45. Pearson WR (2000) Flexible sequence similarity searching with the FASTA3 program package. Methods Mol Biol 132:185–219
46. Gonnet GH, Cohen MA, Benner SA (1992) Exhaustive matching of the entire protein sequence database. Science 256:1443–1445
47. Thompson JD, Koehl P, Ripp R et al (2005) BAliBASE 3.0: latest developments of the multiple sequence alignment benchmark. Proteins 61:127–136
48. Sammeth M, Heringa J (2006) Global multiple-sequence alignment with repeats. Proteins 64:263–274

Chapter 17

PROMALS3D: Multiple Protein Sequence Alignment Enhanced with Evolutionary and Three-Dimensional Structural Information

Jimin Pei and Nick V. Grishin

Abstract

Multiple sequence alignment (MSA) is an essential tool with many applications in bioinformatics and computational biology. Accurate MSA construction for divergent proteins remains a difficult computational task. The constantly increasing protein sequences and structures in public databases could be used to improve alignment quality. PROMALS3D is a tool for protein MSA construction enhanced with additional evolutionary and structural information from database searches. PROMALS3D automatically identifies homologs from sequence and structure databases for input proteins, derives structure-based constraints from alignments of three-dimensional structures, and combines them with sequence-based constraints of profile–profile alignments in a consistency-based framework to construct high-quality multiple sequence alignments. PROMALS3D output is a consensus alignment enriched with sequence and structural information about input proteins and their homologs. PROMALS3D Web server and package are available at http://prodata.swmed.edu/PROMALS3D.

Key words Multiple sequence alignment, Database searches, Three-dimensional structural alignment, Consistency-based scoring, Probabilistic model of profile–profile alignment

1 Introduction

Multiple sequence alignment (MSA) is fundamentally important for a variety of tasks in bioinformatics and computational biology, including homology-based structure modeling, prediction of structural properties, sequence similarity searches, phylogenetic reconstruction, and identification of functionally important sites. For a set of protein sequences, MSA construction involves placement of gap characters in sequences so that each position (column) contains evolutionarily or structurally equivalent amino acid residues. Such a biologically meaningful representation of multiple sequences not only facilitates their visualization and inspection, but also helps extraction of valuable information such as sequence conservation and residue preferences on a positional basis.

David J. Russell (ed.), *Multiple Sequence Alignment Methods*, Methods in Molecular Biology, vol. 1079,
DOI 10.1007/978-1-62703-646-7_17,

Accurate and fast MSA construction has been under extensive research with significant progress made in the last decade [1–5]. Dynamic programming algorithms [6, 7] are effective in aligning a pair of sequences (pairwise alignment), while such techniques are too time-consuming and memory-consuming to align a large number of sequences [8, 9]. Many MSA methods resort to a heuristic, the progressive alignment technique [10, 11], that reduces the task of aligning multiple sequences to a hierarchical series of pairwise alignments of sequence subsets. In early progressive methods, aligning two subsets of sequences only used information from these two subsets, and mistakes introduced in this process were fixed and propagated to later steps. One way to improve the alignment quality is through refinements after MSA assembly, often conducted by repeatedly dividing the MSA into sub-alignments and realigning the sub-alignments [12, 13]. Another popular alignment technique uses consistency-based scoring functions [14–16] to improve alignment quality by exploring information from the entire set of sequences when aligning subsets of sequences.

While various MSA methods generally produce high-quality alignments when sequence similarity is high (e.g., sequence identity above 40 %), it is still difficult to achieve accurate results for distantly related proteins. It is not uncommon for evolutionarily related proteins to have highly divergent sequences (e.g., sequence identity below 20 %) while maintaining similar structures and related functions. Alignments constructed with information from divergent sequences themselves are often prone to mistakes. Additional evolutionary information from homologous sequences is useful to enhance alignment quality. First, a protein sequence can be augmented with information from its homologs by using sequence profile, a numerical representation of positional amino acid usage. Profile-to-profile alignment is generally more accurate than sequence-to-sequence alignment [17, 18]. Secondly, positional structural properties such as secondary structures and solvent accessibilities can be predicted from sequence profile, and scoring functions incorporating predicted structural information can lead to better alignment quality [19, 20]. As protein spatial structures are generally more conserved than sequences [21], comparison of available three-dimensional (3D) structures can offer high-quality alignment constraints for MSA construction [22–24].

PROMALS3D [23, 25] is a tool for MSA construction that integrates various sources of evolutionary and structural information, such as sequence profile derived from database homologs, predicted secondary structures, and available 3D structures. PROMALS3D combines profile-derived alignment constraints and structure-derived alignment constraints within a consistency-based framework to produce protein MSAs of improved quality.

2 Methods

PROMALS3D is a progressive multiple protein sequence alignment tool. A key feature of PROMALS3D is that it uses different strategies to align subsets of sequences with different levels of difficulty to properly balance alignment accuracy and speed. Relatively similar sequences are aligned by a fast algorithm to form pre-aligned groups without retrieving additional information from databases. To align the relatively divergent pre-aligned groups, PROMALS3D resorts to more elaborate alignment techniques and uses additional information from homologous sequences and structures found by database searches. Such a method of using different strategies at different aligning stages allows PROMALS3D to align thousands of protein sequences in manageable time, since the time-consuming and memory-consuming steps of database search and consistency computation are only applied to a subset consisting of representative sequences of the pre-aligned groups instead of the entire set of input sequences. The flowchart of PROMALS alignment procedure is shown in Fig. 1.

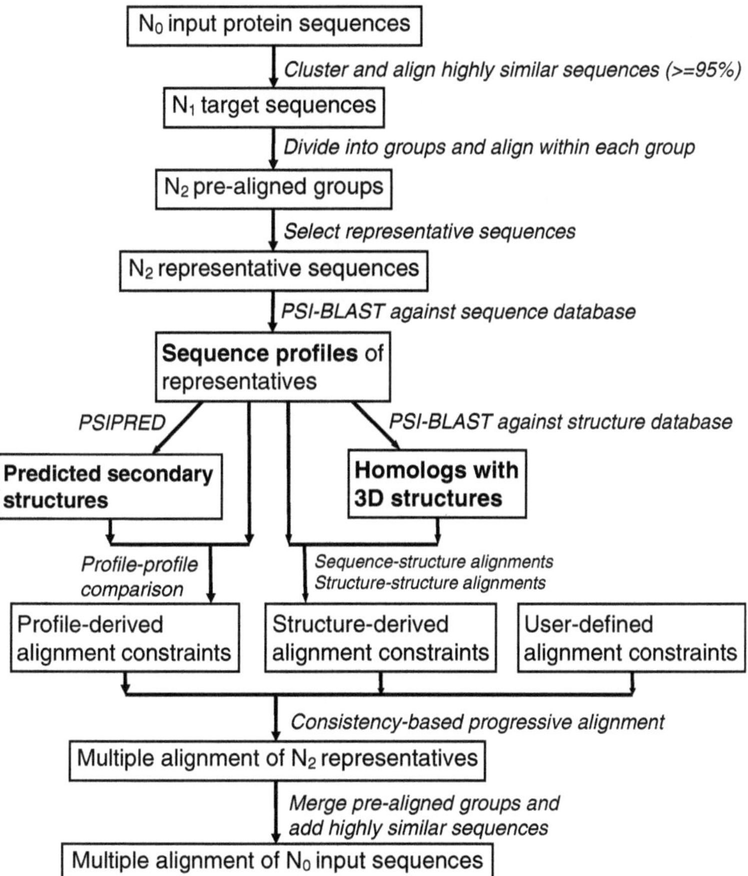

Fig. 1 Flowchart of the PROMALS3D method

2.1 Initial Clustering and Reducing Sequence Redundancy

For an input set with N_0 sequences, PROMALS3D first rapidly clusters sequences using the program CD-HIT [26] with a sequence identity cutoff of 95 % (-c option) and alignment coverage for the longer sequence of 0.95 (-aL option). This initial step results in N_1 clusters ($N_1 \leq N_0$) of highly similar sequences. Clusters with more than one sequence are individually aligned in a fast way by MAFFT (with –auto option) [27]. This step could significantly reduce computation for datasets with a large number of near-identical sequences. One target sequence is selected from each cluster. The N_1 target sequences after initial filtering of highly similar sequences are subject to further alignment steps described below.

2.2 Dividing Target Sequences to Groups and Obtaining Pre-aligned Groups

1. PROMALS3D divides the N_1 nonredundant target sequences into a set of N_2 groups ($N_2 \leq N_1$) and aligns each group without information from sequence and structure databases. Two methods are used to obtain the groups. If N_1 is no more than 200, PROMALS3D uses the UPGMA method to build a tree based on a crude measure of distances (k-mer counting) [12] among the sequences. Given a distance cutoff (-id_thr option, default: 0.6) the tree is divided into a set of subtrees, and the sequences in each subtree form a group [28]. If the number of formed groups is larger than the maximum number of groups set by PROMALS3D (-max_group_number option, default: 60), PROMALS3D automatically adjusts the distance cutoff so that the number of formed groups is the same as the maximum number of groups allowed.
2. We observed that the UPGMA method for deducing groups can produce one or more very large groups when the input dataset is large (e.g., thousands of sequences). These large groups may not be properly aligned without using additional information. Thus, for large sequence input datasets, instead of UPGMA we used a different method based on K-center clustering to divide the target sequences into a number of groups when the number of target sequences is more than 200. Our K-center approach does not allow any group to have more than 200 sequences. This method begins by randomly selecting K target sequences as the centers of K groups. Then the method makes iterations of the following two steps. Step (1) is to assign each target sequence to a group so that its distance to the center of this group is the smallest among its distances to all the group centers. Step (2) is to update the center for each group by selecting a target sequence with the minimum sum of distances to other target sequences in the same group. Our modification of this K-center method to control the maximum size of any group is that any group with 200 target sequences will not accept new members during Step (1).
3. After dividing the target sequences into N_2 groups, each group is aligned, resulting in N_2 pre-aligned groups. We have previously

used a progressive method with the sum-of-pairs BLOSUM62 [29] scores to align sequences within each group. Such an approach does not perform as well as some recent alignment methods. In the later development of PROMALS3D, we used MAFFT (options: –maxiterate 1000 –localpair) to perform alignment within each group to obtain better alignment quality for each pre-aligned group.

2.3 Aligning Pre-aligned Groups Enhanced with Evolutionary and Structural Information

1. The core steps of the PROMALS3D method use advanced techniques to align the relatively divergent pre-aligned groups with additional information from sequence and structure databases. First, a representative sequence is selected from each pre-aligned group, giving rise to N_2 representatives. Instead of using the longest sequence as the representative as in our original PROMALS method, we select the representative sequence that has the highest average similarity to other sequences in the same pre-aligned group.
2. Each representative sequence is subject to PSI-BLAST [30] iterations against the UniRef90 database [31] to retrieve sequence homologs. The sequence profile of PSI-BLAST searches is used to predict secondary structures by PSIPRED [32].
3. For each pair of representative sequences, we used a probabilistic model to obtain posterior profile–profile alignment probabilities for each position pair via the forward–backward algorithm. Strictly speaking, our probabilistic model for profile–profile comparison is not a hidden Markov model (HMM) as originally proposed [19], but a Conditional Random Field (CRF) [33], since we allowed observation-dependent transitions between hidden states. In our model, the transition probabilities depend on predicted secondary structures, which are used as a type of observations. Like that in HMMs, the forward–backward algorithm is applicable to CRFs to obtain posterior alignment probabilities, which serve as profile-derived alignment constraints.
4. PSI-BLAST profile is used to search a sequence database with known structures to retrieve homologs with 3D structures (homolog3Ds). Multiple homolog3Ds could be identified and used for one representative sequence, e.g., if it contains several distinct domains with known spatial structures. Structure-derived alignment constraints for two representative sequences are deduced from profile-based representative-to-homolog3D alignments and structure-based homolog3D-to-homolog3D alignments [23].
5. Profile-derived alignment constraints and structure-derived alignment constraints are combined for all pairs of

representatives. These constraints are subject to consistency measure to derive consistency-based scoring function.

6. The N_2 representatives are then progressively aligned by the consistency-based scoring function, with the aligning order following a UPGMA tree estimated for the representative sequences.
7. The pre-aligned groups are merged to the MSA of the N_2 representatives to form an MSA of N_1 target sequences. Finally, the clusters with highly similar sequences obtained at the initial clustering step are merged to the MSA of N_1 target sequences to form the MSA of all input sequences.

3 PROMALS3D Usage and Practical Issues

1. PROMALS3D is available as a Web server as well as a downloadable package at http://prodata.swmed.edu/PROMALS3D.
2. PROMALS3D Web server allows input of both sequences and structures. The Web server extracts sequences from input structures and combines them with input sequences to form the final input sequence set. The Web server also prepares structural alignments for input structures and feeds them as structural constraints to the PROMALS3D program. On the other hand, the PROMALS3D downloadable package currently only takes sequences as input.
3. If only structures are input to the PROMALS3D Web server, the final alignment is a consistency-based multiple structure alignment that integrates both structural information and homolog-derived sequence information.
4. Input sequences should be in FASTA format and should not have identical names. Certain characters in sequence names are changed to "_", including space, tab, and *?'‘";&\|/{})(][$, but "." (dot) and "–" are kept.
5. PROMALS3D Web server [25] offers various options of customization of the final alignment output, such as displaying the alignment with sequences colored by predicted secondary structures and showing a consensus sequence and positional conservation indices [34].
6. The UPGMA tree built for target sequences is reported. Since it is based on a very crude measurement of evolutionary distances, it would not serve well for phylogenetic purposes.
7. The structure database is regularly updated in an automatic fashion. The structure database contains a nonredundant set of structures from the PDB database. The CD-HIT program is used to cluster sequences with 3D structures at the 70 %

identity level. Within each cluster, one representative structure is selected. X-ray structures are preferred over NMR or CryoEM structures. Among the X-ray structures, the one with the lowest resolution is selected as the representative. The representative structures that are classified in the SCOP database are further split into structural domains according to SCOP domain definitions.

8. The PROMALS3D Web server also offers an option to use PROMALS [19] to align within each pre-aligned group instead of MAFFT. This option is currently not available in the downloadable package. PROMALS uses sequence homologs and predicted secondary structures, and thus often produces better alignment results. This is helpful to achieve better overall alignment quality when pre-aligned groups themselves contain divergent sequences. However, due to database searching PROMALS is more time-consuming compared to MAFFT.
9. Three options of structural alignments and their combinations are offered: DaliLite [35], FAST [36], and TM-align [37]. DaliLite gives slightly better results than FAST and TM-align [23]. Using combinations of them also provides slight improvement of alignment accuracy [23]. The default option of the PROMALS3D Web server is the combination of FAST and TMalign. DaliLite is computationally intensive when the structures are large (e.g., with more than 500 residues).
10. In addition to input sequences and structures, PROMALS3D also allows input of alignment constraints (user-defined constraints).
11. While PROMALS3D compares favorably to a number of other methods on an average basis [23], it does not mean it can outperform any method for any individual alignment case. For regions with uncertainty, inspection of results produced by other methods could be helpful to manually improve alignment quality.
12. The advantage of PROMALS3D is the incorporation of information from homologous sequences and structures. However, mistakes may be introduced in the process. For example, PSI-BLAST may find nonhomologous sequences (profile corrupt), and the PSI-BLAST alignment between the query and its hits may contain errors that could lead to inferior profile or wrong profile–profile alignment. The PSI-BLAST results of sequence and structure database searches are kept and can be accessed from the PROMALS3D Web server.
13. Alignment mistakes could also be caused by wrong secondary structure predictions. While PSIPRED secondary structure prediction accuracy is on average about 70–80 %, it is more difficult to obtain accurate predictions for beta-strands and in cases where few homologous sequences exist.

14. PROMALS3D method generally works best when sequences are of similar lengths and do not contain large nonhomologous regions (e.g., inserted nonhomologous domains).
15. Difficult cases that PROMALS3D may not perform well on include sequences with repeats, duplications or circular permutations, sequences with many disordered regions or low complexity regions, and sequences with predicted transmembrane segments.
16. Input datasets with many long sequences (e.g., >1,000 amino acid residues) may cause memory crash. In these cases, reduction of the number of pre-aligned groups is recommended, which can be done by setting lower distance cutoff (-id_thr option) or setting lower maximum number of pre-aligned groups allowed (-max_group_number option).

Acknowledgments

The work is supported in part by the National Institutes of Health (GM094575 to NVG) and the Welch Foundation (I-1505 to NVG).

References

1. Do CB, Katoh K (2008) Protein multiple sequence alignment. In: Walker J (ed) Methods Mol Biol, vol 484, 1st edn. Humana, Totowa, pp 379–413
2. Pei J (2008) Multiple protein sequence alignment. Curr Opin Struct Biol 18(3):382–386
3. Notredame C (2007) Recent evolutions of multiple sequence alignment algorithms. PLoS Comput Biol 3(8):e123
4. Edgar RC, Batzoglou S (2006) Multiple sequence alignment. Curr Opin Struct Biol 16(3):368–373
5. Wallace IM, Blackshields G, Higgins DG (2005) Multiple sequence alignments. Curr Opin Struct Biol 15(3):261–266
6. Needleman SB, Wunsch CD (1970) A general method applicable to the search for similarities in the amino acid sequence of two proteins. J Mol Biol 48(3):443–453
7. Smith TF, Waterman MS (1981) Identification of common molecular subsequences. J Mol Biol 147(1):195–197
8. Lipman DJ, Altschul SF, Kececioglu JD (1989) A tool for multiple sequence alignment. Proc Natl Acad Sci USA 86(12):4412–4415
9. Wang L, Jiang T (1994) On the complexity of multiple sequence alignment. J Comput Biol 1 (4):337–348
10. Feng DF, Doolittle RF (1987) Progressive sequence alignment as a prerequisite to correct phylogenetic trees. J Mol Evol 25(4):351–360
11. Thompson JD, Higgins DG, Gibson TJ (1994) CLUSTAL W: improving the sensitivity of progressive multiple sequence alignment through sequence weighting, position-specific gap penalties and weight matrix choice. Nucleic Acids Res 22(22):4673–4680
12. Edgar RC (2004) MUSCLE: multiple sequence alignment with high accuracy and high throughput. Nucleic Acids Res 32(5): 1792–1797
13. Katoh K, Misawa K, Kuma K, Miyata T (2002) MAFFT: a novel method for rapid multiple sequence alignment based on fast Fourier transform. Nucleic Acids Res 30(14):3059–3066
14. Notredame C, Higgins DG, Heringa J (2000) T-Coffee: a novel method for fast and accurate multiple sequence alignment. J Mol Biol 302 (1):205–217
15. Do CB, Mahabhashyam MS, Brudno M, Batzoglou S (2005) ProbCons: probabilistic consistency-based multiple sequence alignment. Genome Res 15(2):330–340
16. Pei J, Grishin NV (2006) MUMMALS: multiple sequence alignment improved by using hidden Markov models with local structural

information. Nucleic Acids Res 34(16): 4364–4374

17. Sadreyev R, Grishin N (2003) COMPASS: a tool for comparison of multiple protein alignments with assessment of statistical significance. J Mol Biol 326(1):317–336
18. Soding J (2005) Protein homology detection by HMM-HMM comparison. Bioinformatics 21(7):951–960
19. Pei J, Grishin NV (2007) PROMALS: towards accurate multiple sequence alignments of distantly related proteins. Bioinformatics 23(7): 802–808
20. Deng X, Cheng J (2011) MSACompro: protein multiple sequence alignment using predicted secondary structure, solvent accessibility, and residue-residue contacts. BMC Bioinformatics 12:472
21. Chothia C, Lesk AM (1986) The relation between the divergence of sequence and structure in proteins. EMBO J 5(4):823–826
22. Armougom F, Moretti S, Poirot O, Audic S, Dumas P, Schaeli B, Keduas V, Notredame C (2006) Expresso: automatic incorporation of structural information in multiple sequence alignments using 3D-Coffee. Nucleic Acids Res 34(Web Server issue):W604–W608
23. Pei J, Kim BH, Grishin NV (2008) PROMALS3D: a tool for multiple protein sequence and structure alignments. Nucleic Acids Res 36(7):2295–2300
24. Zhou H, Zhou Y (2005) SPEM: improving multiple sequence alignment with sequence profiles and predicted secondary structures. Bioinformatics 21(18):3615–3621
25. Pei J, Tang M, Grishin NV (2008) PROMALS3D web server for accurate multiple protein sequence and structure alignments. Nucleic Acids Res 36(Web Server issue): W30–W34
26. Li W, Godzik A (2006) Cd-hit: a fast program for clustering and comparing large sets of protein or nucleotide sequences. Bioinformatics 22(13):1658–1659
27. Katoh K, Kuma K, Toh H, Miyata T (2005) MAFFT version 5: improvement in accuracy of multiple sequence alignment. Nucleic Acids Res 33(2):511–518
28. Pei J, Sadreyev R, Grishin NV (2003) PCMA: fast and accurate multiple sequence alignment based on profile consistency. Bioinformatics 19(3):427–428
29. Henikoff S, Henikoff JG (1992) Amino acid substitution matrices from protein blocks. Proc Natl Acad Sci USA 89(22): 10915–10919
30. Altschul SF, Madden TL, Schaffer AA, Zhang J, Zhang Z, Miller W, Lipman DJ (1997) Gapped BLAST and PSI-BLAST: a new generation of protein database search programs. Nucleic Acids Res 25(17):3389–3402
31. Suzek BE, Huang H, McGarvey P, Mazumder R, Wu CH (2007) UniRef: comprehensive and non-redundant UniProt reference clusters. Bioinformatics 23(10):1282–1288
32. Jones DT (1999) Protein secondary structure prediction based on position-specific scoring matrices. J Mol Biol 292(2):195–202
33. Lafferty J, McCallum A, Pereira F (2001) Conditional random fields: Probabilistic models for segmenting and labeling sequence data. Proceedings of the 18th international conference on machine learning, pp 282–289
34. Pei J, Grishin NV (2001) AL2CO: calculation of positional conservation in a protein sequence alignment. Bioinformatics 17(8): 700–712
35. Holm L, Park J (2000) DaliLite workbench for protein structure comparison. Bioinformatics 16(6):566–567
36. Zhu J, Weng Z (2005) FAST: a novel protein structure alignment algorithm. Proteins 58(3): 618–627
37. Zhang Y, Skolnick J (2005) TM-align: a protein structure alignment algorithm based on the TM-score. Nucleic Acids Res 33(7): 2302–2309

Chapter 18

MSACompro: Improving Multiple Protein Sequence Alignment by Predicted Structural Features

Xin Deng and Jianlin Cheng

Abstract

Multiple Sequence Alignment (MSA) is an essential tool in protein structure modeling, gene and protein function prediction, DNA motif recognition, phylogenetic analysis, and many other bioinformatics tasks. Therefore, improving the accuracy of multiple sequence alignment is an important long-term objective in bioinformatics. We designed and developed a new method MSACompro to incorporate predicted secondary structure, relative solvent accessibility, and residue–residue contact information into the currently most accurate posterior probability-based MSA methods to improve the accuracy of multiple sequence alignments. Different from the multiple sequence alignment methods that use the tertiary structure information of some sequences, our method uses the structural information purely predicted from sequences. In this chapter, we first introduce some background and related techniques in the field of multiple sequence alignment. Then, we describe the detailed algorithm of MSACompro. Finally, we show that integrating predicted protein structural information improved the multiple sequence alignment accuracy.

Key words Multiple sequence alignment, Bioinformatics, Secondary structure, Solvent accessibility, Residue–residue contact information, Posterior probability-based

1 Introduction

Multiple sequence alignment methods are central to many challenging bioinformatics problems, such as protein function prediction, protein homology identification, protein structure prediction, protein interaction study, mutagenesis analysis, and phylogenetic tree construction. Since a few decades ago, a number of methods and tools have been developed for multiple sequence alignment, which facilitated the development of the bioinformatics field.

Well-established techniques, such as iterative alignment [1], progressive alignment [2], alignment based on profile hidden Markov models [3], and posterior alignment probability transformation [4, 5] have been widely adapted in state of art multiple sequence alignment methods to enhance alignment accuracy. Besides, known 3D structure information is also used by some alignment methods,

David J. Russell (ed.), *Multiple Sequence Alignment Methods*, Methods in Molecular Biology, vol. 1079,
DOI 10.1007/978-1-62703-646-7_18, © Springer Science+Business Media, LLC 2014

such as 3D-Coffee [6]. However, although such a technique improves multiple sequence alignment, it cannot be applied to the majority of protein sequences without known tertiary structures. Aiming at overcoming such a problem, we have developed MSA-Compro, a new multiple sequence alignment method, which effectively utilizes predicted secondary structure, relative solvent accessibility, and residue–residue contact map together with posterior alignment probabilities produced by both pair hidden Markov models and partition function as in MSAProbs [4]. Moreover, applying predicted relative solvent accessibility and residue–residue contact map to multiple sequence alignment is novel, although a few attempts had been made to use predicted secondary structure information [7–12].

Following the basic scheme in MSAProbs [4], MSACompro has five main steps: (1) construct the pairwise posterior alignment probability matrices based on both pair-HMM and partition function, utilizing the similarity in amino acids, secondary structure, and relative solvent accessibility; (2) generate the pairwise distance matrix from both the pairwise posterior probability matrices constructed in the first stage and the newly introduced pairwise contact map similarity matrices; (3) build up a guide tree according to pairwise distance matrix, and calculate sequence weights; (4) adapt a weighting scheme to transform all the pairwise posterior matrices; (5) perform a progressive alignment by computing the profile–profile alignment from the probability matrices of all sequence pairs, and then an iterative alignment to refine the results from progressive alignment. Different from MSAProbs, our method considers secondary structure and solvent accessibility information in the calculation of the posterior residue–residue alignment probabilities and computes the pairwise distance matrix in light of predicted residue–residue contact information.

2 Materials

We released the latest version of the software MSACompro1.2.0 at the MULTICOM toolbox Web site http://sysbio.rnet.missouri.edu/multicom_toolbox/tools.html. To install MSACompro, users need to download Pspro2.0 package at the same Web site first, and then install MSACompro according to the instruction in the package.

The simplest way to use MSACompro1.2.0 is to go to ./script directory of the software package and directly run the script auto_run_msacompro.pl which automatically predicts secondary structure, solvent accessory and contact map information during the alignment process. The standard command for auto_run_msacompro.pl is:

```
./auto_run_msacompro.pl arg1 arg2 arg3
```

The three inputs are respectively the path for Pspro bin directory, the given file of the target multiple sequences in fasta format and the output multiple sequence alignment file for target proteins. An example command is as follows:

```
./auto_run_msacompro.pl
/storage/shared/pspro2/bin/
../test/BB40004.fasta ../test/BB40004.msa
```

There are ways for users to run MSACompro based on their own structural information data gathered in advance, which are described in the Readme.txt file.

3 Methods

3.1 Calculation of Pairwise Posterior Probability Matrices Integrating the Predicted Structural Information

Fig. 1 shows the workflow of MSACompro method. Given input multiple protein sequences, pairwise posterior probability matrices are first generated based on a partition function which integrates the predicted structural information and pairwise Hidden Markov Model. Then pairwise distance matrices between the proteins are constructed by combining both the posterior probability matrices and newly introduced contact map similarity matrices. Based on the distances between the protein pairs, a guide tree is built up and the posterior probability matrices are transformed by a weighting scheme. Finally, a progressive alignment and iterative alignment refinement were performed to get a final multiple sequence alignment. More details are discussed as follows.

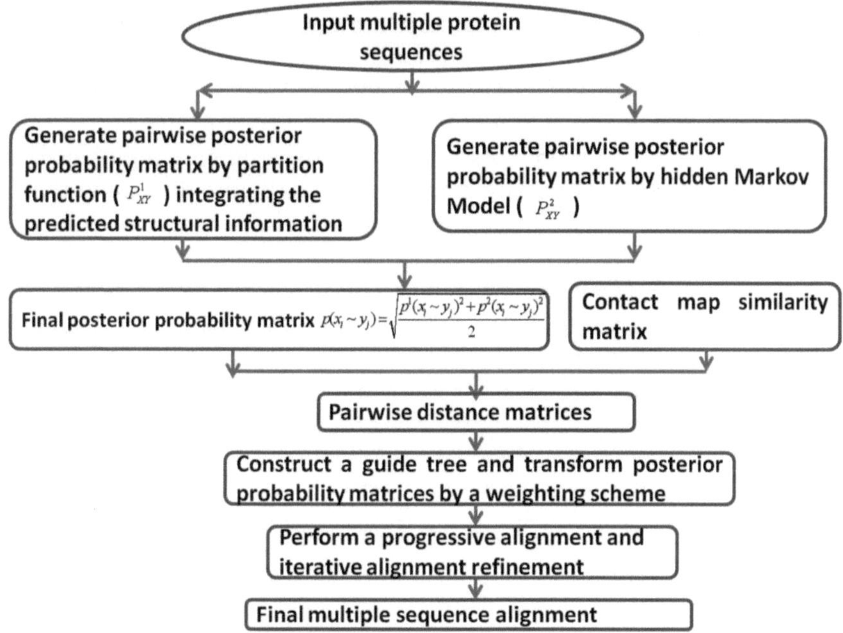

Fig. 1 The workflow of MSACompro method

Suppose we are going to align a protein sequence group S, in which protein sequences X and Y are considered as two representatives. The sequences of X and Y are denoted as $X=(x_1, x_2, \ldots, x_{n1})$, $Y=(y_1, y_2, \ldots, y_{n2})$, where $x_1, x_2, \ldots, x_{n1}$ and $y_1, y_2, \ldots, y_{n2}$ are lists of the residues in X and Y, respectively. n_1 and n_2 are the length of sequence X and Y, respectively. x_i is the i-th amino acid in sequence X, and y_j is the j-th amino acid in sequence Y. We let aln represent a global alignment between X and Y, ALN the set of all the possible global alignments of X and Y, and aln* ∈ ALN the true pairwise alignment of X and Y. Following MSAProbs, the posterior probability that the i-th residue in X (x_i) is aligned to the j-th residue (y_j) in Y in aln* is defined as:

$$p(x_i \sim y_j \in \text{aln}^*|X, Y) = \sum_{\text{aln}\in\text{ALN}} P(\text{aln}|X, Y) I\{x_i \sim y_j \in \text{aln}\} \quad (1 \le x_i \le n_1,\ 1 \le y_j \le n_2) \tag{1}$$

$$I\{x_i \sim y_j \in \text{aln}\} = \begin{cases} 1, & \text{if}(x_i \sim y_j \in \text{aln})\text{true} \\ 0, & \text{otherwise} \end{cases}$$

$P(\text{aln}|X, Y)$ is the posterior probability that aln is the true alignment aln*. Thus, the $n_1 \times n_2$ posterior probability matrix P_{XY} is a matrix including all the values $p(x_i \sim y_j \in \text{aln}^*|X, Y)$ ($p(x_i \sim y_j)$ for short) for $1 \le x_i \le n_1,\ 1 \le y_j \le n_2$. The calculation process of the pairwise posterior probability matrix is described as below.

The pairwise posterior probability matrix in MSACompro is combination of two types of pairwise posterior probability matrices (P^1_{XY} and P^2_{XY}) calculated by two different methods (a pair hidden Markov model and a partition function) respectively. The first kind of pairwise probability matrix P^1_{XY} is calculated by a partition function (F) of alignments based on dynamic programming. $F(i, j)$ represents the probability of all partial global alignments of X and Y ending at position (i, j). Before discussing the calculation of $F(i, j)$, three other probabilities are introduced: $F_M(i, j)$, the probability of all partial global alignments with x_i aligned to y_j; $F_Y(i, j)$, the probability of all partial global alignments with y_j aligned to a gap; $F_X(i, j)$, the probability of all partial global alignments with x_i aligned to a gap. Accordingly, $F(i, j)$ can be calculated recursively as follows:

$$\begin{aligned} F_M(i,j) &= F(i-1, j-1)\mathrm{e}^{W_1\beta s(x_i, y_j) + W_2 \mathrm{SS}(\mathrm{ss}(x_i), \mathrm{ss}(y_j)) + W_3 \mathrm{SA}(\mathrm{sa}(x_i), \mathrm{sa}(y_j))} \\ F_Y(i,j) &= F_M(i, j-1)\mathrm{e}^{\beta \mathrm{gap}} + F_Y(i, j-1)\mathrm{e}^{\beta \mathrm{ext}} \\ F_X(i,j) &= F_M(i-1, j)\mathrm{e}^{\beta \mathrm{gap}} + F_X(i-1, j)\mathrm{e}^{\beta \mathrm{ext}} \\ F(i,j) &= F_M(i,j) + F_Y(i,j) + F_X(i,j) \end{aligned} \tag{2}$$

$$W_1 + W_2 + W_3 = 1$$

In the formula (Eq. 2), $s(x_i, y_j)$ is the amino acid similarity score between x_i and y_j, which is an element at the i-th row and j-th column of the $n_1 \times n_2$ amino acid substitution matrix s. Similarly, $\mathrm{SS}(\mathrm{ss}(x_i), \mathrm{ss}(y_j))$ is the similarity score between the secondary structure $(\mathrm{ss}(x_i))$ of residue x_i in protein X and that of residue y_j in protein Υ according to the secondary structure similarity matrix SS, $\mathrm{SA}(\mathrm{sa}(x_i), \mathrm{sa}(y_j))$ is the similarity score between the relative solvent accessibility $(\mathrm{sa}(x_i))$ of residue x_i in protein X and that of residue y_j in protein Υ according to the solvent accessibility similarity matrix SA. W_1, W_2, W_3 are weights for the amino acid similarity score, secondary structure similarity score and solvent accessibility similarity score. The secondary structure and solvent accessibility can be automatically predicted by PSpro2.0 [13] (http://sysbio.rnet.missouri.edu/multicom_toolbox/) using a multi-threading technique implemented in MSACompro, or alternatively be provided by a user. The three weights W_1, W_2, W_3 are set to 0.4, 0.5, and 0.1 by default, and can be adjusted by users as well. Referring to MSAprobs, β is a parameter measuring the deviation between suboptimal and optimal alignments, $\mathrm{gap}(\mathrm{gap} \leq 0)$ is the gap open penalty, and $\mathrm{ext}(\mathrm{ext} \leq 0)$ is the gap extension penalty. We set these three parameters the same as the values used in MSAprobs.

Gonnet 160 matrix was used as a substitution matrix to generate the similarity scores between two amino acids in proteins [14]. In addition, we designed a simple 3×3 secondary structure similarity matrix SS, containing the similarity scores of three kinds of secondary structures (E, H, C) as follows:

$$\mathrm{SS} = \begin{bmatrix} 100 \\ 010 \\ 001 \end{bmatrix},$$

where two identical secondary structures receive a score of 1 and otherwise receive a score of 0.

Similarly, we also came up with a 2×2 solvent accessibility similarity matrix SA, consisting of the similarity scores of two types of relative solvent accessibilities (e, b) as follows:

$$\mathrm{SA} = \begin{bmatrix} 10 \\ 01 \end{bmatrix},$$

where two identical solvent accessibility receive a score of 1 and different ones a score of 0. Applying more advance scoring matrices defined in [15] may lead to further improvement.

Each posterior residue–residue alignment probability element in the first kind of posterior probability matrix $(P^1_{X\Upsilon})$ can be calculated from the partition function as:

$$p^1(x_i \sim y_j) = \frac{F_M(i-1,j-1)F'_M(i+1,j+1)}{F} \cdot e^{W_1 \beta s(x_i,y_j)+W_2\mathrm{SS}(\mathrm{ss}(x_i),\mathrm{ss}(y_j))+W_3\mathrm{SA}(\mathrm{sa}(x_i),\mathrm{sa}(y_j))}, \quad (3)$$

where $F'_M(i,j)$ is the partition function of all the reverse alignments from the ending position (n_1, n_2) till position (i, j) with x_i aligned to y_j.

The second kind of pairwise probability matrix P^2_{XY} is calculated by a pair hidden Markov model (HMM) combining both Forward and Backward algorithm [4, 5, 16]. State emissions and transitions are used in pair HMM to calculate the pairwise probabilities. No secondary structure and solvent accessibility information is used to generate the second type of pairwise probability matrix P^2_{XY}.

Based on both P^1_{XY} and P^2_{XY}, the final posterior probability matrix P_{XY} is calculated as the root mean square of them:

$$p(x_i \sim y_j) = \sqrt{\frac{p^1(x_i \sim y_j)^2 + p^2(x_i \sim y_j)^2}{2}} \quad (4)$$

where $P^1(x_i \sim y_j)$ and $P^2(x_i \sim y_j)$ denote a posterior probability element in two kinds of posterior probability matrices (P^1_{XY} and P^2_{XY}), respectively.

3.2 Calculation of Pairwise Distance Matrices from Both Pairwise Posterior Probabilities and Pairwise Contact Map Scores

After the posterior probability matrix P_{XY} is built by dynamic programming in the previous stage, an optimal sub-alignment score matrix AS is calculated based on P_{XY} in terms of the Eq. 5 below. The optimal global alignment score Opt(X, Y) of the global alignment is computed according to matrix AS. The optimal sub-alignment score AS(i, j) represents the score of the optimal sub-alignment ending at residues i and j in X and Y. The AS matrix is recursively calculated as:

$$\mathrm{AS}(i,j) = \max\begin{cases} \mathrm{AS}(i-1,j-1) + P_{XY}(x_i \sim y_j) \\ \mathrm{AS}(i-1,j) \\ \mathrm{AS}(i,j-1) \end{cases} \quad (5)$$

AS(n_1, n_2) is the optimal score of the full global alignment between X and Y, which is denoted as Optscore(X, Y).

Consequently, an optimal pairwise alignment of X and Y is generated by tracing back through the matrix AS. We also introduce a contact map score, CMscore(X, Y), for the optimal pairwise alignment of X and Y, since it is believed that the spatially neighboring residues of two aligned residues have higher possibility to be aligned together. CMscore(X, Y) is calculated from the contact map correlation score matrix CMap$_{XY}$ based on the residue–residue contact map matrices CMap$_X$ and CMap$_Y$ of X and Y.

Suppose we get the optimal global alignment of X and Y by tracing back through AS as follows:

$$x_1 x_2 \ldots - x_m \ldots x_p \ldots x_{n1}$$
$$y_1 - \ldots y_k y_{k+1} \ldots - \ldots y_{n2}$$

For the purpose of calculating CMscore(X, Y), a new alignment is generated after removing the pairs containing gaps:

$$x_1 \ldots x_m \ldots x_{n1}$$
$$y_1 \ldots y_{k+1} \ldots y_{n2}$$

We also denote the new alignment as:

$$x'_1 x'_2 \ldots x'_n$$
$$y'_1 y'_2 \ldots y'_n,$$

where n is the length of the new alignment without gaps.

From this alignment, we can construct two contact map matrices, CMap_X and CMap_Y, which consist of predicted contact probability scores for sequences of X and Y respectively, as follows:

$$\text{CMap}_X = \begin{bmatrix} x'_{11} x'_{12} \cdots x'_{1n} \\ x'_{21} x'_{22} \cdots x'_{2n} \\ \vdots \\ x'_{n1} x'_{n2} \cdots x'_{nn} \end{bmatrix} \quad (6)$$

$$\text{CMap}_Y = \begin{bmatrix} y'_{11} y'_{12} \cdots y'_{1n} \\ y'_{21} y'_{22} \cdots y'_{2n} \\ \vdots \\ y'_{n1} y'_{n2} \cdots y'_{nn} \end{bmatrix}$$

x'_{ij} is the predicted contact probability score between amino acid x'_i and x'_j in protein sequence X, and similarly, y'_{ij} is the predicted contact probability score between amino acid y'_i and y'_j in protein sequence Y. The residue–residue contact probability scores introduced above are predicted from the protein sequence by NNcon [17] (http://sysbio.rnet.missouri.edu/multicom_toolbox/). The contact map correlation score matrix CMap_{XY} is designed in our MSACompro as the multiplication of CMap_X and CMap_Y:

$$\text{CMap}_{XY} = \text{CMap}_X \times \text{CMap}_Y$$
$$= \begin{bmatrix} xy'_{11} xy'_{12} \cdots xy'_{1n} \\ xy'_{21} xy'_{22} \cdots xy'_{2n} \\ \vdots \\ xy'_{n1} xy'_{n1} \cdots xy'_{nn} \end{bmatrix} \quad (7)$$

$$\mathrm{CMscore}(X, \Upsilon) = \frac{1}{n^2}\sum_{i=1}^{n} \mathrm{CMap}_{X\Upsilon}(i, i)$$
$$= \frac{1}{n^2}\sum_{i=1}^{n} xy'_{ii} = \frac{1}{n^2}\sum_{i=1}^{n}\sum_{j=1}^{n} x'_{ij}y'_{ji}$$

We only need to calculate the diagonal values in $\mathrm{CMap}_{X\Upsilon}$ in implementation, so as to speed up the program. What's more, the pairwise distance between sequences *X* and *Υ* can be calculated as:

$$\mathrm{d}(X, \Upsilon) = 1 - \frac{W_4\mathrm{Optscore}(X, \Upsilon)}{\min\{n_1, n_2\}} - W_5\mathrm{CMscore}(X, \Upsilon) \quad (8)$$

The sum of W_4 and W_5 is 1. They are used to control the influence of sequences *X* and *Υ*.

3.3 Construction of Guide Tree and Transformation of Posterior Probability

A guide tree is constructed by the UPGMA method based on the linear combinatorial strategy [18]. Referring to the calculation scheme adapted in MSAProbs, the distance between a new cluster *Z* formed by merging clusters *X* and *Υ*, and another cluster *W* is:

$$\mathrm{d}(W, Z) = \frac{\mathrm{d}(W, X) \times \mathrm{Num}(X) + \mathrm{d}(W, \Upsilon) \times \mathrm{Num}(\Upsilon)}{\mathrm{Num}(X) + \mathrm{Num}(\Upsilon)} \quad (9)$$

where Num(*X*) is the number of leafs in cluster *X*.

After constructing the guide tree, we adapted the sequence weighting scheme in [4]. Based on the weights, we transformed the original posterior probability in terms of the equation below, so as to reduce the bias of sampling similar sequences:

$$P'_{X\Upsilon} = \frac{1}{wN}\left((w_X + w_\Upsilon)P_{X\Upsilon} + \sum_{Z\in S, Z\neq X,\Upsilon} w_z P_{XZ}P_{Z\Upsilon})\right) \quad (10)$$

w_X and w_Υ are the weight of sequences *X* and *Υ*, *Z* represents any sequence other than *X* or *Υ* in the given sequence group, w_Z is the weight of *Z*, and *wN* is the sum of sequence weights in dataset *S*.

3.4 Combination of Initial Progressive Alignment and Final Iterative Alignment Refinement

We use the guide tree to create a multiple sequence alignment by progressively aligning two clusters of the most similar sequences together. During the progressive alignment process, we apply a weighted profile–profile alignment to align two clusters of sequences according to the weighting scheme introduced in the previous step. The posterior alignment probability matrix of two clusters/profiles is averaged from the probability matrices of all sequence pairs (*X*, *Υ*), in which *x* and *y* are from two different clusters. In the earlier stage, global profile–profile alignment is generated based on the posterior alignment probability matrices of the profiles in terms of the equation (5). Furthermore, we use a randomized iterative alignment to refine the initial alignment, so as to improve the alignment accuracy. As the iterative refinement process, the given sequence group A is randomly separated into two groups, and further profile–profile alignment is performed on the two groups. In our program, the refinement iteration time is set to be 10 by default, or can also be set by users.

The final progressive alignment is performed along the guide tree from closely related to distantly related. Then a final iterative alignment is applied to refine the results from progressive alignment. In addition, a multi-threading technology based on OPENMP is also used to improve the time efficiency of the program [19].

4 Case Study

We used BAliBASE 3.0 [20] data sets as training sets, and SABmark 1.65 [21] and OXBENCH [22] data sets as testing sets, for the parameter optimization. The alignment results were evaluated according to sum-of-pairs (SP) score and true column (TC) scores [4, 23, 24]. We used the application bali_score provided by BAliBASE 3.0 to calculate these scores. Users can download the protein sequence alignment benchmark datasets from http://www.drive5.com/bench/.

Fig. 2 shows an example input multiple sequence file BB12003 from BAliBASE 3.0 database and its output by running

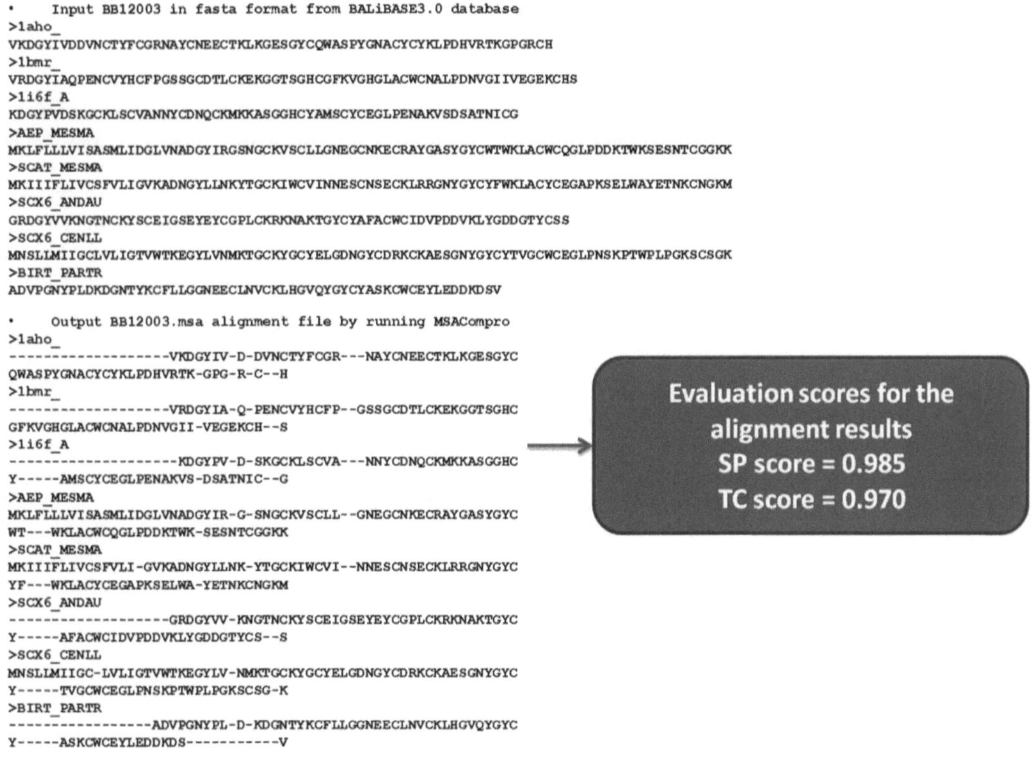

```
•    Input BB12003 in fasta format from BALiBASE3.0 database
>1aho_
VKDGYIVDDVNCTYFCGRNAYCNEECTKLKGESGYCQWASPYGNACYCYKLPDHVRTKGPGRCH
>1bmr_
VRDGYIAQPENCVYHCFPGSSGCDTLCKEKGGTSGHCGFKVGHGLACWCNALPDNVGIIVEGEKCHS
>1i6f_A
KDGYPVDSKGCKLSCVANNYCDNQCKMKKASGGHCYAMSCYCEGLPENAKVSDSATNICG
>AEP_MESMA
MKLFLLLVISASMLIDGLVNADGYIRGSNGCKVSCLLGNEGCNKECRAYGASYGYCWTWKLACWCQGLPDDKTWKSESNTCGGKK
>SCAT_MESMA
MKIIIFLIVCSFVLIGVKADNGYLLNKYTGCKIWCVINNESCNSECKLRRGNYGYCYFWKLACYCEGAPKSELWAYETNKCNGKM
>SCX6_ANDAU
GRDGYVVKNGTNCKYSCEIGSEYEYCGPLCKRKNAKTGYCYAFACWCIDVPDDVKLYGDDGTYCSS
>SCX6_CENLL
MNSLLMIIGCLVLIGTVWTKEGYLVNMKTGCKYGCYELGDNGYCDRKCKAESGNYGYCYTVGCWCEGLPNSKPTWPLPGKSCSGK
>BIRT_PARTR
ADVPGNYPLDKDGNTYKCFLLGGNEECLNVCKLHGVQYGYCYASKCWCEYLEDDKDSV

•    Output BB12003.msa alignment file by running MSACompro
>1aho_
-------------------VKDGYIV-D-DVNCTYFCGR---NAYCNEECTKLKGESGYC
QWASPYGNACYCYKLPDHVRTK-GPG-R-C--H
>1bmr_
-------------------VRDGYIA-Q-PENCVYHCFP--GSSGCDTLCKEKGGTSGHC
GFKVGHGLACWCNALPDNVGII-VEGEKCH--S
>1i6f_A
--------------------KDGYPV-D-SKGCKLSCVA---NNYCDNQCKMKKASGGHC
Y-----AMSCYCEGLPENAKVS-DSATNIC--G
>AEP_MESMA
MKLFLLLVISASMLIDGLVNADGYIR-G-SNGCKVSCLL--GNEGCNKECRAYGASYGYC
WT---WKLACWCQGLPDDKTWK-SESNTCGGKK
>SCAT_MESMA
MKIIIFLIVCSFVLI-GVKADNGYLLNK-YTGCKIWCVI--NNESCNSECKLRRGNYGYC
YF---WKLACYCEGAPKSELWA-YETNKCNGKM
>SCX6_ANDAU
-------------------GRDGYVV-KNGTNCKYSCEIGSEYEYCGPLCKRKNAKTGYC
Y-----AFACWCIDVPDDVKLYGDDGTYCS--S
>SCX6_CENLL
MNSLLMIIGC-LVLIGTVWTKEGYLV-NMKTGCKYGCYELGDNGYCDRKCKAESGNYGYC
Y-----TVGCWCEGLPNSKPTWPLPGKSCSG-K
>BIRT_PARTR
-----------------ADVPGNYPL-D-KDGNTYKCFLLGGNEECLNVCKLHGVQYGYC
Y-----ASKCWCEYLEDDKDS-----------V
```


Fig. 2 The example input file BB12003 and its output generated by MSACompro

MSACompro. The example command to perform multiple sequence alignment by MSACompro is as follows:

```
./auto_run_msacompro.pl
/storage/shared/pspro2/bin/  ../test/BB12003  ../
test/BB12003.msa
```

The SP and TC scores for the output alignments respectively generated by running MSACompro and MSAProbs are illustrated on the right of the figure, as well.

Acknowledgment

This work was supported by an NIH grant (1R01GM093123) to JC.

References

1. Barton GJ, Sternberg M (1987) A strategy for the rapid multiple alignment of protein sequences. Confidence levels from tertiary structure comparisons. J Mol Biol 198(2):327
2. Feng DF, Doolittle RF (1987) Progressive sequence alignment as a prerequisitetto correct phylogenetic trees. J Mol Evol 25(4):351–360
3. Krogh A, Brown M, Mian IS, Sjolander K, Haussler D (1994) Hidden Markov models in computational biology: applications to protein modeling. J Mol Biol 235(5):1501–1531
4. Liu Y, Schmidt B, Maskell DL (2010) MSA-Probs: multiple sequence alignment based on pair hidden Markov models and partition function posterior probabilities. Bioinformatics 26(16):1958–1964
5. Do CB, Mahabhashyam MSP, Brudno M, Batzoglou S (2005) ProbCons: probabilistic consistency-based multiple sequence alignment. Genome Res 15(2):330–340
6. Poirot O, Suhre K, Abergel C, O'Toole E, Notredame C (2004) 3DCoffee@ igs: a web server for combining sequences and structures into a multiple sequence alignment. Nucleic Acids Res 32(Suppl 2):W37–W40
7. Heringa J (1999) Two strategies for sequence comparison: profile-preprocessed and secondary structure-induced multiple alignment. Comput Chem 23(3):341–364
8. Kim NK, Xie J (2006) Protein multiple alignment incorporating primary and secondary structure information. J Comput Biol 13(9): 1615–1629
9. Subramanian AR, Hiran S, Steinkamp R, Meinicke P, Corel E, Morgenstern B (2010) DIALIGN-TX and multiple protein alignment using secondary structure information at GOBICS. Nucleic Acids Res 38(Suppl 2): W19–W22
10. Zhou H, Zhou Y (2005) SPEM: improving multiple sequence alignment with sequence profiles and predicted secondary structures. Bioinformatics 21(18):3615–3621
11. Pei J, Grishin NV (2006) MUMMALS: multiple sequence alignment improved by using hidden Markov models with local structural information. Nucleic Acids Res 34(16):4364–4374
12. Pei J, Grishin NV (2007) PROMALS: towards accurate multiple sequence alignments of distantly related proteins. Bioinformatics 23(7): 802–808
13. Cheng J, Randall A, Sweredoski M, Baldi P (2005) SCRATCH: a protein structure and structural feature prediction server. Nucleic Acids Res 33(Web Server Issue):W72–W76
14. Gonnet GH, Cohen MA, Benner SA (1992) Exhaustive matching of the entire protein sequence database. Science 256(5062): 1443–1445
15. Kawabata T, Nishikawa K (2000) Protein structure comparison using the Markov transition model of evolution. Proteins 41(1): 108–122
16. Durbin R, Eddy SR, Krogh A, Mitchison G (1998) Biological sequence analysis: probabilistic models of proteins and nucleic acids. Cambridge University Press, Cambridge, MA
17. Tegge AN, Wang Z, Eickholt J, Cheng J (2009) NNcon: improved protein contact map prediction using 2D-recursive neural networks. Nucleic Acids Res 37(Suppl 2): W515–W518

18. Sneath PHA, Sokal RR (1973) Numerical taxonomy. The principles and practice of numerical classification. Freeman, San Francisco, CA
19. Barney B (2011) OpenMP tutorial
20. Thompson JD, Koehl P, Ripp R, Poch O (2005) BAliBASE 3.0: latest developments of the multiple sequence alignment benchmark. Proteins 61(1):127–136
21. Van Walle I, Lasters I, Wyns L (2004) Align-m—a new algorithm for multiple alignment of highly divergent sequences. Bioinformatics 20(9):1428–1435
22. Raghava GPS, Searle SMJ, Audley PC, Barber JD, Barton GJ (2003) OXBench: a benchmark for evaluation of protein multiple sequence alignment accuracy. BMC Bioinformatics 4(1):47
23. Thompson JD, Plewniak F, Poch O (1999) A comprehensive comparison of multiple sequence alignment programs. Nucleic Acids Res 27(13):2682–2690
24. Deng X, Cheng J (2011) MSACompro: protein multiple sequence alignment using predicted secondary structure, solvent accessibility, and residue-residue contacts. BMC Bioinformatics 12:472

INDEX

A

David J. Russell (ed.), *Multiple Sequence Alignment Methods*, Methods in Molecular Biology, vol. 1079,
DOI 10.1007/978-1-62703-646-7,

S

T

MIX
Papier aus verantwortungsvollen Quellen
Paper from responsible sources
FSC® C105338

If you have any concerns about our products,
you can contact us on
ProductSafety@springernature.com

In case Publisher is established outside the EU,
the EU authorized representative is:
Springer Nature Customer Service Center GmbH
Europaplatz 3, 69115 Heidelberg, Germany

Printed by Libri Plureos GmbH
in Hamburg, Germany